HVAC

heating, ventilating, and air conditioning Second Edition

AMERICAN TECHNICAL PUBLISHERS, INC.
HOMEWOOD, ILLINOIS 60430

S. Don Swenson

© 1995 by American Technical Publishers, Inc.
All rights reserved

2 3 4 5 6 7 8 9 – 95 – 9 8 7 6 5 4 3

Printed in the United States of America

Library of Congress Cataloging-in-Publication Data

Swenson, S. Don, 1928-
    HVAC heating, ventilating, and air conditioning / S. Don Swenson.
    -- 2nd ed.
        p.   cm.
    Includes index.
    ISBN 0-8269-0675-3
    1. Heating.   2. Ventilation.   3. Air conditioning.   I. Title.
TH7012.S94   1995
697--dc20   94-46551
                                                                         CIP

Contents

Acknowledgments

The author and publisher are grateful to the following companies and organizations for providing technical information and assistance. A bullet (•) preceding a chart or table indicates that it has been reset for educational purposes.

Air Conditioning Contractors of America
ALCO Controls Division Emerson Electric Company
Alnor Instrument Company
ASHRAE 1989 Handbook — Fundamentals
Bacharach, Inc.
R.W. Beckett Corp.
Bethlehem Steel Corporation
Carrier Corporation
Cleaver-Brooks, Division of Aqua-Chem, Inc.
Copeland Corporation
Crane Valves
Crosby Valve & Gage Company
Dunham-Bush
Du Pont Co.
Dwyer Instruments Inc.
Eaton Corp., Cutler-Hammer Products
Henry Valve Co.
Honeywell Inc.
ITT Bell & Gossett
Johnson Controls
Lau, a Division of Tomkins Industries
Lennox Industries Inc.
Parker Hannifin Corp.
Ranco Inc.
Rheem Air Conditioning Division
Robinair Div. Sealed Power Corp.
Sporlan Valve Company
Superior Valve Company Division of Amcast Industrial Corporation
Super Radiator Coils
Tecumseh Products Company
The Trane Company
Turbotec Products, Inc.

Introduction

HVAC is an introduction to the areas of heating, ventilating, and air conditioning that emphasizes the fundamentals of HVAC systems. The components of forced-air and hydronic heating and air conditioning systems are covered in detail.

Information on heat transfer, psychrometrics, refrigeration principles, pressure-enthalpy diagrams, load calculations, equipment selection, troubleshooting, and maintenance is included and explained in detail. Equations are presented in a step-by-step format and are followed by examples to help reinforce the information. Each chapter concludes with subjective review questions, which cover the information presented in the chapter.

The Appendices contain useful tables and charts used in the HVAC trade. An illustrated Glossary provides easy-to-find definitions for key terms. The Index is cross-referenced so information can be found quickly by using key words.

Photos of HVAC equipment and detailed line illustrations help clarify technical material. The following charts are provided to clarify symbols used in the illustrations in the text.

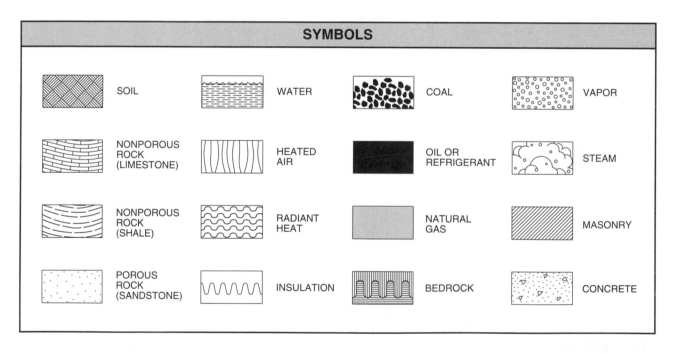

Comfort

Chapter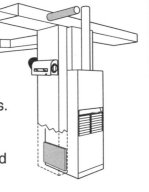

Heating, ventilating, and air conditioning (HVAC) systems used in residential, commercial, and industrial buildings provide comfort to occupants. A feeling of comfort results when temperature, humidity, circulation, filtration, and ventilation of the air are controlled. HVAC systems are designed to fulfill comfort requirements with maximum efficiency.

HVAC systems contain mechanical, electrical, and chemical components. Mechanical components include all moving parts. Electrical components include blowers, compressors, and controls. Chemical components include fuels and refrigerants. Precautions should be taken and safety rules followed when working with HVAC systems.

COMFORT

Comfort is the condition that occurs when a person cannot sense a difference between themselves and the surrounding air. Comfort occurs when no differences exist or undesirable conditions have been corrected. The five requirements for comfort are proper temperature, humidity, circulation, filtration, and ventilation. See Figure 1-1. *Discomfort* is the condition that occurs when a person can sense a difference between themselves and the surrounding air. It can occur when any of the five requirements for comfort are not met.

Temperature

Controlling the temperature of the human body is an important physiological function. *Physiological functions* are the natural physical and chemical functions of an organism. The body produces energy by digesting food. Some energy is used for normal living processes, some is stored, and some is used as thermal energy (heat). Physiological systems regulate body temperature to maintain comfort.

Normal body temperature is 98.6°F. The body has natural heating and cooling systems to maintain this temperature. These systems control heat output by responding to the conditions of the air according to the internal temperature of the body. The body responds by controlling blood flow at the surface of the skin, radiating heat from body surfaces, or using evaporation of perspiration from skin. *Evaporation* is the process that occurs when a liquid changes to a vapor by absorbing heat. When the body is clothed, the body's temperature control system provides comfort at an air temperature of approximately 75°F. If the air temperature varies much above or below 75°F, the body begins to feel uncomfortably warm or uncomfortably cool.

Signals (electrical impulses) from different points in the body are sent through a network of nerves to the hypothalamus, a gland in the brain. The hypothalamus regulates body temperature by controlling blood flow to capillaries (tiny blood vessels) located in the skin. Capillaries regulate perspiration flow to the surface of the skin. If the body temperature rises, blood flow to the skin increases. Blood carries heat to the skin, where it is given off to the air. A person who becomes overheated also becomes flushed because the blood flow to the skin increases. If signals indicate the body is cooling off, the system reduces the cooling effect by allowing less blood flow to the skin. In cold weather fingers and toes become cold before the rest of the body because blood flow to these areas is reduced. Cold areas also appear pale because of the reduced blood

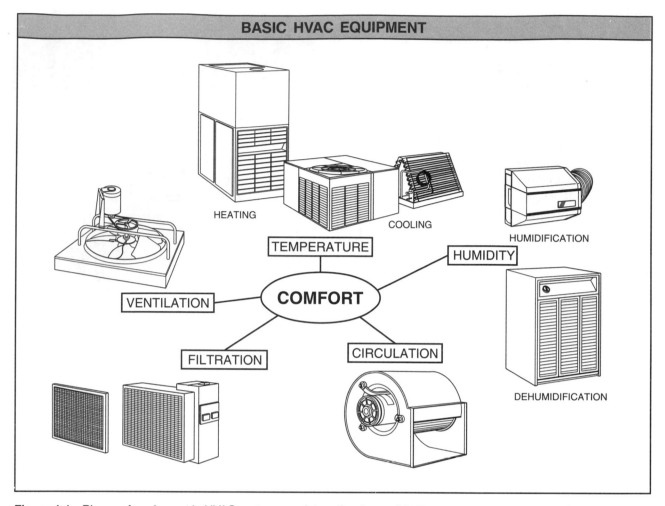

Figure 1-1. Pieces of equipment in HVAC systems work together to provide the proper temperature, humidity, circulation, filtration, and ventilation required for comfort.

flow. Temperature control in buildings is provided by warm air from heating equipment or cool air from air conditioning equipment. Heating equipment supplies the proper amount of heat in cold weather to offset heat loss from a building.

Heating equipment is rated in Btu per hour (Btu/hr). A *Btu* (*British thermal unit*) is the quantity of heat required to raise the temperature of 1 lb of water 1°F. Air conditioning equipment supplies the proper amount of cooling in hot weather to offset heat gain to a building. Air conditioning equipment is rated in Btu per hour or ton of cooling. A *ton of cooling* is the amount of heat required to melt a ton of ice (2000 lb) over a 24-hour period. One ton of cooling equals 288,000 Btu per 24-hour period or 12,000 Btu/hr ($12,000 \times 24 = 288,000$). See Figure 1-2.

The operation of heating and air conditioning equipment is a function of temperature control in

the HVAC system. For maximum comfort, temperature control equipment maintains air temperature in a building within 1°F or 2°F of the temperature necessary for comfort.

Humidity

Humidity is the amount of moisture (water vapor) in air. Humidity is always present in air. A low humidity level indicates dry air that contains little moisture. A high humidity level indicates damp air that contains a significant amount of moisture.

Relative Humidity. *Relative humidity* is the amount of moisture in air compared to the amount of moisture the air would hold at the same temperature if it were saturated (full of water). Relative humidity is always expressed as a percentage. For example, air at 50% relative humidity holds one-half of the amount of moisture it would hold at the same temperature if it were saturated. Air at 60% relative humidity holds 60% of the amount of moisture it would hold at the same temperature if it were saturated. The amount of moisture required to saturate the air changes as the dry bulb temperature changes. In addition, the relative humidity and the capacity to hold moisture change as the dry bulb temperature changes. Humidity is important in determining comfort.

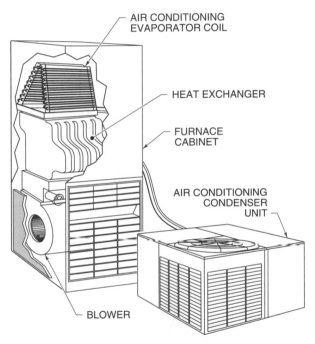

Figure 1-2. Heating and air conditioning equipment control the temperature in a building.

Humidity affects comfort because it determines how slowly or rapidly perspiration evaporates from the body. The flow of perspiration is controlled by the cooling system in the body, which regulates body temperature. Evaporation of perspiration cools the body. The higher the relative humidity, the slower the evaporation rate. The lower the relative humidity, the faster the evaporation rate. For example, with no temperature change and an increase in relative humidity, a person feels warmer because of the slower evaporation rate. With no temperature change and a decrease in relative humidity, a person feels cooler because of the faster evaporation rate. A steam room is an example of an area with high temperature and high relative humidity. These conditions cause the body to have a high perspiration rate with a slow evaporation rate.

Comfort is usually attained at normal cooling and heating temperatures with a relative humidity of about 50%, which is 50% saturated. If the humidity is too low, a higher air temperature is required to feel comfortable. If the humidity is too high, a lower air temperature is required for the same feeling of comfort.

In some cases the humidity in a building may be too low or too high for comfort. Humidity level is controlled by humidifiers and dehumidifiers. See Figure 1-3.

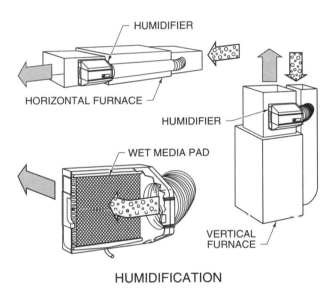

HUMIDIFICATION

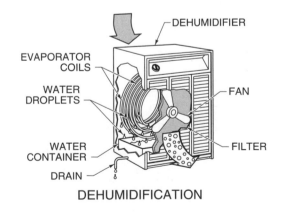

DEHUMIDIFICATION

Figure 1-3. Humidifiers and dehumidifiers control the level of humidity in a building.

A *humidifier* is a device that adds moisture to air by causing water to evaporate into air. In cold climates and dry climates, the humidity in a building may be too low for comfort. For example, in cold climate areas such as Chicago in the winter, the relative humidity is low. In dry climate areas such as southern Arizona, relative humidity is also low. In these cases, humidifiers are used in buildings to add moisture to the air and to maintain a comfortable humidity level.

A *dehumidifier* is a device that removes moisture from air by causing moisture to condense. *Condensation* is the formation of liquid (condensate) as moisture or other vapor cools below its dew point. *Dew point* is the temperature below which moisture begins to condense. In buildings with swimming pools or a large number of potted plants, the humidity level may be too high for comfort.

Circulation

Circulation is the movement of air. Air in a building must be circulated continuously to provide maximum comfort. A total comfort system always includes gently circulating air that is clean and fresh. In a building where there is improper circulation, air rapidly becomes stagnant and uncomfortable. *Stagnant air* is air that contains an excess of impurities and lacks the oxygen required to provide comfort. When air is improperly circulated, temperature stratification occurs. *Temperature stratification* is the variation of air temperature in a building space that occurs when warm air rises to the ceiling and cold air drops to the floor. Temperature stratification in a building space causes discomfort. See Figure 1-4.

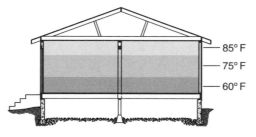

Figure 1-4. Temperature stratification occurs when there is improper circulation in a building space.

Air velocity is the speed at which air moves from one point to another. Air velocity is measured in feet per minute (fpm). If air is not moving, heat cannot be carried away from skin surfaces. If heat is carried to the surface of the skin but is not carried

away by the air, the body will overheat. Air circulation is important because it helps cool the body by causing perspiration to evaporate. An increase in air velocity increases the rate of evaporation of perspiration from the skin, which causes a person to feel cool. A decrease in air velocity or no air velocity reduces the evaporation rate, which causes a person to feel warm. Air movement is used to supply clean and fresh air and to remove stagnant air. Air movement at 40 fpm is nearly ideal.

Supply air ductwork and registers are used to distribute air to different building spaces. Supply air ductwork is sized and located for maximum efficiency. The type, location, and size of registers determine the amount of supply air introduced into each building space and the distribution pattern of the air.

Return air ductwork and grills are used to move air from building spaces back to the blower. Return air ductwork and grills, like supply air ductwork and registers, are sized and located for maximum efficiency. See Figure 1-5.

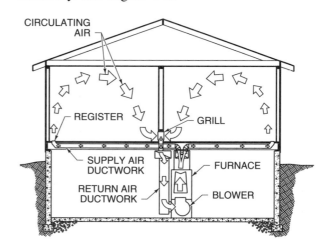

Figure 1-5. Supply and return air ductwork are sized and located to provide efficient flow of air through building spaces.

Filtration

Filtration is the process of removing particles and contaminants from air that circulates through an air distribution system. Circulated air in a building must be clean and includes both return air that is recirculated and fresh air that is used for ventilation. The air in a forced-air heating system is automatically cleaned by filters that are placed in the return air ductwork where the air enters the heating or cooling equipment. See Figure 1-6.

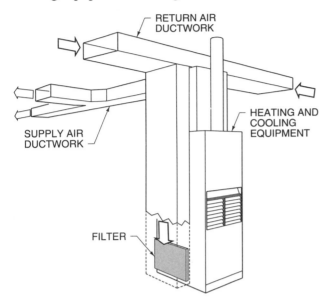

Figure 1-6. Circulated air is cleaned by filters placed in the return air ductwork near heating and cooling equipment.

The primary consideration in air filter selection is the level of clean air desired. Filters are commercially available in low-, medium-, and high-efficiency levels. In one-family dwellings or small commercial buildings, low-efficiency filters are sufficient for normal filtering applications. Medium- and high-efficiency filters are used according to the degree of filtration required. Some filters are designed for disposal after use, while others may be cleaned and reused.

Ventilation

Ventilation is the process of introducing fresh air into a building. Ventilation and air circulation are necessary for comfort inside a building. Offensive odors and vapors are created in closed living spaces. The best method for keeping the air in living spaces comfortable is to dilute the contaminated air. This is accomplished by introducing fresh air from outdoors with recirculated (filtered) air.

Some odors that originate in buildings are simply offensive, while others may be dangerous. Offensive odors are eliminated by bringing fresh air into a system from outdoors and introducing it into the return air portion of the HVAC system. See Figure

1-7. The amount of ventilation necessary for comfort is determined by the number of occupants and the kind of activity taking place inside the building. If odors or contaminants originating in a building are offensive or potentially toxic, the air in the building may require continuous exhaust with no recirculation. The building would require 100% makeup air for ventilation. *Makeup air* is air that is used to replace air that is lost to exhaust.

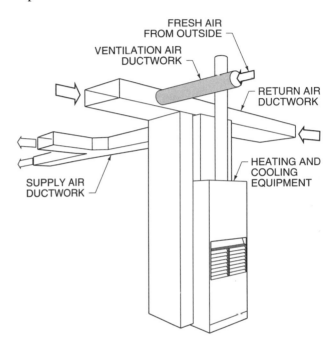

Figure 1-7. Ventilation is the process of introducing fresh air into a building.

HVAC SYSTEM SAFETY

HVAC systems use equipment consisting of mechanical, electrical, and chemical components. Each of these components has characteristics that could be hazardous if proper safety precautions are not followed. Mechanical hazards are related to mass, motion, or the condition of the HVAC equipment. Electrical hazards relate to the potential of electrical shock or fires. Chemical hazards relate to the flammability, caustic action, or high temperature of chemicals used in HVAC systems. See Figure 1-8.

Mechanical

Mechanical components of HVAC systems include moving parts such as motors, blower wheels,

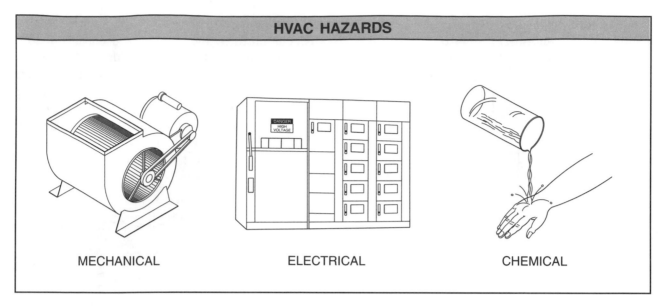

Figure 1-8. HVAC systems contain mechanical, electrical, and chemical components that could be hazardous.

pulleys, and belts. Loose clothing, neckties, and jewelry should not be worn. Clothing can become entangled in rotating shafts and serious injury may result. Caution must be taken whenever working around moving parts.

All equipment should be installed according to manufacturers' specifications. A complete inspection should be conducted on all equipment after installation or servicing before turning the system ON. Heavy and bulky equipment should be moved using lift trucks or cranes. To prevent personal injury, heavy objects should be lifted using the legs rather than the back.

Electrical

HVAC systems use electricity to turn blower and compressor motors and actuate controls. During installation, troubleshooting, service, and repair procedures, many electrical connections are exposed. These electrical connections should be avoided to prevent serious injury or death. Always shut power to HVAC equipment OFF when working on the equipment. Never work in locations that are wet or damp, which could cause electrical shock. A short circuit in part of an electrical system can cause current to flow through the equipment itself. Current flow can generate heat and possibly start a fire. All electrical equipment must be installed by a licensed electrician according to applicable provisions of the National Electrical Code®.

Combustible materials and fuels used in heating equipment burn readily and require special handling. By maintaining careful control of combustible materials and fuels, the danger of a fire hazard is reduced. Fires are classified by the combustible material involved. The three most common classes of fires are Class A, Class B, and Class C. Class A includes fires that burn wood, paper, textiles, and other ordinary combustible materials containing carbon. Class B includes fires that burn oil, gas, grease, paint, or other liquids that convert to gas when heated. Class C include electrical, motor, or transformer fires.

Class D is a rare, specialized class of fires including fires caused by combustible metals such as zirconium, titanium, magnesium, sodium, and potassium. A special powder is applied to put out this class of fire.

Chemical

Chemicals present in HVAC systems include fuels, refrigerants, and chemicals used for cleaning and water treatment. When properly contained in fuel lines, tanks, or other vessels, the hazards of fuels are reduced. Danger occurs when fuels leak out of contained areas and mix with air. If this happens, the fuel could ignite and explode. All fuels, regardless of quantity, must be handled with care.

Refrigerants commonly used in HVAC systems can cause injuries if handled improperly. Ammonia, used in large systems, is dangerous. Other refrigerants are dangerous if exposed to open flames or high temperatures. When this occurs, the refrigerant compounds change chemically and toxic gases form. Most refrigerants boil at very low temperatures. If they come in contact with skin, severe frostbite can result. Protective clothing should always be worn to prevent injury from leaks. Always refer to manufacturers' recommended safety procedures when handling chemicals used in HVAC systems.

Safety Precautions

An HVAC technician must develop safety habits to prevent personal injury, injury to others, and damage to equipment. Safety procedures vary depending on the type and size of the equipment. However, basic safety rules are common to all equipment.

Wear approved clothing and shoes at all times.

Wear gloves when handling hot lines and refrigerant system charging equipment.

Wear appropriate eye protection and respirators.

Wear a hard hat when working where there is a possibility of head injury.

Do not use hands to stop moving equipment.

Store all oily rags or waste in approved containers to prevent fires caused by spontaneous combustion.

Use approved safety cans to store combustible liquids.

Do not use unsafe ladders or substitutes for ladders.

Do not carry tools in back pockets.

Do not throw tools.

Use the proper tool for the job.

Do not use defective tools.

Precheck all equipment for starting hazards.

Only authorized personnel should repair electrical equipment.

Never remove high-voltage fuses without the proper tools.

Never enter an enclosed area that has toxic or flammable gas without adequate ventilation.

Check automatic combustion controls routinely for proper operation.

Review Questions

1. Define *comfort* as it relates to heating and air conditioning systems.

2. List the five requirements for comfort.

3. At what temperature do most individuals feel comfortable?

4. At what relative humidity do most individuals feel comfortable?

5. At what velocity should air in a building space be moving for most individuals to feel comfortable?

6. Define *humidity* as it relates to a comfort system.

7. What piece of equipment is used to add humidity to building spaces?

8. Define *condensation*.

9. Define *relative humidity*.

10. In a comfort system, what are supply ductwork and registers used for?

11. In a comfort system, what are return air ductwork and grills used for?

12. What device is used to clean the air as it circulates in a heating or air conditioning system?

13. What is used to dilute contaminated air?

14. List the three categories of system components that require safety precautions.

15. How do moving parts such as sheaves, belts, and pulleys create hazards?

16. What should be done to minimize hazards when lifting heavy and bulky equipment?

17. What should be done to reduce electrical hazards when working on heating and air conditioning equipment?

18. What should be done to reduce chemical hazards of fuels or refrigerants used in heating and air conditioning systems?

19. How do humidifiers add moisture to air?

20. Define *temperature stratification* and explain what causes it.

Thermodynamics and Heat

Chapter 2

Thermodynamics is the science of the relationships between heat and mechanical action. Heat is the form of energy that is transferred because of a temperature difference. Heat is transferred by conduction, convection, and radiation. In each method of heat transfer, there are certain factors and variables that affect the process of heat transfer. Heat is produced by combustion, electrical energy, or alternate heat sources. Laws of thermodynamics apply to heating and air conditioning systems.

THERMODYNAMICS

Thermodynamics is the science of thermal energy (heat) and how it transforms to and from other forms of energy. Thermodynamics relates to events that occur when air is heated or cooled in a building space. The two laws of thermodynamics that apply to heating and air conditioning systems are the first law of thermodynamics and the second law of thermodynamics. Both laws apply in operating heating and air conditioning systems.

First Law of Thermodynamics

The *first law of thermodynamics,* the law of conservation of energy, states that energy cannot be created or destroyed but may be changed from one form to another. An example of this law is the combustion process. As fuel burns, the elements hydrogen and carbon, which are found in all fuels, combine with oxygen in the air. As these three elements combine, a chemical reaction occurs, and chemical energy in the elements is released in the form of thermal energy. The hydrogen, carbon, and oxygen recombine to form new compounds, which are found in the products of combustion. Two of the new compounds are carbon dioxide (CO_2) and water vapor (H_2O). During the process, energy has changed from chemical energy

to thermal energy, but no energy has been created or destroyed. See Figure 2-1.

Another example of the first law of thermodynamics is a compressor operating in an air conditioning system. An electric motor uses electrical energy to drive the compressor. The electrical energy is converted to mechanical energy in the motor. The mechanical energy compresses the refrigerant in the system. As the compressed refrigerant expands, it produces a cooling effect. Most of the electrical energy used to drive the compressor results in the cooling effect from the system. Some of the mechanical energy is converted to thermal energy because of friction, but no energy is created or destroyed in the process.

Second Law of Thermodynamics

The *second law of thermodynamics* states that heat always flows from a material at a high temperature to a material at a low temperature. This flow of heat is natural and does not require energy. See Figure 2-2. The second law of thermodynamics applies to all cases of heat transfer. For example, air in a furnace is heated by the products of combustion. Heat flows from the warm burner flame to the cool air. Air conditioning systems use energy to control the movement of heat.

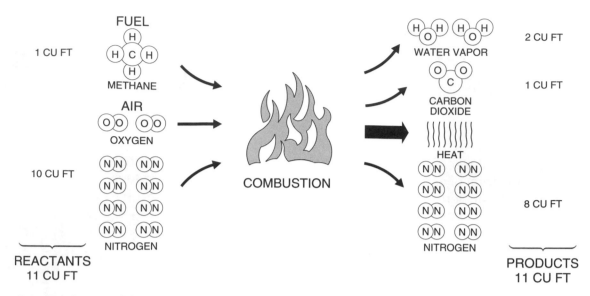

Figure 2-1. The first law of thermodynamics states that energy cannot be created or destroyed but may be changed from one form to another.

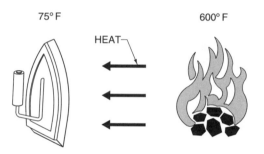

Figure 2-2. The second law of thermodynamics states that heat always flows from a material at a high temperature to a material at a low temperature.

TEMPERATURE

Temperature is the measurement of the intensity of heat. A measurement of 70°F indicates that there is enough heat in the air to register a 70°F reading on a thermometer. The temperature of 70°F indicates intensity of heat, not quantity of heat. Quantity of heat is expressed in Btu.

Two scales used for measuring temperature are the Fahrenheit (°F) and the Celsius (°C) scales. The Fahrenheit scale is used with the U.S. system of measurements. Atmospheric pressure is measured in pounds per square inch (psi). On the Fahrenheit scale, 32°F is the freezing point and 212°F is the boiling point of water at normal atmospheric pressure (14.7 psi). The Celsius scale is used with the metric system of measurements. On this scale, 0°C is the freezing point and 100°C is the boiling point of water at normal atmospheric pressure.

Temperature Conversion

The relationships between the freezing and boiling points on the Fahrenheit and Celsius scales make it possible to convert from one scale to the other. To make a conversion between the two scales, the difference between the bases and the ratio of the difference between the bases are used. On the Fahrenheit scale, 32°F is the base. On the Celsius scale, 0°C is the base. Therefore, 32° is the difference between the bases. To find the difference between the bases apply the formula:

Base difference = Base F − Base C

where

Base difference = difference between bases

Base F = base of Fahrenheit scale (32)

Base C = base of Celsius scale (0)

Example: Finding the Difference Between Bases

Base difference = Base F − Base C

Base difference = 32 − 0

Base difference = **32°**

The ratio of the difference between the two scales is determined by the difference between the freezing point and the boiling point of water on each scale. There is a range of 180°F between 32°F and 212°F on the Fahrenheit scale and 100°C between 0°C and 100°C on the Celsius scale. The ratio for this conversion is found by dividing 180 by 100. To find the ratio between the bases, apply the formula:

$$Ratio = \frac{Range\ F}{Range\ C}$$

where

Ratio = ratio of the ranges of the temperature scales

Range F = difference between Fahrenheit freezing and boiling points

Range C = difference between Celsius freezing and boiling points

Example: Finding the Ratio of the Ranges

$$Ratio = \frac{Range\ F}{Range\ C}$$

$$Ratio = \frac{(212 - 32)}{(100 - 0)}$$

$$Ratio = \frac{180}{100}$$

$$Ratio = \mathbf{1.8}$$

There is 1.8°F for every 1.0°C. See Figure 2-3. Both of these numbers are used in the formulas for converting from one scale to the other. To convert a Fahrenheit reading to Celsius, subtract 32 from the Fahrenheit reading and divide by 1.8. To convert Fahrenheit to Celsius, apply the formula:

$$°C = \frac{(°F - 32)}{1.8}$$

where

°*C* = degrees Celsius

°*F* = degrees Fahrenheit

32 = difference between bases

1.8 = ratio between bases

Example: Converting Fahrenheit to Celsius

Convert 70°F to Celsius.

$$°C = \frac{(°F - 32)}{1.8}$$

$$°C = \frac{(70 - 32)}{1.8}$$

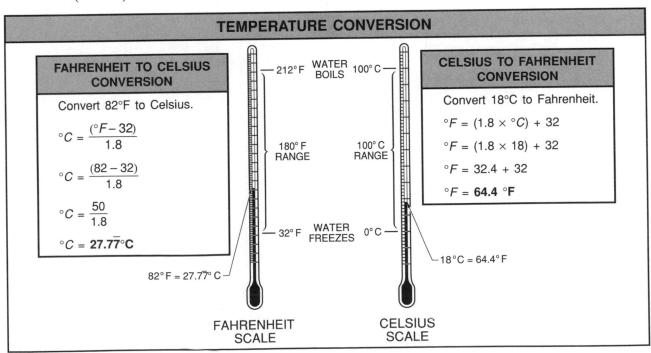

Figure 2-3. The difference of 32 and the ratio of 1.8 are used when converting between Fahrenheit and Celsius temperature scales.

$$°C = \frac{38}{1.8}$$

$$°C = \textbf{21.1}\overline{\textbf{1}} \; °C$$

To convert a Celsius reading to Fahrenheit, multiply 1.8 by the Celsius reading and add 32. To convert Celsius to Fahrenheit, apply the formula:

$$°F = (1.8 × °C) + 32$$

where

$°F$ = degrees Fahrenheit

1.8 = ratio between bases

$°C$ = degrees Celsius

32 = difference between bases

Example: Converting Celsius to Fahrenheit

Convert 26°C to Fahrenheit.

$$°F = (1.8 × °C) + 32$$

$$°F = (1.8 × 26) + 32$$

$$°F = 46.8 + 32$$

$$°F = \textbf{78.8} \; °\textbf{F}$$

A temperature of 0°F is entirely different from 0°C. Either scale can be used, but the difference between the two becomes significant when using mathematical calculations, which require a common base. A common base is also required in pressure-temperature calculations and pressure-heat diagrams.

To work calculations with a common base, degrees Fahrenheit must be converted to absolute temperature in degrees Rankine (°R) and degrees Celsius must be converted to absolute temperature in degrees Kelvin (°K). The Kelvin and Rankine scales use absolute zero as the common base. *Absolute zero* is a theoretical condition at which no heat is present. On the Rankine scale, absolute zero (0°R) is 460° below 0°F. On the Kelvin scale, absolute zero (0°K) is 273° below 0°C. See Figure 2-4.

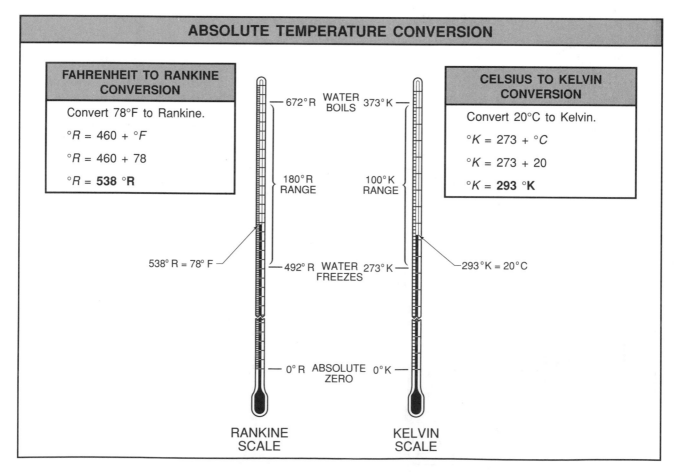

ABSOLUTE TEMPERATURE CONVERSION

FAHRENHEIT TO RANKINE CONVERSION

Convert 78°F to Rankine.

$°R = 460 + °F$

$°R = 460 + 78$

$°R = \textbf{538} \; °\textbf{R}$

CELSIUS TO KELVIN CONVERSION

Convert 20°C to Kelvin.

$°K = 273 + °C$

$°K = 273 + 20$

$°K = \textbf{293} \; °\textbf{K}$

672°R WATER BOILS 373°K

180°R RANGE — 100°K RANGE

538°R = 78°F

492°R WATER FREEZES 273°K

293°K = 20°C

0°R ABSOLUTE ZERO 0°K

RANKINE SCALE

KELVIN SCALE

Figure 2-4. The base on the Rankine and Kelvin scales is absolute zero. The Rankine scale is used with degrees Fahrenheit, and the Kelvin scale is used with degrees Celsius.

To convert a Fahrenheit temperature to Rankine, add 460° to the Fahrenheit temperature. To convert Fahrenheit to Rankine, apply the formula:

$$°R = 460 + °F$$

where

$°R$ = degrees Rankine

460 = difference between bases

$°F$ = degrees Fahrenheit

Example: Converting Fahrenheit to Rankine

Convert 96°F to Rankine.

$$°R = 460 + °F$$
$$°R = 460 + 96$$
$$°R = \mathbf{556 \ °R}$$

To convert a Celsius temperature to Kelvin, add 273° to the Celsius temperature. To convert Celsius to Kelvin, apply the formula:

$$°K = 273 + °C$$

where

$°K$ = degrees Kelvin

273 = difference between bases

$°C$ = degrees Celsius

Example: Converting Celsius to Kelvin

Convert 34°C to Kelvin.

$$°K = 273 + °C$$
$$°K = 273 + 34$$
$$°K = \mathbf{307 \ °K}$$

HEAT

Heat is the form of energy identified by a temperature difference or a change of state. All substances exist in either a solid, liquid, or gas state. *Change of state* is the process that occurs when enough heat is added to a substance to change it from one physical state to another, such as from ice to water or water to steam.

Sensible heat is heat measured with a thermometer or sensed by a person. Sensible heat does not involve a change of state. *Latent heat* is heat identified by a change of state and no temperature change. Latent heat is heat added to ice that changes it to water, or

heat added to water that changes it to steam. See Figure 2-5. Heat transferred in heating systems is usually sensible heat. Heat transfer in air conditioning systems includes both sensible heat and latent heat.

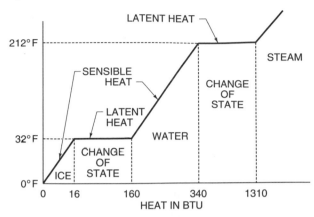

Figure 2-5. Sensible heat changes the temperature of a substance without changing the state of the substance. Latent heat changes the state of a substance without changing the temperature of the substance.

Heat Measurement

In the U.S. system of measurements, quantity of heat is measured in Btu (British thermal units). A Btu is the quantity of heat required to change the temperature of 1 lb of water 1°F. Large quantities of heat are measured in thousands of Btu per hour (Mbh). For example, 2000 Btu/hr equals 2 Mbh. See Figure 2-6.

1 BTU = 1° F TEMPERATURE CHANGE

Figure 2-6. A Btu is the quantity of heat required to change the temperature of 1 lb of water 1°F.

In the metric system of measurements, quantity of heat is measured in calories. A *calorie (cal)* is the quantity of heat required to change the temperature of 1 g (grams) of water 1°C. One calorie equals .004186 J. A *joule (J)* is a unit of measure that expresses the quantity of heat. Joules are much smaller than calories. Large quantities of heat are measured in kilocalories (kcal) and kilojoules (kJ). The prefix kilo indicates thousands. For example, 1 kJ equals 1000 J. One kilojoule equals .9486 Btu. One kilocalorie equals 3.968 Btu. See Figure 2-7.

UNITS OF ENERGY			
Energy	Btu	J	kcal
1 British thermal unit		1055	.2520
1 joule	9.481×10^{-4}		2.390×10^{-4}
1 kilocalorie	3.968	4.184	

Figure 2-7. The Btu, joule, and kilocalorie are units of measure for heat energy.

Heat Transfer

Heat transfer is the movement of heat from one material to another. Heat is transferred by conduction, convection, and radiation, or any combination of the three. The second law of thermodynamics states that heat always flows from a material at a high temperature to a material at a low temperature. This flow of heat is natural and does not require energy. This law applies to all methods of heat transfer. The quantity of heat involved in heat transfer is a function of weight, specific heat, and temperature difference.

Weight is the force with which a body is pulled downward by gravity. In heat calculations weight is used instead of volume because weight, unlike volume, remains constant despite changes in temperature. The volume of most materials changes with a change in temperature. In heat transfer calculations, weight is expressed in pounds or grams.

Specific heat is the ability of a material to hold heat. Specific heat is expressed as the ratio of the quantity of heat required to raise the temperature of a material 1°F compared to the quantity required to raise the temperature of an equal mass of water 1°F. Specific heat is used in calculations and normally given in Btu per pound.

The specific heat of water is 1.0 and is used as the standard for calculating specific heat. The specific heat of most materials is less than the specific heat of water because water holds large quantities of heat. Each substance has a constant value for specific heat that can be found on tables and charts. See Figure 2-8. See Appendix.

SPECIFIC HEAT*					
Solid		Liquid		Gas	
Aluminum	.214	Alcohol	.615	Air	.24
Brass	.09	Ammonia	1.099	Butane	.377
Coal	.3	Kerosene	.5	CO_2	.20
Concrete	.156	Mineral oil	.5	Chlorine	.117
Glass	.18	Petroleum	.4	Helium	1.241
Gold	.031	R-22	.26	Methane	.520
Ice	.487	R-502	.255	Neon	.246
Iron	.12	Saltbrine	.745	Oxygen	.218
Rubber	.48	Turpentine	.42	Propane	.375
Wood	.45	Water	1.00	Steam	.48

* in Btu/lb

Figure 2-8. Specific heat is the ability of material to hold heat. Values of specific heat are constants that are given in tables and charts.

Temperature difference is the difference between the temperatures of two materials, the temperatures on both sides of a material, or the initial and final temperatures of a material through which heat has been transferred. An example of temperature difference is the difference between the inside and outside temperatures of a heat exchanger. A *heat exchanger* is anything that transfers heat from one substance to another without allowing the substances to mix. A temperature difference must exist between two materials or at least two locations within a material for heat transfer to occur.

The Greek letter delta (Δ) represents a difference. Temperature difference (ΔT) is found by applying the formula:

$$\Delta T = T_1 - T_2$$

where

ΔT = temperature difference

T_1 = initial temperature

T_2 = final temperature

Example: Finding Temperature Difference

What is the temperature difference when a piece of metal heated to 600°F cools to 57°F?

$$\Delta T = T_1 - T_2$$

$$\Delta T = 600 - 57$$

$$\Delta T = \textbf{543°F}$$

Heat is transferred, for example, from liquid in a glass to an ice cube in the liquid. The quantity of heat involved in the process is a function of the weight of ice involved, the specific heat of the ice, and the temperature difference between the time when the ice starts to melt and when it is completely melted. The amount of time, however, is not included in this calculation. The quantity of heat transferred is found by applying the formula:

$$Q = wt \times c \times \Delta T$$

where

Q = quantity of heat (in Btu)

wt = weight (in lb)

c = specific heat (constant)

ΔT = temperature difference (in °F)

Example: Finding Quantity of Heat Transferred

Find the quantity of heat involved when 36 lb of iron is heated from a temperature of 46°F to 208°F.

$$Q = wt \times c \times \Delta T$$

$$Q = 36 \times .12 \times (208 - 46)$$

$$Q = 36 \times .12 \times 162$$

$$Q = \textbf{699.84 Btu}$$

Conduction. *Conduction* is heat transfer that occurs when molecules in a material are heated and the heat is passed from molecule to molecule through the material. Conduction occurs when one end of a metal rod is heated in a flame. The heat travels by conduction from molecule to molecule until it reaches the opposite end of the rod. When heat is transferred by conduction, there is no flow of material. See Figure 2-9.

Conduction also occurs between two different materials that are in direct contact. The process of heat transfer is the same, but the rate of heat transfer differs depending on the materials. Conduction factors rep-

resent the rates of heat transfer through different materials per degree Fahrenheit difference between the materials. The quantity of heat transfer through materials is usually measured in Btu per hour.

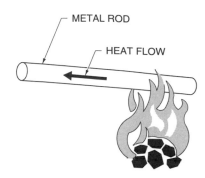

Figure 2-9. Conduction is heat transfer that occurs when molecules in a material are heated and the heat is passed from molecule to molecule through the material.

Because conduction takes time, a temperature gradient (variation in temperature) can occur within a material. The surface of the hot side of the material becomes nearly the same temperature as the source of heat, while the surface of the cool side of the material remains nearly the same as the original temperature. The temperature at any point in the conducting material falls in the range between the two surface temperatures. The temperature at any point in the conducting material depends on the distance from the surface and the resistance of the material to heat flow. See Figure 2-10.

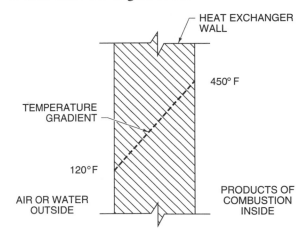

Figure 2-10. The temperature through a material varies whenever the temperature on each side of the material is different.

Surface area is the number of square feet (sq ft) of a material through which heat flows. A *square foot* equals 144 sq in. The area of a flat surface is found by multiplying the length by the width. Surface area is found by applying the formula:

$$A = l \times w$$

where

A = area

l = length

w = width

Example: Finding Surface Area — Flat Surface

A piece of sheet metal has a length of 6′ and a width of 4′. What is the surface area?

$$A = l \times w$$

$$A = 6 \times 4$$

$$A = \textbf{24 sq ft}$$

The surface area of a cylinder is found by applying the formula:

$$A = \pi D \times l$$

where

A = area

π = 3.14 (constant)

D = diameter

l = length

Example: Finding Surface Area — Cylinder

A cylinder has a length of 18″ and a diameter of 10″. What is the surface area?

$$A = \pi D \times l$$

$$A = 3.14 \times 10 \times 18$$

$$A = \textbf{565.20 sq in.}$$

$$A = \frac{565.20}{144}$$

$$A = \textbf{3.93 sq ft}$$

A *thermal transmission factor* is a numerical value that represents the amount of heat that passes through a material when there is a temperature difference across the material. Transmission factors are based on the thickness of a material because thickness affects heat transfer. Tables of thermal properties of materials, such as transmission factors, have been prepared for use in heat transfer calculations. These tables include conductivity and conductance factors for different materials. See Figure 2-11.

The *conductivity (k) factor* is the amount of heat transferred through 1 sq ft of material that is 1″ thick in Btu per hour, per 1°F difference through the material. Conductivity through a material 1″ thick is found by applying the formula:

$$Q = A \times k \times \Delta T$$

where

Q = quantity of heat (in Btu/hr)

A = surface area (in sq ft)

k = conductivity factor (constant)

ΔT = temperature difference (in °F)

CONDUCTION FACTORS

Materials	k	C	Materials	k	C
Aluminum	1536.0	—	Gypsum board		
Brass	936.0	—	.5″	—	2.22
Brick	4.8	—	Hardwoods	1.10	—
Cast iron	331.2	—	Lead	241.2	—
Cement plaster			Slate		
.75″		6.66	.5″	—	20.00
Coal	1.176	—	Softwoods	.80	—
Concrete	6.48	—	1.5″	—	.53
8″	—	.90	2.5″	—	.32
12″	—	.81	Steel	314.4	—
Glass	.35	—	Tin	450.0	—

Figure 2-11. Conductivity (k) factors and conductance (C) factors are used to calculate the quantity of heat transferred through a material.

Example: Finding Quantity of Heat — Conductivity Factor

The surface area of a 1″ thick steel boiler is 25.13 sq ft. The conductivity factor for the steel is 314.4 per inch of thickness. The temperature of the water in the boiler is 200°F, and the temperature of the surrounding air is 67°F. Find the amount of heat that will pass through the boiler walls in 1 hour.

$$Q = A \times k \times \Delta T$$

$$Q = 25.13 \times 314.4 \times (200 - 67)$$

$$Q = 25.13 \times 314.4 \times 133$$

$$Q = \textbf{1,050,815.98 Btu/hr}$$

The *conductance (C) factor* is the amount of heat transferred through 1 sq ft of area of the surface of a material of a given thickness in Btu per hour, per 1°F difference through the material. The mathematical formula for calculating heat transfer by conductance is the same as that for conductivity except the C factor is used for a given thickness of the material. Conductance through a material of a given thickness is found by applying the formula:

$$Q = A \times C \times \Delta T$$

where

Q = quantity of heat (in Btu/hr)

A = surface area (in sq ft)

C = conductance factor (constant)

ΔT = temperature difference (in °F)

Example: Finding Quantity of Heat — Conductance Factor

A wall is built of 8″ thick concrete blocks. The surface area of the wall is 144 sq ft. The conductance factor for concrete blocks is .90. The temperature difference through the concrete blocks is 30°F. Find the amount of heat that will pass through the blocks in 1 hour.

$$Q = A \times C \times \Delta T$$

$$Q = 144 \times .90 \times 30$$

$$Q = \textbf{3888 Btu/hr}$$

Convection. *Convection* is heat transfer that occurs when currents circulate between warm and cool regions of a fluid. A *fluid* is a substance that takes the shape of its container. Liquids and gases are fluids. Heat is transported by currents that circulate between warm and cool regions in a fluid. See Figure 2-12. For example, convection occurs when a pan of water is heated. A flame from below heats the pan. The water nearest the flame gains heat rapidly. The warming happens so quickly that conduction cannot move heat away from the area fast enough. The water nearest the heat becomes warmer, expands, and becomes less dense. This water begins to rise to the surface and is replaced by cooler, denser water, which falls to the bottom near the heat.

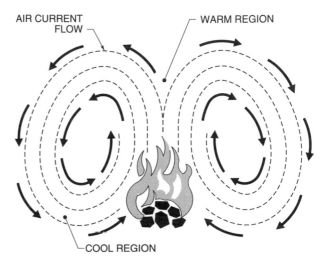

Figure 2-12. Convection is heat transfer that occurs when currents circulate between warm and cool regions of a fluid.

Heat is transferred by convection and conduction in hydronic heating systems. Heat is transferred by convection as hot water circulates through piping to terminal devices where the heat is transferred through the metal by conduction. The heat is then transferred from the metal to the air by convection. See Figure 2-13.

Heat transfer by convection is calculated using the weight of the material carrying the heat, the specific heat of the material, and the temperature difference between the beginning and end of the heat transfer process. Heat transfer by convection is found by applying the formula:

$$Q = wt \times c \times \Delta T$$

where

Q = quantity of heat (in Btu/hr)

wt = weight (in lb)

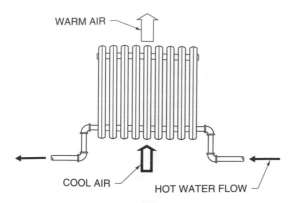

Figure 2-13. Heat is transferred by convection as hot water is pumped through a radiator in a hydronic heating system.

c = specific heat (constant)

ΔT = temperature difference (in °F)

Example: Finding Quantity of Heat — Convection

As the blower in a gas furnace moves air through the heat exchanger in the furnace, the air temperature is increased 80°F. The blower moves 4167 lb of air through the furnace in 1 hour. The specific heat of air is .24 Btu/lb/°F. Find the amount of heat added to the air as it passes through the furnace.

$Q = wt \times c \times \Delta T$

$Q = 4167 \times .24 \times 80$

$Q = \textbf{80,006.40 Btu/hr}$

Radiation. *Thermal radiation* is heat transfer that occurs as radiant energy (electromagnetic) waves transport energy from one object to another. All objects transmit radiant energy. The amount of heat transmitted depends on the intensity of heat (temperature) in the material. Heat from an object becomes waves that are transmitted from the object. These waves are radiant energy and not heat energy while they are transmitted. The waves move through space and transparent materials without producing heat. Heat is produced when the energy waves contact an opaque object. *Opaque* means that light cannot pass through an object. Radiant energy when it strikes an opaque object becomes thermal energy. See Figure 2-14. Rays of the sun are an example of heat transferred by radiation. The rays travel through space between the earth

and the sun, but the space is not heated. Heat is produced only when the waves contact the atmosphere and surface of the earth.

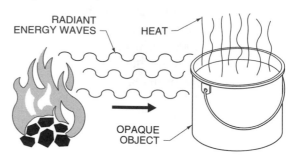

Figure 2-14. Radiation is the transfer of energy in the form of radiant energy waves. Heat is produced when the energy waves contact an opaque object.

According to the second law of thermodynamics, heat transferred by radiation, like heat transferred by conduction and convection, flows from a high temperature to a low temperature. An example is a person facing a fire, who feels heat even though the temperature of the air does not change. If an object is placed between the person and the fire, the person will not feel heat.

Heat transfer by radiation is measured using the surface temperature of the radiant surface, the temperature difference between the radiant surface and the receiving surface. The space, shape, size, and surface conditions of the radiant surface and receiving surface are also considered.

Surface temperature is the temperature of a radiant surface. The higher the temperature of a radiant surface, the more radiation flows from the surface. The sun radiates large amounts of energy because of its high surface temperature, approximately 11,000°F. To produce a heating effect, radiant heating systems commonly have radiation surface temperatures of 50°F to 150°F. Some specialized radiant heating system applications require temperatures above 150°F.

Heat transfer by radiation does not warm the air within a space. Only opaque objects in direct line of sight from the source are heated by radiation. Radiant heating systems are often used to heat small areas instead of entire buildings. In radiant heating systems, some heat transfer by convection occurs from the radiant surface and from heated objects.

The condition of the radiant surface and the receiving surface affects the amount of heat transferred.

Energy is radiated from the surface area of an object, which may include curves, recesses, and other irregularities. *Emissivity* is the measure of the ability of a surface to absorb or emit heat by radiation. The receiving surfaces absorb and/or reradiate heat depending on the condition of the surfaces. See Figure 2-15. Calculating the amount of radiation involves many variables. Consult the *ASHRAE Handbook Series* for additional information on design data for various types of radiant heating systems. *ASHRAE (American Society of Heating, Refrigerating and Air Conditioning Engineers)* is a professional organization that is dedicated to the advancement of the HVAC profession and associated industries.

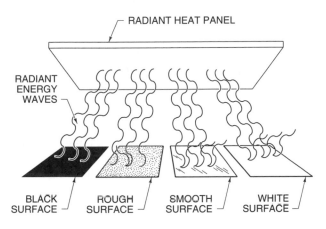

Figure 2-15. A material absorbs and/or reradiates heat depending on the condition of the surface of the material.

Heat Production

Heat is produced by combustion, electricity, or alternate sources. Combustion is the most common method of producing heat. In the combustion process, fuel combines with oxygen at a temperature high enough to start a chemical reaction. The reaction produces heat. Some heating systems use electricity to produce heat. In this process, heat is produced when a wire conductor resists the flow of electricity. Alternate sources, such as solar and geothermal heating systems, are sometimes used to produce heat or supplement other heating systems.

Combustion Heat. *Combustion heat* is heat produced during combustion. To start combustion, elements in the fuel combine with oxygen in the air at the ignition temperature. *Ignition temperature*

is the intensity of heat required to start a chemical reaction (combustion). The combustion process requires fuel, oxygen, and heat. See Figure 2-16. Fuels commonly used for combustion include coal, oil, and natural gas. All fuels contain hydrogen and carbon.

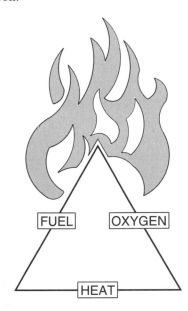

Figure 2-16. The three requirements for combustion are fuel, oxygen, and enough heat to start a chemical reaction.

When fuel combines with oxygen at the ignition temperature, the hydrogen and carbon in the fuel oxidize (chemically react with the oxygen in the air). During the reaction, the chemical energy in the fuel is converted to thermal energy. The result is products of combustion that are much hotter than the air or the fuel. *Products of combustion* are the heat and gases produced when chemicals react and recombine to form new compounds. See Figure 2-17. The heat produced in the combustion process is the amount of thermal energy that changed from chemical energy. Forced-air and hydronic heating systems use combustion heat to warm air or water that is then distributed to building spaces.

Electric Heat. *Electric heat* is electrical energy that has changed to thermal energy. When electricity passes through a conductor, resistance to the flow of electricity generates heat. The type of conductor determines the amount of resistance. A heating element is formed by using a conductor with the proper resistance and physical strength to resist

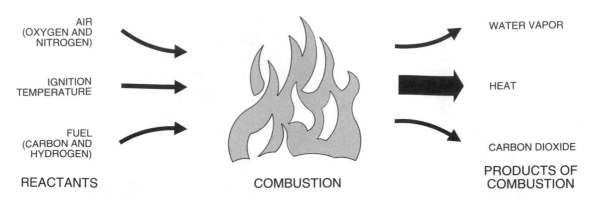

Figure 2-17. The products of combustion are much hotter than the reactants.

burning. Nichrome, a nickel-chromium alloy, is commonly used for resistance heating elements. Heating systems that use electrical energy include radiant panel, unit, duct, and baseboard heaters.

Alternate Heat Sources. *Alternate heat sources* are natural sources of heat that do not require the combustion process or electricity to produce heat. Solar energy, wind power, and geothermal heat are three of the most common alternate heat sources.

Solar energy is the energy transmitted from the sun by radiation. *Solar heat* is the heat created by the visible (light) and invisible (ultraviolet) energy rays of the sun. These rays travel through space and the earth's atmosphere. The rays provide energy in the form of radiant waves.

Passive and active systems are the two basic types of solar heating systems. See Figure 2-18. *Passive heating systems* are systems that use no mechanical means for bringing heat into a building or for storing heat. In a passive system, the building is designed and constructed so that it acts as both a collector and a storage unit. *Active heating systems* are systems that use mechanical components for bringing heat into a building and for storing heat. In an active system, special collectors and storage units are used. Active systems may use a fluid, usually water, as the collecting medium. When water is used as the collecting medium, it is heated by the solar rays and stored in tanks until the heat is needed. Solar rays are also used to heat rocks or salt bags, which then heat the air.

Solar energy collection and storage systems are built to collect and store the maximum amount of solar energy. The amount of energy received at the surface of the earth can exceed 200 Btu/hr per sq ft of surface depending on the angle of the rays and the position of the collector. Solar energy is converted to thermal energy and used to heat building spaces.

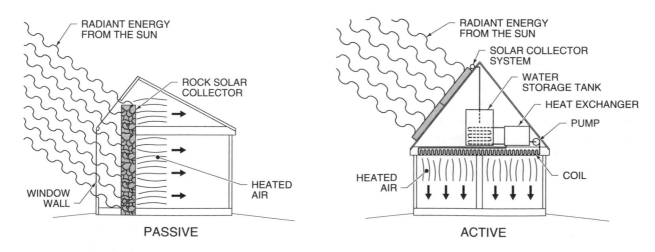

Figure 2-18. Solar energy systems convert radiant energy from the sun into thermal energy used to heat air and water.

Solar energy can also be converted to electrical energy by collectors. *Collectors* are either panels containing a substance that is heated by the sun's rays or are made up of photovoltaic cells that convert solar energy directly to electrical energy. If collector cells such as water or air collectors are used, the heated material is used as a heat source. If photovoltaic cells are used, the electrical energy produced is often used for heating and cooling buildings.

Wind power is energy created by the movement of wind. Wind power is used to generate electricity in many parts of the world where constant winds of sufficient speed are common. Wind power turns propellers that drive electric generators. The assemblies are mounted on towers. The electricity generated by the windmills is fed to the local electric company and may be used as heat energy.

Geothermal heat is heat that results when magma from deep within the earth's crust comes in contact with groundwater. *Magma* is molten rock. *Groundwater* is water that sinks into soil and subsurface rocks. When groundwater comes in contact with magma, the groundwater is heated to high temperatures, and it returns to the surface as hot water springs and steam geysers.

In some areas this supply of hot water and steam is used for heating purposes. This process occurs naturally in some places. In others, deep wells and pumps are used to bring hot water to the surface.

Review Questions

1. Define *thermodynamics.*

2. State the first law of thermodynamics.

3. Give an example of the first law of thermodynamics.

4. State the second law of thermodynamics.

5. Give an example of the second law of thermodynamics.

6. List the two temperature scales that are most commonly used.

7. What temperature scale is used with the U.S. system of measurements?

8. What temperature scale is used with the metric system of measurements?

9. What is the base temperature of the Kelvin and Rankine temperature scales?

10. What does heat indicate?

11. Define *change of state.*

12. List and explain the two types of heat in heating and air conditioning applications.

13. What is the equivalent of 1 Btu of heat?

14. Define *specific heat.*

15. List and describe the three ways in which heat is transferred.

16. Define *thermal conduction factor.*

17. What is the difference between a conductivity factor and a conductance factor?

18. Describe the combustion process.

19. List three alternate heat sources used to collect energy for heating and air conditioning applications.

20. Explain the difference between an active solar heating system and a passive solar heating system.

Combustion and Fuels

Chapter  3

Combustion and electrical resistance are two methods of producing heat for buildings. Combustion is a chemical reaction between elements in fuel and oxygen. During the reaction, chemical energy is converted to thermal energy. If enough oxygen is available for complete combustion, the products of combustion are harmless. If enough oxygen is not available for complete combustion, the products of combustion can be dangerous. The quantity of heat produced by burning a fuel is directly related to the elements in a fuel.

COMBUSTION

Combustion is the chemical reaction that occurs when oxygen reacts with the elements hydrogen (H) and carbon (C) in a fuel at an ignition temperature. A *chemical reaction* is the chemical change of a substance or substances. An *element* is pure matter that cannot be broken down into simpler substances. Fuel, oxygen, and ignition temperature are the three requirements for combustion. Fuel provides chemical energy, oxygen reacts with the fuel, and the ignition temperature starts the reaction.

Combustion produces heat and light as elements in the fuel and oxygen recombine to form different substances. When combustion is used to heat a building, it takes place in a furnace or a boiler. Combustion in a furnace or boiler must be controlled for the heat to be used efficiently.

Fuels and Heating Values

A *fuel* is any material that is burned to produce heat. Fuels are classified by physical characteristics or chemical composition. Physical characteristics identify a fuel and determine the type of equipment required for transporting, storing, and burning the fuel. The most important physical characteristic of a fuel is its state. Fuels are available in a solid, liquid, or gas state. *Chemical composition* is all of the elements that make up a compound. A *compound* is a pure substance that can be broken down into simpler substances (elements). Each compound has a unique chemical composition. Most fuels are compounds. The unit of measure used for a fuel depends on the state of the fuel. Solid fuels are measured in pounds, liquid fuels in gallons, and gas fuels in cubic feet (cu ft).

All fuels contain hydrocarbons. *Hydrocarbons* are compounds that contain hydrogen and carbon. Some fuels also include small quantities of other elements. Each element burns at a different rate. Hydrogen burns quickly, and carbon, the major element in most fuels, burns slowly.

Each element has a specific heating value. *Heating value* is the quantity of heat in Btu that is released during combustion of one unit of an element. When a given quantity of an element is burned, a fixed amount of heat is produced. If the elements in a fuel are known, the heating value of the fuel can be calculated. It is possible to calculate the quantity of heat that will be produced when a fuel is burned because each element has a unique heating value. See Figure 3-1.

The quantity of heat produced by burning a fuel is determined by the heating value and the amount of fuel used in a given time. Because fuels are compounds, different fuels have different heating values. However, the heating value for a fuel can be determined if a chemical analysis of the fuel is available and the amount of oxygen combined with the fuel is known.

HEATING VALUES OF ELEMENTS AND COMPOUNDS		
Substance	Molecular Symbol	Heating Value*
Acetylene	C_2H_2	2355
Butane	C_4H_{10}	3444
Carbon (to CO)	C	308
Carbon (to CO_2)	C	1733
Carbon monoxide	CO	339
Ethylene	C_2H_4	1641
Hydrogen	H_2	315
Methane	CH_4	1016
Propane	C_3H_8	2411
Propylene	C_3H_6	2477

* in Btu/cu ft

Figure 3-1. Heating value is the quantity of heat in Btu that is produced when one unit of fuel is burned.

Heating values are measured in Btu per pound (Btu/lb) for solid fuels, Btu per gallon (Btu/gal.) for liquid fuels, and Btu per cubic foot (Btu/cu ft) for gas fuels. See Figure 3-2. For example, when 1 lb of hydrogen is burned, 51,000 Btu is produced. When 1 lb of carbon is burned, 14,093 Btu is produced. If a sample of a fuel containing 1 lb of hydrogen and 1 lb of carbon is burned, the heat output would be 14,093 Btu plus 51,000 Btu, or 65,093 Btu. This assumes that combustion is 100% efficient.

FUEL HEATING VALUES*		
Solids (Btu/lb)	Liquids (Btu/gal.)	Gases (Btu/cu ft)
Wood 8600	Fuel oil 140,000	Natural gas 1000
Coal 14,000		Propane 2500
		Butane 3200

* standard heating values for commercial fuels

Figure 3-2. The unit of measure used for a fuel depends on the state of the fuel.

Fuels are used depending on regional availability and cost. Because the prices of fuels and electricity

vary, selecting a fuel for a specific job is a difficult and important task. Petroleum-base fuels are popular, but are subject to availability. Electricity is used more often for heating, but the cost of electricity is rising. The availability and relatively stable cost of natural gas make it a popular fuel for heating. Coal and wood are used for residential heating in areas where they are inexpensive and readily available.

Coal. The most common solid fuel used for heating is coal. *Coal* is a solid black fuel that formed when organic material hardened in the earth over millions of years. Water, soil, and rock covered organic material. Pressure and heat changed the organic material to coal. Because of the way it forms, coal is usually found in layers or seams. Depending on the location of a seam, coal is mined either deep underground or close to the surface of the earth. Underground coal is mined in tunnels, shafts, or galleries that are dug in the earth. See Figure 3-3. Surface coal is mined in strip mines. Soil and rock that cover the coal is stripped away and the coal is dug out.

Bethlehem Steel Corporation

Figure 3-3. A coal mining machine rips coal from a seam deep underground.

Coal contains a large percentage of carbon but also contains hydrogen, ash, and water with trace amounts of oxygen (O), nitrogen (N), and sulfur

(S). Coal is classified according to grade and rank. *Grade* identifies the size, heating value, and ash content of the coal. *Rank* identifies the hardness of coal, which varies among the different kinds of coal. Coals usually used for heating include anthracite and bituminous coal. See Appendix. *Anthracite coal* is a hard, high luster coal that yields a small amount of volatile matter when it burns. *Volatile matter* is the gas (usually oil and tar) given off when coal burns. Anthracite coal is a clean burning coal. It is found inland along the East Coast of the United States and in various other places in the world. *Bituminous coal* is a softer coal that yields a large amount of volatile matter when it burns. Bituminous coal burns well, but does not burn as clean as anthracite coal. Bituminous coal is found in many places in the United States. At the present time, it is plentiful and used extensively. See Figure 3-4.

Different kinds of coal vary in size and weight but a standard sample weighs 90 lb/cu ft. A *standard sample* is a sample of coal that consists of different kinds of coal. A standard sample of coal from an area is used to represent all of the coal in that area. Coal is removed from mines by mine cars or conveyor belts. It is then cleaned, sorted, and graded. The graded coal is shipped either to a distributor or directly to the point of use. At one time, coal was used for all heating applications. Today it is used for commercial and industrial heating and for producing electricity. Because of the abundant supply of coal, research is being conducted to find ways to reduce coal to either a liquid or gas form to make it easier to transport and burn.

The standard heating value for coal is 12,000 Btu/lb, which is slightly less than the heating value of the hydrogen and carbon alone. Coal heating values vary with the chemical content of the coal. Heating values range from 6900 Btu/lb to 14,350 Btu/lb. See Figure 3-5.

COAL HEATING VALUES	
Coal	**Heating Values***
Anthracite	12,700–13,600
Bituminous	11,000–13,800
Medium-volatile bituminous	14,000 (standard)

* in Btu/lb

Figure 3-5. Heating values of coal vary with the chemical content of the coal.

Fuel Oil. *Fuel oil* is one of the petroleum-base products made from crude oil. Different grades of fuel oil are available. *Crude oil* is a mixture of semi-solids, liquids, and gases formed from the remains of organic materials that have been changed by pressure and heat over millions of years. Crude oil deposits are found under the surface of the earth in various parts of the world.

Geologists, scientists who study rock formations, use various devices and techniques to locate structures underground that could hold oil. One method of finding these structures is by detonating an explosive charge on the surface of the earth. A seismograph is used to listen to the echoes that bounce back from the various layers of rock in the earth. A *seismograph* is a device that measures and records

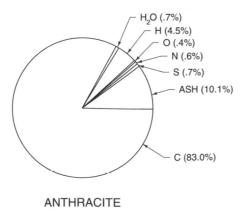

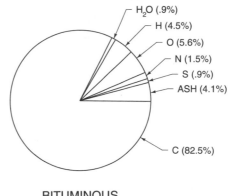

Figure 3-4. Coals used for heating contain carbon with small amounts of hydrogen and other elements.

vibrations in the earth. The seismograph readings are used to determine if an underground structure may hold oil. A pilot well is then drilled at the site. If crude oil is found, more wells are drilled to collect the oil. The crude oil either flows from a well by natural pressure or it is pumped. See Figure 3-6.

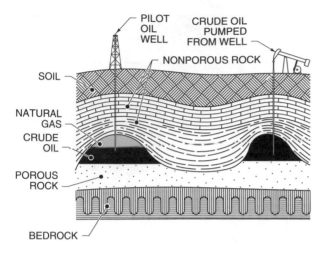

Figure 3-6. Crude oil flows from deposits deep underground by natural pressure or it is pumped.

Once the crude oil is collected, it is transported to a refinery. A *refinery* is a plant where crude oil is separated into petroleum products. Heat and pressure vaporize (change to a gas) crude oil in one part of the refinery. Different petroleum products condense out of the vapor at different temperatures. The products are collected and stored separately. Classifications of petroleum products are distillates (light oils) and residuals (heavy oils). Distillates are refined into fuels such as liquefied petroleum gas, jet and automobile gasoline, kerosene, light heating oils, and diesel fuels. Residuals are refined into industrial fuel oils and other commercial and industrial products. See Figure 3-7.

The grade of a fuel oil is based on the weight and viscosity of the oil. *Viscosity* is the ability of a liquid or semiliquid to resist flow. As the numbers of the grades of fuel oil increase, so does the viscosity. Each grade of oil has properties required for specific applications. Grade No. 1 is a light distillate fuel oil used in open pot burners. It is almost as light as diesel oil, which is burned in automobiles and trucks. An *open pot burner* is a burner used in space heaters and other small heating units. Oil flows into an open pot and begins to vaporize. Combustion air is introduced

around the pot and combustion starts when the oil vapors ignite at the top of the pot.

Grade No. 2 fuel oil is slightly heavier than Grade No. 1. Grade No. 2 is usually used for heating with pressure atomizing burners. A *pressure atomizing burner* is a burner that sprays oil into the combustion chamber through an atomizing nozzle. *Atomization* is the process of breaking a liquid into small droplets. Air mixes with the oil droplets and an electric spark ignites the air-oil mixture.

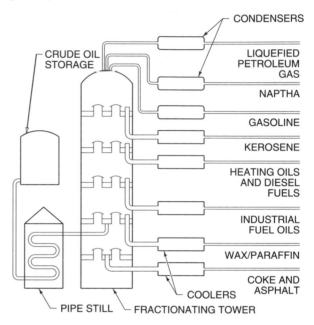

Figure 3-7. Heat and pressure vaporize crude oil. Petroleum products condense at different temperatures and are collected separately.

Grade No. 4 fuel oil is an intermediate weight fuel oil used for commercial and industrial burners. Grade No. 5 (light) is a residual fuel oil that is heavier than Grade No. 4. Grade No. 5 (heavy) is a residual fuel oil heavier than Grade No. 5 (light). Preheating is usually required when using Grade No. 5 and heavier oils in cold climates. Fuel tanks and fuel lines are heated when using heavier fuel oils to lower the viscosity. Grade No. 6 is the heaviest fuel oil. It is a residual oil usually used for heating. It is sometimes called bunker C and requires heating in the tank, in the lines, and before it enters the burner.

Fuel oil is transported from a refinery to a distributor by ships, barges, railroad tank cars, or trucks. Tank trucks transport fuel oil from the distributor to the consumer. The fuel oil is deposited in a storage tank that is piped to the furnace or boiler inside a building. A pump on the burner draws oil from the storage tank to the burner.

Heating values of fuel oil range from 132,900 Btu/gal. to 155,900 Btu/gal. for different grades. See Figure 3-8. Grade No. 2 is used for residential heating. The standard heating value used for Grade No. 2 is 140,000 Btu/gal. Higher grades of fuel oil have higher Btu/gal. heating values.

FUEL OIL HEATING VALUES	
Grade	**Heating Value***
No. 1	132,900–137,000
No. 2	137,000–141,800
No. 4	143,100–148,100
No. 5 (light)	146,800–150,000
No. 5 (heavy)	149,400–152,000
No. 6	151,300–155,900

* in Btu/gal

Figure 3-8. The various grades of fuel oils have different heating values.

Gas. The gas fuels used for heating are natural, liquefied petroleum (LP), and manufactured gas. Natural gas and LP gas are the two gas fuels commonly used for heating. These three fuels are stored and transported differently but all burn in the gas state. This assures clean combustion with little or no residue.

Natural gas is a colorless and odorless gas fuel. Because it is colorless and odorless, odorants are added to the gas to make it noticeable. Natural gas is in the gas state at all times. It is found near crude oil deposits, but is sometimes found where there is no oil. See Figure 3-9. When it is found, natural gas may contain varying amounts of methane (CH_4), ethane (C_2H_6), water vapor, hydrogen sulfide (H_2S), helium (He), propane (C_3H_8), and butane (C_4H_{10}). Before it is used as a fuel, natural gas is scrubbed (cleaned) to remove LP gas, gasoline, and noncombustible material. The product that is distributed as fuel is composed of 70% to 96% methane and 1% to 14% ethane. Small percentages (.4% to 17%) of propane, butane, pentane, hexane, carbon dioxide, oxygen, and nitrogen are also found in natural gas. Natural gas for distribution has a

specific gravity of .660 to .708. *Specific gravity* of a gas is the weight of 1 cu ft of the gas compared to the weight of 1 cu ft of dry air at 29.9″ Hg (inches of mercury) and 60°F. Because natural gas is lighter than air, it will dissipate into the atmosphere if released. Natural gas is available for heating in most urban areas.

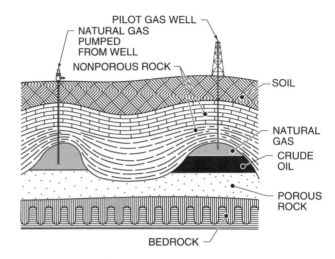

Figure 3-9. Natural gas is found near crude oil deposits and is sometimes found where there is no oil.

The heating value for natural gas varies as the chemical composition of the gas varies. The heating value for gas delivered by the gas utility is about 950 Btu/cu ft to 1050 Btu/cu ft. The natural gas supplied by a utility is composed of about 25% hydrogen and 75% carbon, which burns efficiently. See Figure 3-10. For general use, the standard heating value of natural gas is 1000 Btu/cu ft. Gas is sold to consumers by the therm. A *therm* is the quantity of gas required to produce 100,000 Btu of heat.

GAS HEATING VALUES	
Fuel	**Heating Value***
Natural gas	950–1050
Manufactured gas	500–600

Liquid petroleum
  Propane-Butane mixtures
    (depends on mixutres)
  Propane
  Butane
	2500–3200
	
	2500
	3200

* in Btu/cu ft

Figure 3-10. Heating values of gas fuels range between 500 Btu/cu ft and 3200 Btu/cu ft.

Liquefied petroleum (LP) gas is fuel refined from crude oil or natural gas. It is the second most popular gas fuel used for heating. LP gas is used in rural areas where natural gas pipelines are not available. LP gas is collected as a gas, but is reduced to liquid form by pressure and cooling for transportation in tank trucks. When the fuel is used, the liquid vaporizes when it leaves the tank. LP gas consists of propane, butane, and propane-butane mixtures.

Propane is fuel that contains 90% to 95% propane gas and 5% to 10% propylene (C_3H_6). Propane gas has a specific gravity of 1.5, which means that it is heavier than air. Because propane is heavier than air, care must be taken in storage and piping systems to prevent leaks. If propane leaks, it will accumulate in any low area and create a potential for fire or explosion. Propane can be used as a fuel in most climates because it has a low boiling point. The poiling point is −43.7°F.

Butane is fuel that contains 95% to 99% butane gas and up to 5% butylene (C_4H_8). Butane has a specific gravity of 2.0, which is heavier than propane and air. Butane will also settle in low areas and must be handled with care. Butane cannot be used where the storage temperature would drop below freezing because it has a boiling point of 32°F.

A *propane-butane mixture* is fuel used when the properties of propane or butane alone are not acceptable. Butane is stored as a liquid. If the temperature falls below 32°F, it will not vaporize. By mixing butane with propane, which has a boiling point of −43.7°F, the mixture will vaporize at winter temperatures in most climates. Properties of propane-butane mixtures vary as the ratio of the two gases vary.

LP gases have higher heating values than natural gas. Heating values for LP gases are 2500 Btu/cu ft for propane and 3200 Btu/cu ft for butane. LP gas is often abbreviated as LPG.

Manufactured gas is any gas produced from coal, oil, liquid petroleum, or natural gas. Most of these gas fuels are by-products of a manufacturing process and are used for industrial operations or as specialty fuels. At one time many cities had gas plants where gas was manufactured for lighting and heating. Today most manufactured gas systems are local and do not provide fuel for many gas heating systems.

Alternate Fuels. *Alternate fuels* are combustible materials other than coal, oil, and gas. Some alternate fuels are wood pellets, sawdust, refuse, and waste gas from manufacturing processes. Most alternate fuels are not used as often as primary fuels and are therefore not readily available in large quantities for commercial use. Even when alternate fuels are available, the unusual form of the fuel may make it difficult to use for heating.

Wood is the alternate fuel used most often for residential heating. Wood burns most efficiently in airtight fireplaces or stoves. Wood is not used extensively for heating, but its popularity has increased as the cost and availability of other fuels have changed. Heating values of various woods and efficiencies of different stoves vary, which makes it difficult to evaluate the savings of using wood as a fuel. Wood is cost-effective where it can be obtained locally at low cost.

Determining the heating values of different alternate fuels is difficult. Firing methods, which affect combustion efficiency, densities, and moisture contents, vary among alternate fuels and affect heating values. For example, heating values of wood vary depending on the density and moisture content of the wood. Heating values are about 5980 Btu/lb for softwoods, which are from the cottonwood family, and about 9500 Btu/lb for hardwoods, which are oak and hickory. See Appendix.

Combustion Air

Combustion air is air provided at a burner for proper combustion of fuel. The precise amount of air for proper combustion must be available. The actual amount of air depends on the fuel used. If enough air is not available, dangerous products of combustion will be generated.

Each burner in a heating system must have the proper amount of combustion air. Combustion air can be taken from inside a building if the air can get to the burner without any obstruction. If the heating equipment is isolated from the rest of the building by doors or walls, louvers may be installed in doors or through walls to provide a passage for combustion air. If the building area is small or combustion air cannot be acquired from inside the building, outside air must be ducted directly to the heating equipment area.

Oxygen. In a typical heating unit, oxygen for combustion comes from atmospheric air. Air consists of approximately 20.95% oxygen and 78.08% nitrogen. During combustion, oxygen reacts with elements in fuel. Nitrogen goes through combustion unchanged. Different fuels require different air-fuel ratios for combustion. If too little or too much air mixes with the fuel, the mixture will not burn. For example, a carburetor on a gasoline engine mixes air with fuel to fire the engine. If the carburetor air adjustment is set too lean or too rich, the engine will not run. Adjusting for lean or rich mixtures is done by changing the amount of air that is mixed with fuel. The air-fuel ratio for burning natural gas is 96% to 86% air and 4% to 14% gas. At 4% gas, the mixture is lean. At 14% gas, the mixture is rich. Combustion takes place readily when the mixture is near the center of the range. See Figure 3-11.

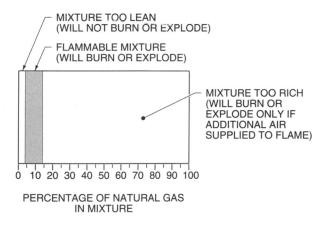

Figure 3-11. The air-fuel ratio for natural gas must be between 4% and 14% for combustion to take place.

Ignition Temperature. Ignition temperature is the intensity of heat required to start a chemical reaction between the elements in a fuel and oxygen. Ignition temperature is provided by a pilot light, electric spark, or some other means. After combustion has started, the temperature must remain at or above the ignition temperature or combustion will stop. The heat produced by combustion usually maintains the ignition temperature.

Flames

Flames are generated as fuel burns. Flames are the result of the chemical reaction between fuel and oxygen. Combustion is a chemical reaction where compounds break apart into elements and release chemical energy in the form of heat and light. The elements recombine into the products of combustion.

Flame Parts. Flames consist of either two or three parts. Each part of a flame relates to the chemical reaction taking place in that part and the time required for the reaction to take place. The *inner mantle* is the center part of the flame. The inner mantle is dark to light blue in color and is produced as hydrogen reacts with oxygen. Hydrogen burns quickly. The *outer mantle* is the light blue area of the flame that surrounds the inner mantle. The outer mantle is produced as carbon reacts with oxygen. The carbon reaction takes place farther away from the burner face because it occurs after the hydrogen reaction. Carbon does not burn as quickly as hydrogen. The outermost tips of a flame are yellow in color. The yellow tips of a flame consist of incandescent particles of heated carbon that give off heat but are not yet consumed by the reaction. See Figure 3-12.

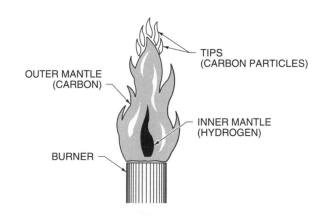

FLAME PARTS	
Tip	Yellow
Outer mantle	Light blue
Inner mantle	Dark blue

Figure 3-12. The different burning rates of the chemical elements in a fuel cause the three parts of a flame.

Flame Types. The size and color of a flame and the products of combustion depend on the kind of fuel burned, how the combustion air mixes with the fuel, and the kind of burner used. When combustion air mixes with fuel before the mixture reaches the burner face, a blue flame is produced. The *burner face* is the point at which the flame is established. A blue flame is usually produced on a gas fuel burner. Fuel in the gas state mixes with air before the air-fuel mixture enters the burner. Flames produced by natural gas are hot blue flames.

A yellow flame is produced when combustion air mixes with a fuel at the burner face. Solid or liquid fuels cannot mix efficiently with combustion air before they burn. Burning liquid and solid fuels is less efficient than burning a gas fuel. Coal and fuel oil fires produce yellow flames for this reason. Modern oil burners are built for high efficiency, but the fuel oil flames are still yellow.

Products of Combustion

During combustion, elements react with oxygen and form new compounds. These new compounds are the products of combustion. Complete combustion produces carbon dioxide and water vapor, which are not dangerous. Incomplete combustion produces carbon monoxide (CO) and aldehydes (CHO), which are dangerous. Complete combustion requires enough combustion air to provide enough oxygen for the process. If there is not enough combustion air, there will not be enough oxygen.

Complete Combustion. *Complete combustion* is ideal combustion that occurs when enough oxygen is supplied by combustion air to combine with all of the elements in a fuel. In complete combustion, hydrogen, carbon, and oxygen recombine to form carbon dioxide and water vapor. Nitrogen in the combustion air passes through the process unchanged. Chemical energy in the fuel is changed to thermal energy, but there is no loss or gain of energy. There are the same number of molecules of chemicals in the products of combustion as there were in the air and fuel. See Figure 3-13.

The quantity of air for complete combustion varies with the type of fuel used. Coal requires approximately 135 cu ft of air per pound of coal burned. Grade No. 2 fuel oil requires about 1400 cu ft of air for every gallon of oil burned. Natural gas requires 10 cu ft of air, butane requires 30 cu ft of air, and propane requires 24 cu ft of air for every cubic foot of gas burned.

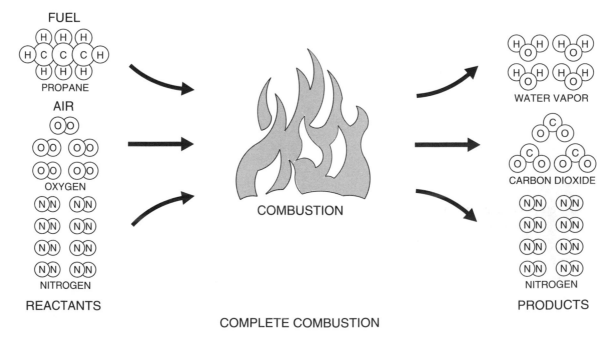

Figure 3-13. Complete combustion produces carbon dioxide and water vapor. No matter is created or destroyed.

Incomplete Combustion. *Incomplete combustion* is improper combustion that occurs when not enough oxygen is supplied by combustion air to combine with all of the elements in a fuel. Incomplete combustion produces carbon monoxide, water vapor, and aldehydes. Carbon monoxide is a dangerous gas. *Aldehydes* are chemical compounds that contain hydrogen, carbon, and oxygen. They have a harsh odor and can burn the eyes and nose. When incomplete combustion occurs, the elements in the fuel and oxygen recombine in a different arrangement. Incomplete combustion is caused by a lack of combustion air and rapid cooling of the flame. See Figure 3-14.

Combustion Efficiency

Combustion efficiency is the evaluation of how well the chemical energy in a fuel is converted to thermal energy during combustion. When combustion efficiency is high, most of the chemical energy is converted to thermal energy. If combustion efficiency is low, most of the chemical energy is wasted and dangerous products of combustion form.

If a chemical analysis of a fuel is available, the heating value of the fuel can be calculated by adding the heating value of each of the elements in the fuel. The heating value of hydrogen and carbon, which are the main elements of all fuels, are 61,500 Btu/lb and 14,550 Btu/lb respectively. Using these figures, it is possible to calculate the heating value of a fuel containing 1 lb each of hydrogen and carbon. For 1 lb of hydrogen to burn completely, 8 lb of oxygen is required. This reaction produces 9 lb of water vapor and 61,500 Btu. For 1 lb of carbon to burn completely, $2\frac{2}{3}$ lb of oxygen is required. This reaction produces $3\frac{2}{3}$ lb of carbon dioxide and 14,550 Btu.

If 100% combustion efficiency is achieved, 76,050 Btu will be produced by burning a fuel containing 1 lb each of hydrogen and carbon in the presence of $10\frac{2}{3}$ lb of oxygen (amount required for complete combustion). Ten pounds of products of combustion would be produced. Combustion is seldom 100% efficient. The two main causes of inefficiencies are incomplete combustion and excess air.

Hydrogen in fuel burns rapidly and therefore cleanly. Carbon burns slowly. If there is not enough oxygen for complete combustion, some carbon particles will leave the process unburned. When this happens, smoke and soot form. *Smoke* is unburned particles of carbon that are carried away from the flame by the convection currents generated by the heat of the flame. *Soot* is unburned particles of carbon

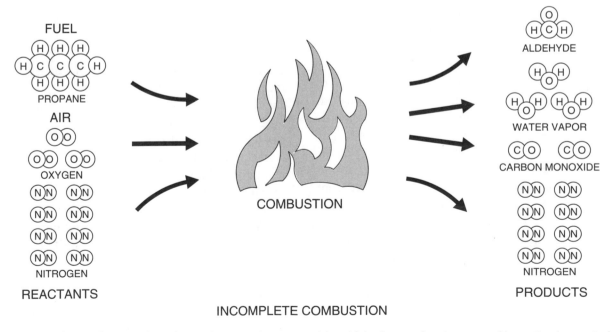

FUEL

PROPANE

AIR

OXYGEN

NITROGEN

REACTANTS

COMBUSTION

ALDEHYDE

WATER VAPOR

CARBON MONOXIDE

NITROGEN

PRODUCTS

INCOMPLETE COMBUSTION

Figure 3-14. Incomplete combustion produces carbon monoxide, aldehydes, and water vapor. No matter is created or destroyed.

that collect on the burner, in the combustion chamber, or in the flue when the surface temperature of these parts is lower than the ignition temperature of the carbon. The lower temperature prevents the carbon particles from burning. Soot causes a smoky fire and acts as an insulator when it collects on the inside of a furnace. A coat of soot on the surface of a heat exchanger reduces efficiency because it slows the heat transfer through the exchanger.

Burners and combustion air blowers should be adjusted to provide more air than required for complete combustion. For maximum efficiency, 5% to 50% excess air is required for combustion. If too much combustion air is introduced into a burner, the air that is not required for combustion passes through the process unchanged except that it is heated. The excess air takes heat from combustion. This reduces combustion efficiency because fuel is used to heat air that goes up the flue. The combustion efficiency is directly related to the amount of excess air present during combustion.

Efficiency Tests. Combustion efficiency is tested by measuring the flue-gas temperature and amount of carbon dioxide in flue gas. A theoretical maximum percentage of carbon dioxide (ultimate carbon dioxide), is produced when precisely enough air is supplied to burn all of the fuel. If more air is supplied than is required for combustion, the excess air will dilute the carbon dioxide. Measuring the percentage of carbon dioxide in flue gas is a method of determining the amount of excess air in combustion. The proper amount of carbon dioxide will ensure clean complete combustion, with a safety margin for variations in fuel, draft, or burner conditions. Recommended carbon dioxide values for natural gas burners are from 8% to 9.5%. Grade No. 2 fuel oil burners should have from 10% to 12.5% carbon dioxide. Bituminous coal furnaces should have from 13% to 15% carbon dioxide. These values are approximate percentages for equipment that is in good condition. The recommendations of equipment manufacturers should be followed at all times.

A *carbon dioxide analyzer* is an instrument used to measure the percentage of carbon dioxide found in flue gas. See Figure 3-15. When using a carbon dioxide analyzer, a metal sampling tube is inserted through a small hole in the flue pipe leading from

a heating unit. The hole must be located just outside the unit and ahead of any device that mixes air with the flue gas, such as a barometric damper. Flexible rubber tubing connects the metal sampling tube to a transparent vessel that contains a fluid that absorbs carbon dioxide. A respirator located on the rubber tube pumps flue gas into the vessel. Once the flue gas is inside the analyzer vessel, the vessel is shaken to mix the gas with the absorbent fluid. The volume of the absorbent fluid increases in proportion to the amount of carbon dioxide present in the flue gas. A calibrated scale on the side of the vessel shows the percentage of carbon dioxide in the gas.

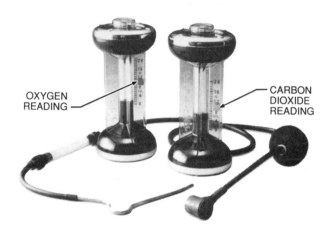

OXYGEN READING

CARBON DIOXIDE READING

Bacharach, Inc.

Figure 3-15. Combustion efficiency is determined using flue-gas analyzers.

The temperature of the sample of flue gas is also measured. The temperature of the flue gas is measured with a stack thermometer. A *stack thermometer* is a thermometer designed to measure high flue-gas temperatures. It is inserted in a hole in the flue or attached to the flue near the furnace. The temperature of the flue (net flue temperature) is found by subtracting the flue-gas temperature from the room temperature. The carbon dioxide reading and net flue temperature are used with appropriate tables to determine the actual combustion efficiency. See Figure 3-16.

If the carbon dioxide reading is not as high as desired, the quantity of combustion air should be reduced. The carbon dioxide and flue-gas temperature readings are used to adjust a burner so it will operate as efficiently as possible. The combustion air is adjusted based on the carbon dioxide level and flue-gas temperatures.

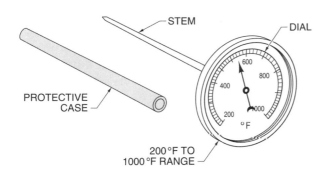

Figure 3-16. A stack thermometer is used to measure the temperature of flue gases.

When testing the combustion efficiency of fuels, levels of oxygen and carbon monoxide should also be tested. Oxygen and carbon monoxide analyzers are similar to carbon dioxide analyzers. Carbon dioxide analyzers and complete system flue-gas analyzers are available in conventional readout or electronic digital readout models. Conventional models require several steps. Electronic models, however, may require only one sample. Some analyzers are permanently attached to a heating unit while others are hand-held. The amounts of oxygen, carbon monoxide, and carbon dioxide must all be used to determine the combustion efficiency of a fuel. Follow the recommendations of heating equipment manufacturers about the percentage of oxygen allowed. There should never be more than a trace of carbon monoxide present.

Combustion has been used throughout history as a source of heat. Today there are concerns about the increasing amounts of carbon dioxide in the atmosphere. Carbon dioxide is a product of combustion. Industrial plants and combustion engines that burn fossil fuels emit most of the carbon dioxide. *Fossil fuels* are fuels, such as coal, crude oil, and natural gas, derived from plant and other organic matter that has decayed beneath the surface of the earth over millions of years. The United States is responsible for 24% of the total carbon emissions from fossil fuels. If carbon dioxide emissions continue to increase, the earth's atmosphere will become

warmer. To reduce carbon dioxide emissions, alternate energy sources such as nuclear energy, hydroelectric energy, and wind power might be used. These alternate energy sources produce electrical energy without burning fossil fuels.

ELECTRICAL ENERGY

Electrical energy is used to produce heat for residential, commercial, and industrial use. Electrical energy is the most popular form of energy used in noncombustion heating units. Other forms of energy such as solar energy are used in limited applications.

Electrical Energy Production

Many different sources of mechanical energy are used to produce electricity. Water power and steam power are the most common sources.

Water is used for producing electricity in hydroelectric power generating plants. A *hydroelectric power generating plant* is a power plant where energy from moving water is changed to electrical energy. See Figure 3-17. Water is piped from a reservoir that is at a higher elevation than the plant. The reservoir is fed from a river or lake. The force of the weight of the water falling through large pipes (penstocks) creates pressure, which turns turbines in the power generating plant. The mechanical energy of the turbines turns the generators, which changes the mechanical energy to electrical energy. Because of the high costs of building dams and hydroelectric power generating plants, these facilities are constructed by the central government in most countries.

Combustion power generating plants use steam to produce electricity. When steam is used to produce electricity, water must be changed into steam. The most common method of producing steam is in boilers. Coal, fuel oil, or gas is burned to heat water to produce steam. The steam is then piped at high pressures to turbines, which turn the generators. See Figure 3-18. Steam from other sources can also be used for operating a turbine. Steam geysers with high enough temperatures and pressures can turn turbines. Nuclear energy also generates heat for steam production. The steam is used to turn turbines to produce electricity.

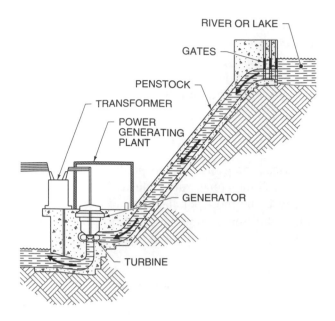

Figure 3-17. Hydroelectric power generating plants use the pressure of falling water to turn turbines. The turbines then turn generators that produce electricity.

Electricity

Electricity is produced at a generating plant at very high voltages. *Voltage* is the force that pushes electricity through an electrical circuit. It is much like water pressure that makes water flow through piping. A *volt* (V) is the unit of measure that expresses the voltage of electric current. High voltage is used for transmitting electricity over long distances. When electricity is produced by a power plant, the voltage is raised by step-up transformers before the electricity enters transmission lines. The voltage is raised from about 18,000 V to 207,000 V. High voltage makes it possible to use smaller cables in transmission lines. The electricity is carried over transmission lines and steel towers. See Figure 3-19.

As transmission lines approach areas where the electricity is needed, the electricity is routed through transmission substations. At this point, step-down transformers reduce the voltage from about 207,000 V to 23,000 V. Smaller distribution lines are routed from transmission substations to distribution substations, where the voltage is reduced from about 23,000 V to 4160 V. As the distribution lines approach buildings, other step-down transformers reduce the voltage to the level needed in a particular building. Most electric companies are publicly or privately owned utilities. The government regulates the pricing and distribution of electricity.

Electrical energy is rated in watts (W) or in kilowatts (kW). A kilowatt is 1000 W. When electricity is used for heating, kilowatts are used to define quantities of heat. When converted to heat, 1 W of electrical energy equals 3.416 Btu. One kilowatt of

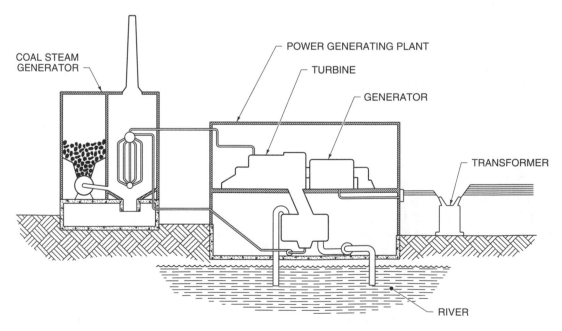

Figure 3-18. Steam used in coal-fired power generating plants is produced by combustion.

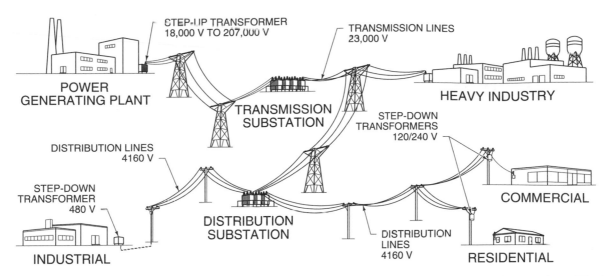

Figure 3-19. High-voltage electricity from power plants is transmitted through electrical lines and substations. The high voltage is reduced by step-down transformers to the voltage needed by the consumer.

electrical energy equals 3416 Btu of heat energy (3.416 × 1000 = 3416). See Appendix.

Compared with fuel oil and gas, coal is used more often for generating electricity. Coal is burned to produce steam that is used to produce electricity. When coal is used, the generating plant is often located near a coal mine, since it is less expensive to transport electricity than it is to transport coal.

Electrical energy produced by nuclear power plants, hydroelectric power plants, and wind power are cleaner forms of energy because they do not emit carbon dioxide or other pollutants to the atmosphere. Hydroelectric power plants and wind power are renewable sources of energy that do not pollute the environment. Nuclear power plants produce large amounts of energy efficiently, but the hazardous waste produced must be handled properly.

Review Questions

1. Define *combustion*.

2. List and explain the three requirements for combustion.

3. List the two main factors considered when selecting a fuel for a particular application.

4. What is the most important physical characteristic of a fuel?

5. What two elements are found in all fuels?

6. How can the heating value for a fuel be found if the chemical composition of the fuel is known?

7. What are the units of measure for solid, liquid, and gas fuels?

8. What two elements make up most of the air?

9. What element in air passes through the combustion process unchanged?

10. How is ignition temperature produced for typical heating units?

11. List the three parts of a flame and explain the process that occurs in each part.

12. What causes a yellow flame?

13. What are the products of combustion when complete combustion occurs?

14. What are the undesirable products of incomplete combustion?

15. What normally causes incomplete combustion?

16. What is soot and what causes it?

17. Explain how the percentage of carbon dioxide in flue gas is used to determine combustion efficiency.

18. What is the most common method of producing electricity for heating and air conditioning purposes?

19. What is the conversion rate for electrical energy to Btu?

20. List and describe three ways of producing electricity without combustion.

Psychrometrics

Psychrometrics is the branch of physics that describes the properties of air and the relationships between them. Properties of air are temperature, humidity, enthalpy, and volume. Psychrometric charts are used to show the relationships between the various properties of air at any condition.

As air is conditioned, one or more of the properties of air change. When one property changes, the others are affected. If any two properties of air are known, the others can be found by using a psychrometric chart.

PROPERTIES OF AIR

Atmospheric air is the mixture of dry air, moisture, and particles. *Dry air* is the elements that make up atmospheric air with the moisture and particles removed. *Moist air* is the mixture of dry air and moisture. The *properties of air* are the characteristics of air, which are temperature, humidity, enthalpy, and volume. The properties of air determine the condition of the air, which is related to comfort.

Temperature

Temperature is the intensity of heat. Temperature is measured with a dry bulb thermometer. A *dry bulb thermometer* is a thermometer that measures dry bulb temperature. *Dry bulb temperature (db)* is the measurement of sensible heat. Temperature controls in HVAC systems turn equipment ON and OFF in response to changes in sensible heat. Dry bulb temperature is expressed in degrees Fahrenheit and degrees Celsius.

Humidity

Humidity is moisture in the air that comes from water that has evaporated into the air. The volume of moisture in the air compared to the total volume of air is small. Humidity is expressed as either relative humidity or humidity ratio.

Relative Humidity. Relative humidity (*rh*) is the amount of moisture in the air compared to the amount it would hold if the air were saturated. Relative humidity is always expressed as a percentage. For example, air that has a relative humidity of 55% holds 55% of the moisture it can hold at the same temperature when it is saturated.

Humidity Ratio. *Humidity ratio (W)* is the ratio of the mass (weight) of the moisture in a quantity of air to the mass of the air and moisture together. Humidity ratio indicates the actual amount of moisture found in the air. Humidity ratio is expressed in grains (gr) of moisture per pound of dry air (gr/lb) or in pounds of moisture per pound of dry air (lb/lb). A *grain* is the unit of measure that equals $\frac{1}{7000}$ lb. For example, 1 lb of air contains 78 gr or .0111 lb of moisture (1 ÷ 7000 × 78 = .0111 lb).

Humidity represents latent heat. Latent heat is heat identified by a change of state and no temperature change. Therefore, latent heat cannot be measured with a thermometer.

A certain quantity of heat is required to change water (liquid) into water vapor (gas). The heat required to make this change is latent heat. Because a certain amount of latent heat is required to evaporate a certain amount of water to water vapor, the amount of moisture (water vapor) in the air represents the amount of latent heat. See Figure 4-1. Heat must be removed from the air for the moisture to condense back to water.

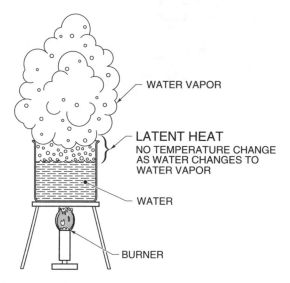

Figure 4-1. Latent heat is identified by the amount of water vapor in the air.

readings indicate the amount of moisture in the air. See Figure 4-2.

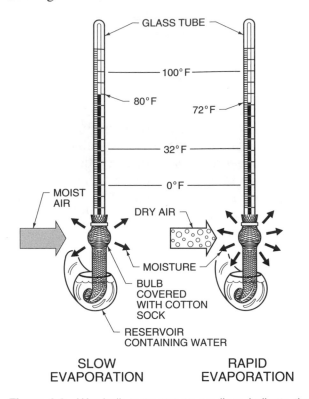

Figure 4-2. Wet bulb temperature readings indicate the amount of moisture in the air. The cotton sock serves as a wick, which keeps the bulb of the thermometer wet.

Wet Bulb Temperature. *Wet bulb temperature* (*wb*) is a measurement of the amount of moisture in the air. Wet bulb temperature is measured with a wet bulb thermometer. A *wet bulb thermometer* is a thermometer that has a small cotton sock placed over the bulb. The end of the sock is placed in a reservoir that contains water. The sock draws water from the reservoir, which keeps the sock and the bulb of the thermometer wet.

A wet bulb thermometer measures humidity by measuring the temperature of the air when the bulb of the thermometer is wet. The bulb is cooled as water evaporates from the sock. Because of this cooling effect, the thermometer reads lower than it would without the sock. The cooling effect is a function of the rate of evaporation of water from the sock. The rate of evaporation is a function of the amount of moisture in the air. The drier the air, the quicker the evaporation. Wet bulb temperature

Dew Point. *Dew point* (*dp*) is the temperature below which moisture in the air begins to condense. Dew point varies with the dry bulb temperature and the amount of humidity in the air. At dew point, air is saturated with moisture and the dry bulb and wet bulb temperatures are the same. Dew point is also known as saturation temperature. On a psychrometric chart, the dew point values are the same as the saturation temperature values.

Humidity Measurement Equipment. A *psychrometer* is an instrument used for measuring humidity that consists of a dry bulb thermometer and a wet bulb thermometer mounted on a common base. A psychrometer measures humidity by comparing the temperature readings on the dry bulb and wet bulb thermometers. *Wet bulb depression* is the difference between the wet bulb and dry bulb temperature readings and is directly related to the amount of moisture in the air. A *sling psychrometer* is an instrument used for measuring humidity that consists of a wet bulb and a dry bulb thermometer mounted on a base. The base is mounted on a handle so it can be rotated rapidly. See Figure 4-3. Air flows over the bulbs of the thermometers when the sling psychrometer is rotated. As the sling psychrometer is rotated, the water on the wet bulb thermometer evaporates. The sling psychrometer is rotated in the air until the temperature reading on each thermometer stabilizes. The readings are taken immediately. Charts or graphs must then be used to find the relative humidity from the two temperature readings. See Appendix.

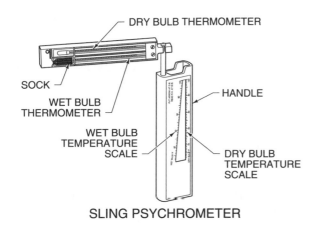

SLING PSYCHROMETER

Figure 4-3. Sling psychrometers measure wet and dry bulb temperature simultaneously. Charts or graphs are required to determine relative humidity.

Humidity can also be measured with a hygrometer. A *hygrometer* is any instrument used for measuring humidity. A psychrometer is one kind of hygrometer. A *dimensional change hygrometer* is a hygrometer that operates on the principle that some materials absorb moisture and change size and shape depending on the amount of moisture in the air. A material in the hygrometer is exposed to the humidity in the air. As the material expands or contracts based on the moisture absorbed or evaporated, a linkage moves an indicator according to the motion of the material. Materials used in dimensional change hygrometers include hair, wood, and plastic.

Electrical impedance hygrometers measure humidity electronically. An *electrical impedance hygrometer* is a hygrometer based on the principle that the electrical conductivity of a substance changes as the amount of moisture in the air

changes. Sensors on the hygrometer are covered with a salt-base substance such as lithium chloride. The electrical conductivity of the substance changes with the amount of moisture in the air. The hygrometer senses the amount of electricity conducted and gives a reading based on that amount. Most hygrometers give direct percentage readings of humidity (relative humidity). See Figure 4-4.

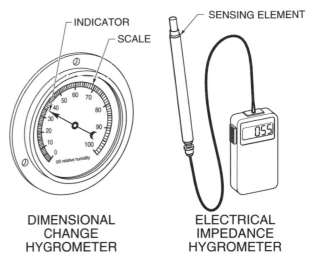

DIMENSIONAL CHANGE HYGROMETER   ELECTRICAL IMPEDANCE HYGROMETER

Figure 4-4. Hygrometers measure and provide readings of relative humidity. Hygrometers are based on the principle that the characteristics of certain materials change with the amount of moisture in the air.

Enthalpy

Any material that has a dry bulb temperature above absolute zero contains heat. *Enthalpy* (*h*) is the total heat contained in a substance, which is the sum of sensible heat and latent heat. Enthalpy of air is expressed in Btu per pound of moist air. The enthalpy of air at different conditions is found on a psychrometric chart.

Volume

Air, like most substances, expands when heated and contracts when cooled. *Volume* (*V*) is the amount of space occupied by a three-dimensional figure. It is expressed in cubic units such as cubic inches or cubic feet. *Specific volume* (*v*) is the volume of a substance per unit of the substance. The specific volume of air is expressed in cubic feet per pound at a given temperature. Because air expands and contracts at different temperatures, air volume varies as temperature varies.

Standard Conditions. The properties of air are related so that a change in one of the properties causes a change in the other three. Properties of air are compared at standard conditions. *Standard conditions* are values used as a reference for comparing properties of air at different elevations and pressures. One pound of dry air and its associated moisture at standard conditions has a pressure of 29.92″ Hg (14.7 psi), temperature of 68°F, volume of 13.33 cu ft/lb, and density of .0753 lb/cu ft. Standard conditions are used when comparing the relationships between the different properties of air such as on a psychrometric chart.

Relationships Between Properties

The properties of moist air are related in such a way that a change in one brings about changes in one or more of the others. The two properties used most often for identifying specific conditions of the air are temperature and humidity.

A change in any property of the air affects other properties. Both dry bulb and wet bulb temperatures must be taken into account when considering the effect of changes in temperature on the other properties of air. Together, dry bulb and wet bulb temperatures affect relative humidity, humidity ratio, dew point, enthalpy, and volume.

A change in wet bulb temperature changes the humidity ratio and the relative humidity of the air. A change in wet bulb temperature indicates that moisture has been added to or removed from the air. Wet bulb temperature, humidity ratio, and relative humidity are all directly related to the amount of moisture in the air. A change in one property changes the others.

A change in both dry bulb and wet bulb temperatures affects enthalpy, which is total heat content. A change in dry bulb temperature indicates a change in sensible heat. A change in wet bulb temperature indicates a change in latent heat.

A change in dry bulb temperature affects the specific volume of the air. If the dry bulb temperature increases, the specific volume increases. If the

dry bulb temperature decreases, the specific volume will decrease.

When considering the effect of changes in humidity on other properties of air, the humidity ratio must be considered. As the humidity ratio increases, the latent heat content of the air also increases. As the humidity ratio decreases, the latent heat content of the air decreases.

PSYCHROMETRIC CHART

A *psychrometric chart* is a graph that defines the condition of the air at various properties. See Figure 4-5. Each of the properties of air is shown on the chart. The properties of air for any condition can be determined by using a psychrometric chart. Psychrometric charts are available for standard conditions and special conditions such as higher- or lower-than-normal pressures or higher- or lower-than-normal temperatures.

Using the Psychrometric Chart

The properties of air found on a psychrometric chart are dry bulb temperature, relative humidity, humidity ratio, wet bulb temperature, dew point, enthalpy, and specific volume. The properties of the air define the condition of the air. See Figure 4-6. See Appendix.

The dry bulb temperature scale is found along the bottom of the chart. Vertical lines on the chart identify the dry bulb temperature of the air at a given point. The dry bulb temperature scale begins at the left side of the chart and increases to the right. On a chart for air at standard conditions, the dry bulb temperature scale begins at about 35°F and extends to about 120°F.

Relative humidity of air is found on the curving lines that run from the bottom left side of the chart up to the right side. Relative humidity is expressed as the percentage of saturation of the air by moisture. The scale begins at 10% at the bottom right of the chart and increases up to the saturation line, which is 100%.

The humidity ratio scale is found on the right side of the chart. The humidity ratio for air is found on the horizontal lines that run across the chart. The humidity ratio scale begins at the bottom of the chart and extends to the top. On a chart for

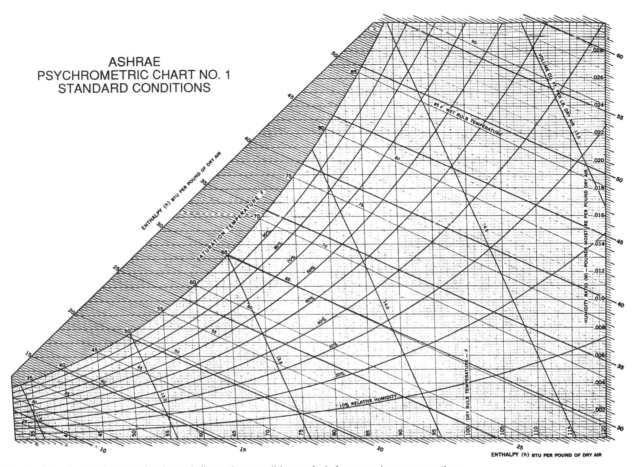

Figure 4-5. A psychrometric chart defines the conditions of air for any given properties.

standard conditions, humidity ratio begins at 0 lb of moisture per pound of dry air and increases to about .030 lb of moisture per pound of dry air.

The wet bulb temperature and dew point scales are found along the curve of the chart. The scale begins at the bottom left and increases as the curve extends up to the right. On a chart for standard conditions, the scale begins at about 25°F and increases to about 90°F. Lines related to wet bulb temperature run from the scale on the curve of the chart diagonally down to the right at about a 35° angle. Any point along one of the wet bulb temperature lines defines the amount of moisture in the air at that point. Lines related to dew point run horizontally from the scale on the curve to the right side of the chart. The scale is the same as that used for wet bulb temperature. The *saturation line* is the curve where the wet bulb temperature and dew point scales begin. Air at any point on this line is saturated with moisture.

The enthalpy scale is located above the saturation line on the curve of the chart. The numbers on the enthalpy scale coincide with lines that extend from the wet bulb temperature lines on the chart. On some charts an enthalpy scale runs below the dry bulb temperature scale, which runs across the bottom of the chart. Enthalpy is found by extending a line through a known point approximately parallel to the wet bulb temperature lines and beyond the saturation line to the enthalpy scale. On a chart for standard conditions, enthalpy begins at about 10 Btu/lb of air at the bottom of the scale and increases to about 60 Btu/lb of air.

Specific volume of air is found on the lines that run at a steep angle from the saturation line to the bottom of the chart. The specific volume scale is shown on the specific volume lines or on an extension of the lines at the bottom of the chart. On a chart for standard conditions, specific volume begins at about 12.5 cu ft/lb of air and increases to about 15.0 cu ft/lb of air.

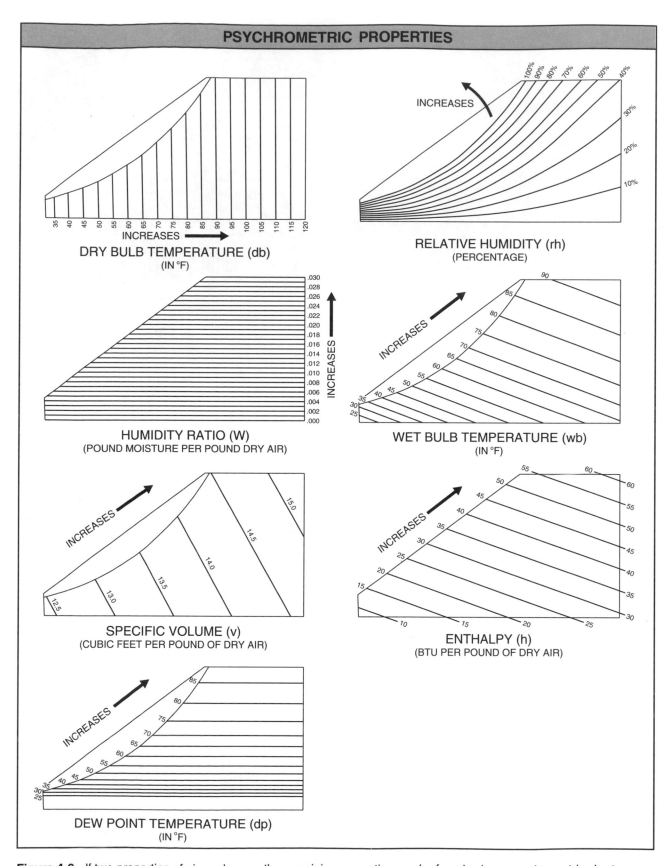

Figure 4-6. If two properties of air are known, the remaining properties can be found using a psychrometric chart.

If any two properties of air are known, the others can be found by locating the point defined by those properties. This is done by applying the procedure:

1. To find the properties of air at 90°F and 40% relative humidity, first locate the 90°F dry bulb temperature line on the scale across the bottom of the chart. Follow this line upward until it crosses the 40% relative humidity line. The point where the two lines intersect defines the condition of the air. This point is used to find the other properties of the air. See Figure 4-7.

PROPERTIES OF AIR

db = Dry bulb temperature
rh = Relative humidity
W = Humidity ratio
wb = Wet bulb temperature
dp = Dew point temperature
h = Enthalpy
v = Specific volume

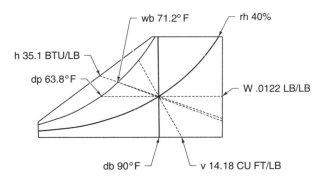

Figure 4-7. All properties of air are found by the intersection of the lines that represent any two properties.

2. To find humidity ratio at this point, follow the horizontal line that runs through the point to the humidity ratio scale on the right side of the chart. The humidity ratio is .0122 lb of moisture per pound of dry air.

3. To find wet bulb temperature, follow the wet bulb temperature line that runs through the point to the wet bulb temperature scale on the curve of the chart. The wet bulb temperature is 71.2°F.

4. To find dew point, follow the horizontal line that runs close to or through the point to the dew point scale on the curve. The dew point is 63.8°F.

5. To find enthalpy, extend the wet bulb temperature line for the point through the curve to the enthalpy scale. The enthalpy at saturation for the point is 35.1 Btu/lb.

6. To find specific volume, locate the specific volume lines on both sides of the point and approximate the specific volume for the point. The specific volume is about 14.18 cu ft/lb.

Example: Finding Properties of Air at a Specific Point

The air in a building has a dry bulb temperature of 75°F and 50% relative humidity. Find the wet bulb temperature, humidity ratio, dew point, enthalpy, and specific volume of the air.

1. Locate the 75°F dry bulb temperature line on the scale across the bottom of the chart. Follow this line upward until it crosses the 50% relative humidity line. The point where the two intersect is the point that defines the condition of the air at the two given properties. See Figure 4-8.

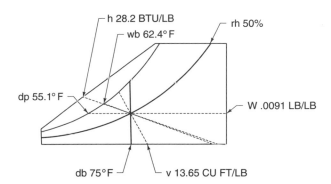

Figure 4-8. Properties of air at 75°F and 50% relative humidity are found at the intersection of the 75°F dry bulb temperature line and the 50% relative humidity line.

2. Find humidity ratio by following the horizontal line that runs through the point to the humidity ratio scale on the right side of the chart. The humidity ratio is .0091 lb of moisture per pound of dry air.

3. Find wet bulb temperature by following the wet bulb temperature line that runs through the point to the wet bulb temperature scale on the curve of the chart. The wet bulb temperature is 62.4°F.

4. Find dew point by following the horizontal line that runs closest to or through the point to the

dew point scale on the curve. The dew point temperature is 55.1°F.

5. Find enthalpy by extending the wet bulb temperature line for the point through the curve to the enthalpy scale. The enthalpy at saturation for the point is 28.2 Btu/lb.

6. Find specific volume by locating the specific volume lines on both sides of the point and approximating the specific volume for the point. The specific volume for the point is approximately 13.65 cu ft/lb.

Applying the Psychrometric Chart

Engineers and air conditioning technicians use psychrometric charts to define properties of moist air during processes that change the properties of air. Heating, cooling, humidification, and ventilation are the most common processes that change the properties of air.

Heating. When moist air is heated, sensible heat is added and there is no change in humidity. If only sensible heat changes, the properties of heated air are determined by horizontal movement between points on the chart. The horizontal movement is the difference between the initial condition and final condition of the air. *Initial condition* is the point that represents the properties of air before it goes through a process. *Final condition* is the point that represents the properties of air after it goes through a process. The point that represents the final condition of the air when sensible heat is added is located directly to the right of the original point on the same humidity ratio line.

The final condition of the air when sensible heat is added is found by applying the procedure:

1. The point that represents the initial condition of the air is located at the intersection of the dry bulb temperature and relative humidity lines on the chart. If air has a dry bulb temperature of 60°F and 40% relative humidity, the intersection of these two lines represents the initial condition of the air. See Figure 4-9.

2. The point that represents the final condition of the air is located directly to the right of the initial

point on the same humidity ratio line where it intersects the final dry bulb temperature line. If sensible heat is added to the air raising it to a temperature of 100°F, the intersection of the humidity ratio line from the initial point and the 100°F dry bulb temperature line represents the final condition of the air.

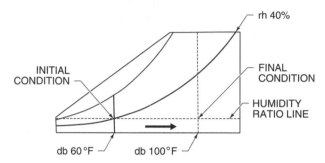

Figure 4-9. The point that represents the condition of air when sensible heat is added is located on the same humidity ratio line at the intersection with the final temperature.

Example: Finding Points — Heated Air

Air enters a furnace at a dry bulb temperature of 75°F and 50% relative humidity and leaves the furnace at a dry bulb temperature of 94°F. Find the point that represents the condition of the air as it leaves the furnace.

1. Locate the point at the intersection of the 75°F dry bulb temperature line and the 50% relative humidity line. This point represents the initial condition of the air. See Figure 4-10.

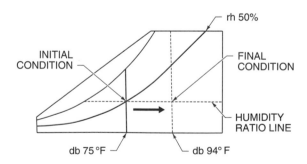

Figure 4-10. To find the point that represents the condition of the heated air, follow the same humidity ratio line to the 94°F dry bulb temperature line.

2. Locate the intersection (to the right) of the humidity ratio line of the initial point with the 94°F dry bulb temperature line. This point represents the final condition of the air.

Cooling. Psychrometric charts are often used to find the total cooling that occurs when air passes through an air conditioning coil. The properties of cooled air are found by locating the initial condition and final condition of the air on the chart and identifying the appropriate properties of the air at each point. The differences between the two conditions indicate the changes made to the air during the process.

Total cooling is found by applying the procedure:

1. The points that represent the initial condition and final condition of the air are located at the intersection of the lines that identify the two given properties. If the air has an initial dry bulb temperature of 80°F and an initial wet bulb temperature of 71°F, the intersection of the two lines (point 1) represents the initial condition of the air. See Figure 4-11. If the air has a final dry bulb temperature of 60°F and a final wet bulb temperature of 50°F, the intersection of the two lines (point 2) represents the final condition of the air.

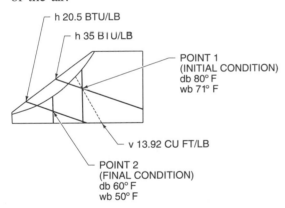

Figure 4-11. Properties of cooled air are found by locating the initial and final enthalpy and initial specific volume of the air on the chart.

2. Enthalpy of the air at the initial condition and final condition is found by using points 1 and 2 and the enthalpy scale on the chart. The enthalpy of the air at the initial condition, point 1, is 35 Btu/lb. The enthalpy at the final condition, point 2, is 20.5 Btu/lb.

3. The amount of heat removed from the air is found by subtracting the enthalpy of the air at the final condition from the enthalpy of the air at the initial condition.

Heat removed = $h_{\text{point 1}} - h_{\text{point 2}}$

Heat removed = 35 − 20.5

Heat removed = **14.5 Btu/lb**

4. Specific volume of the air at the initial condition is found on the chart by locating the specific volume lines on both sides of point 1 and approximating the specific volume for that point. Specific volume at the initial condition is 13.92 cu ft/lb.

5. To find total cooling, first find the weight of the air being cooled per minute (rate of cooling). If the quantity of air being cooled is 11,000 cfm (cubic feet per minute), the weight of the air being cooled per minute is found by dividing 11,000 cfm by the specific volume.

$$Rate\ of\ cooling = \frac{Quantity\ of\ air}{v}$$

$$Rate\ of\ cooling = \frac{11,000}{13.92}$$

Rate of cooling = **790.23 lb/min**

6. Total cooling is found by multiplying the rate of cooling by the amount of heat removed from the air. Because the capacity of cooling equipment is usually rated in Btu per hour, the result is multiplied by 60.

Total cooling = Rate of cooling × Heat removed × 60

Total cooling = 790.23 × 14.5 × 60

Total cooling = **687,500.1 Btu/hr**

Example: Finding Points — Total Cooling

A cooling coil cools 12,500 cfm of air. The air enters the coil at a dry bulb temperature of 74°F and a wet bulb temperature of 63.2°F. The air leaves the coil at a dry bulb temperature of 55°F and a wet bulb temperature of 52.5°F. Find the total cooling as the air passes through the cooling coil.

1. Locate point 1 at the intersection of the 74°F dry bulb temperature line and the 63.2°F wet bulb temperature line. Locate point 2 at the intersection of the 55°F dry bulb temperature line

and the 52.5°F wet bulb temperature line. See Figure 4-12.

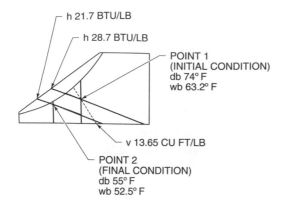

Figure 4-12. Total cooling is the amount of heat removed from air as the air passes through an air conditioning coil.

2. Locate the enthalpy at each point. The enthalpy at point 1 is 28.7 Btu/lb. The enthalpy at point 2 is 21.7 Btu/lb.

3. Subtract the enthalpy of the air leaving the coil from the enthalpy of the air entering the coil. This is the amount of heat removed from the air.

Heat removed $= h_{\text{point 1}} - h_{\text{point 2}}$

Heat removed $= 28.7 - 21.7$

Heat removed $=$ **7.0 Btu/lb**

4. Find the specific volume of the air at initial condition (from the chart). The specific volume is 13.65 cu ft/lb.

5. Find the rate of cooling.

$$Rate \ of \ cooling = \frac{Quantity \ of \ air}{v}$$

$$Rate \ of \ cooling = \frac{12,500}{13.65}$$

Rate of cooling $=$ **915.75 lb/min**

6. Find the total cooling.

Total cooling $=$ *Rate of cooling* $\times$ *Heat removed* $\times$ 60

Total cooling $= 915.75 \times 7.0 \times 60$

Total cooling $=$ **384,615 Btu/hr**

Humidification. The humidity in a building is often too low for comfort or for a process. The quantity of water that must be added to air to increase the relative humidity is found by applying the procedure:

1. The points that represent the initial condition and final condition of the air are located at the intersection of the lines that identify the two given properties. See Figure 4-13. If air has a dry bulb temperature of 68°F and a relative humidity of 45%, the point that represents the initial condition of the air is found at the intersection of the two lines (point 1). To raise the relative humidity to 57%, a certain quantity of water must be added to the air. The point that represents the desired final condition of the air (point 2) is found at the intersection of the 57% relative humidity line and the same dry bulb temperature line (68°F).

2. Humidity ratio at these two points is found by following the horizontal lines that run through the points to the scale on the right side of the chart. The humidity ratio at point 1 is .0066 lb/lb and the humidity ratio at point 2 is .0086 lb/lb.

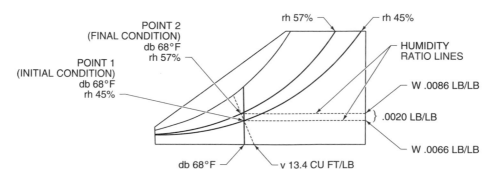

Figure 4-13. Properties of moist air are determined by finding the initial and final humidity ratio and the initial specific volume of the air on the chart.

3. The amount of water that must be added to each pound of air to increase the relative humidity of the air from point 1 to point 2 is found by subtracting the humidity ratio of point 2 from the humidity ratio of point 1.

Humidity ratio difference $= W_{point\ 2} - W_{point\ 1}$

Humidity ratio difference $= .0086 - .0066$

Humidity ratio difference $=$ **.0020 lb/lb**

4. To find the amount of water to be added to the air in a building, first determine the volume of air in the building. If the dimensions of a building are $60' \times 50' \times 10'$, the volume of the building is found by applying the formula:

$V = l \times w \times h$

$V = 60 \times 50 \times 10$

$V =$ **30,000 cu ft**

5. The total weight of the air in the building is found by dividing the volume of the air by the specific volume of the air at point 1 (initial condition). Specific volume at this point is 13.4 cu ft/lb.

Total weight of air $= \dfrac{V}{v_{point\ 1}}$

Total weight of air $= \dfrac{30,000}{13.4}$

Total weight of air $=$ **2238.8 lb**

6. The weight of the water to be added is found by multiplying the total weight of the air by the humidity ratio difference, which is .0020 lb of water per pound of air.

Weight of water to be added =
   Total weight of air × Humidity ratio difference

Weight of water to be added $= 2238.8 \times .0020$

Weight of water to be added $=$ **4.48 lb**

7. To find the number of gallons of water to be added to the air, divide the weight of the water by 8.345, which is the weight of 1 gal. of water.

Gallons of water $= \dfrac{Weight\ of\ water\ be\ added}{8.345}$

Gallons of water $= \dfrac{4.48}{8.345}$

Gallons of water $=$ **.537 gal.**

Example: Finding Points — Humidification

The air in a building has a dry bulb temperature of 72°F and 40% relative humidity. The building is 45′ long, 37′ wide, and has a 10′ high ceiling. Find the quantity of water required to raise the relative humidity to 55%.

1. Locate point 1 at the intersection of the 72°F dry bulb temperature line and the 40% relative humidity line. Locate point 2 at the intersection of the 72°F dry bulb temperature line and the 55% relative humidity line. See Figure 4-14.

2. Locate the humidity ratio of the air at each point. Humidity ratio at point 1 is .0066 lb of moisture per pound of air and the humidity ratio at point 2 is .0092 lb of moisture per pound of air.

3. Find the humidity ratio difference.

Humidity ratio difference $= W_{point\ 2} - W_{point\ 1}$

Humidity ratio difference $= .0092 - .0066$

Humidity ratio difference $=$ **.0026 lb/lb**

4. Find the volume of building.

$V = l \times w \times h$

$V = 45 \times 37 \times 10$

$V =$ **16,650 cu ft**

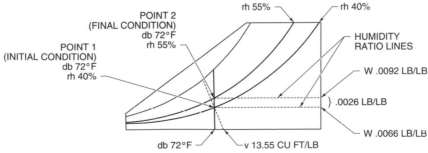

Figure 4-14. The amount of water to be added to the air is the amount that will change the condition of the air to the desired humidity.

5. Find the total weight of the air in building.

$$Total\ weight\ of\ air = \frac{V}{v_{point\ 1}}$$

$$Total\ weight\ of\ air = \frac{16,650}{13.55}$$

Total weight of air = **1228.78 lb**

6. Find the weight of water to be added.

Weight of water to be added =
  Total weight of air × Humidity ratio difference

Weight of water to be added = 1228.78 × .0026

Weight of water to be added = **3.19 lb**

7. Find the number of gallons of water.

Gallons of water =

$$\frac{Weight\ of\ water\ to\ be\ added}{8.345}$$

$$Gallons\ of\ water = \frac{3.19}{8.345}$$

Gallons of water = **.382 gal.**

Air Mixtures. Air conditioning systems use ventilation to mix makeup air from outdoors with return air from building spaces. The condition of the air mixture is found by applying the procedure:

1. The point that represents the condition of the makeup air is located at the intersection of the two given properties. If makeup air is at a dry bulb temperature of 60°F and 65% relative humidity, the intersection of these two lines represents the condition of the makeup air. See Figure 4-15.

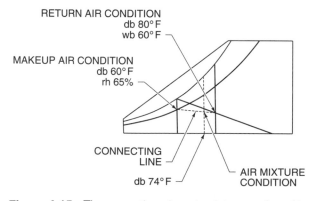

RETURN AIR CONDITION
db 80°F
wb 60°F

MAKEUP AIR CONDITION
db 60°F
rh 65%

CONNECTING
LINE

db 74°F

AIR MIXTURE
CONDITION

Figure 4-15. The properties of an air mixture are found by locating a point on a line drawn between the makeup air condition and the return air condition.

2. The point that represents the condition of the return air is located at the intersection of the two given properties. If return air is at 80°F dry bulb temperature and 60°F wet bulb temperature, the intersection of these two lines represents the condition of the return air.

3. Draw a line on the chart that connects the point that represents the condition of the makeup air and the point that represents the condition of the return air (connecting line).

4. Multiply the percentage of makeup air and return air by the corresponding dry bulb temperature. The resulting values are the temperature ratio of the makeup air and return air. If air consists of 30% makeup air and 70% return air, the temperature ratio of the makeup air and return air is determined by applying the formula:

Temperature ratio = db × Percentage of air

For makeup air:

Temperature ratio = db × Percentage of air

Temperature ratio = 60 × .3

Temperature ratio = **18°F**

and

For return air:

Temperature ratio = db × Percentage of air

Temperature ratio = 80 × .7

Temperature ratio = **56°F**

5. Find the dry bulb temperature of the air mixture by adding the temperature ratios.

Air mixture db = Makeup air temperature ratio
  + Return air temperature ratio

Air mixture db = 18 + 56

Air mixture db = **74°F**

6. The point that represents the condition of the air mixture is found at the intersection of the 74°F dry bulb temperature line and the connecting line. This point represents the condition of the makeup air and the return air mixture.

Example: Finding Points — Air Mixture

Twenty percent makeup air at 90°F dry bulb temperature and 70% relative humidity is mixed with 80% return air at 75°F dry bulb temperature and

65°F wet bulb temperature. Find the final condition of the air mixture.

1. Locate the point that represents the condition of the makeup air at the intersection of the 90°F dry bulb temperature line and the 70% relative humidity line. See Figure 4-16.

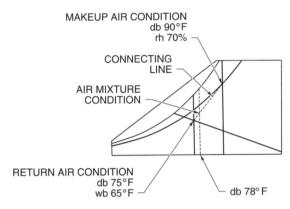

Figure 4-16. The condition of the air mixture is located at the intersection of the 78°F dry bulb temperature line and the connecting line.

2. Locate the point that represents the condition of the return air at the intersection of the 75°F dry bulb temperature line and the 65°F wet bulb temperature line.

3. Draw a line on the chart that connects the point that represents the condition of the makeup air and the point that represents the condition of the return air (connecting line).

4. Find temperature ratio for makeup air and return air.

   For makeup air:

   Temperature ratio = db × Percentage of air

   Temperature ratio = 90 × .2

   Temperature ratio = **18°F**

   For return air:

   Temperature ratio = db × Percentage of air

   Temperature ratio = 75 × .8

   Temperature ratio = **60°F**

5. Find the dry bulb temperature of the air mixture.

   Air mixture db = Makeup air temperature ratio + Return air temperature ratio

   Air mixture db = 18 + 60

   Air mixture db = **78°F**

6. The point that represents the condition of the air mixture is found at the intersection of the 78°F dry bulb temperature line and the connecting line. This point represents the condition of the makeup air and the return air mixture.

Review Questions

1. Define *temperature*.

2. List the four properties of air.

3. List and explain the two ways humidity is expressed.

4. What does wet bulb temperature indicate?

5. Define *sensible heat*.

6. Define *moist air*.

7. What kind of heat does moisture in the air represent? Explain.

8. What is the relative humidity of air that holds 35% of the moisture it can hold at a given pressure and temperature?

9. How does a wet bulb thermometer work?

10. Define *hygrometer*. List and describe two kinds.

11. What effect does an increase in dry bulb temperature have on the specific volume of air?

12. List the seven properties of air identified on a psychrometric chart.

13. What property of air identifies total heat content of air?

14. Where is the dry bulb temperature scale found on a psychrometric chart?

15. What three properties of air are identified by the scale that runs along the curve of a psychrometric chart?

16. How is specific volume found on a psychrometric chart?

17. Where is the enthalpy scale found on a psychrometric chart?

18. What is the wet bulb temperature of the air when the dry bulb temperature is 56°F and the relative humidity is 60%?

19. What is the relative humidity when the dry bulb temperature is 93°F and the wet bulb temperature is 76°F?

20. How much water (in gallons) must be added to air at 82°F dry bulb temperature and 65°F wet bulb temperature to change it to 55% relative humidity at the same dry bulb temperature?

Forced-air Heating Systems

Forced-air heating systems consist of a heating unit, distribution system, and controls. Supply air is heated by a furnace and distributed to building spaces by a blower through supply air ductwork. Air is returned to the furnace through return air ductwork. Forced-air heating systems are used in small- to medium-size buildings where the ductwork is not long. Controls operate the individual components of a heating unit. Some special-purpose heating units do not require ductwork. The air in a special-purpose heating unit is heated and distributed within a specific building space such as a garage or warehouse.

FORCED-AIR HEATING SYSTEMS

Forced-air heating systems use air to carry heat. The air is heated and distributed through a building to control the temperature in the building spaces. The major parts of a forced-air heating system are the heating unit, distribution system, and controls. Different types of forced-air heating systems are used for different applications. The kind of system used for a particular application depends on the amount of heat required and the physical layout of the building. The two main types of forced-air heating systems are the central forced-air and modular forced-air heating system.

Central Forced-air Heating System

A *central forced-air heating system* uses a centrally located furnace to produce heat for a building. Supply air ductwork runs from the heating unit to the building spaces. Return air ductwork runs from the building spaces back to the heating unit.

Modular Forced-air Heating System

A *modular forced-air heating system* uses more than one heating unit to produce heat for a building. Each heating unit produces heat for a module or zone of the building. A *zone* is a specific section of a building that requires separate temperature control. Heating units in modular forced-air heating systems are usually located on roofs.

FURNACES

The furnace is the central element in a forced-air heating system. A *furnace* is a self-contained heating unit that includes a blower, burner(s), heat exchanger or electric heating elements, and controls. See Figure 5-1. Furnaces are available in different sizes and styles for a variety of applications. Forced-air furnaces are categorized by direction of air flow, fuel or energy used, dimensions, and heating capacity.

Furnaces are categorized based on the direction of air flow out of the furnace. The three most common styles of furnaces are upflow, horizontal, and downflow. An *upflow furnace* is a furnace in which heated air flows upward as it leaves the furnace. Return air enters through or near the bottom of the upflow furnace and exits out of the top. Upflow furnaces are used where the ductwork is located above the furnace such as where the furnace is located in a basement. See Figure 5-2.

A *horizontal furnace* is a furnace in which heated air flows horizontally as it leaves the furnace. Return air enters horizontally on one end of the furnace, and supply air exits horizontally on the other end. Horizontal furnaces are used where headroom is limited such as in attics or crawl spaces of buildings. The ductwork of a horizontal furnace is usually located on the same level as the furnace.

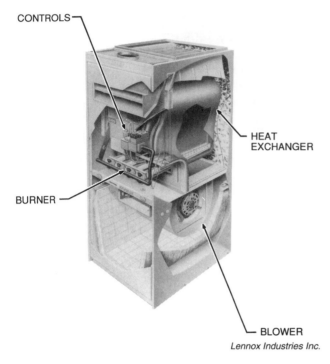

Lennox Industries Inc.

Figure 5-1. A furnace is the central element in a forced-air heating system.

A *downflow furnace* is a furnace in which heated air flows downward as it leaves the furnace. Return air enters through the top of the furnace and supply air exits out of the bottom. Downflow furnaces are used where the ductwork is located below the furnace, where the furnace is located on one floor and the ductwork is installed in the ceiling space of the floor below, or where ductwork is located under a concrete floor.

Furnaces produce heat by either combustion or electrical energy. The heat in a combustion furnace is produced by burning fuel, which may be coal, wood, fuel oil, or gas fuel. Many furnaces, such as coal-burning and wood-burning furnaces, are made for burning a specific kind of fuel, but some furnaces can burn any solid fuel. Fuel oil-burning furnaces usually burn Grade No. 2 fuel oil. Fuel oil-burning furnaces are designed for use in residential and small commercial buildings. Gas fuel-burning furnaces burn natural or LP gas. Some gas fuel-burning furnaces can burn both kinds of gas fuel. Natural gas-burning furnaces are used in areas where natural gas pipelines are in place and the fuel is available. LP gas-burning furnaces are used where the fuel must be transported to the point of use in tanks. The differences between natural gas- and LP gas-burning furnaces are the size of the orifices (openings) on the burner(s) and the pressure of the gas fuel.

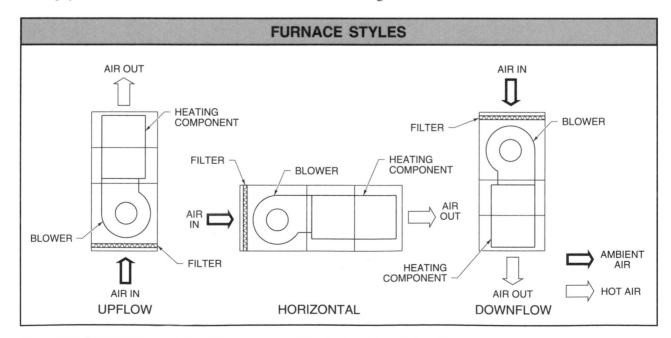

Figure 5-2. Three common styles of furnaces are upflow, horizontal, and downflow.

Electric furnaces produce heat as electricity flows through resistance heating elements. Air is heated as it passes by the hot elements. National safety codes must be followed when installing any kind of furnace. Check jurisdictional codes for further regulations.

The dimensions of a furnace depend on the size and arrangement of the components in the furnace. Furnaces have small components that are arranged to fit close together. A furnace used to heat a residence or small commercial building may be 32″ wide, 28″ deep, and 54″ high. Manufacturer's specification sheets are used to find the characteristics of a furnace, which are dimensions, input rating, output rating, and efficiency. See Figure 5-3.

Input rating is the amount of heat in Btu produced as a fuel burns. Input rating is found by multiplying the heating value of the fuel by the flow rate. Heating value is the amount of heat produced per unit of fuel in Btu per hour. The heating value of a fuel depends on the chemical makeup of the fuel. *Flow rate* is the rate at which a furnace burns fuel. A furnace produces a given amount of heat for each unit of fuel burned. Input rating is found by applying the formula:

$$IR = HV \times Q$$

where

IR = input rating (in Btu/hr)

HV = heating value (in Btu/cu ft)

Q = flow rate (in cu ft/hr)

Example: Finding Input Rating

A natural gas-fired furnace burns 100 cu ft of gas fuel per hour. Find the input rating of the furnace. *Note:* The heating value for natural gas is 1000 Btu/cu ft.

$$IR = HV \times Q$$

$$IR = 1000 \times 100$$

$$IR = \textbf{100,000 Btu/hr}$$

Output rating (heating capacity) of a furnace is the amount of heat in Btu the furnace will produce in 1 hour. Output rating is found by multiplying the input rating by the efficiency rating of the furnace, which is provided by the manufacturer. *Efficiency rating* is the comparison of the furnace input rating with the output rating. Efficiency rating is the evaluation of how well a furnace burns fuel. Output rating is found by applying the formula:

$$OR = IR \times ER$$

where

OR = output rating (in Btu/hr)

IR = input rating (in Btu/hr)

ER = efficiency rating (in percent)

Example: Finding Output Rating

A furnace with an input rating of 100,000 Btu/hr has an efficiency rating of 80%. Find the output rating for the unit.

UPFLOW GAS FURNACE CHARACTERISTICS (STANDING PILOT)							
Model Number	Dimensions*			Input Rating§	Output Rating§	Shipping Weight#	AFUE**
	D	W	H				
005-AX-10	28.5	14	46	40,000	30,000	100	75
010-AX-10	28.5	14	46	44,000	35,000	105	79
050-AX-6	28.5	18	46	50,000	38,000	120	76
070-AX-5	28.5	20	46	75,000	58,000	120	77
100-AX-3	28.5	20	46	100,000	75,000	166	75
200-AX-3	28.5	20	46	125,000	97,000	178	78
300-AX-6	28.5	24	46	132,000	105,000	205	80
400-AX-0	28.5	24	46	150,000	127,000	255	85

* in inches
§ in Btu/hr
# in lb
**Annual fuel utilization efficiency (in percent)

Figure 5-3. Furnace characteristics are found on manufacturer's specification sheets.

$OR = IR \times ER$

$OR = 100,000 \times .80$

$OR = \mathbf{80,000\ Btu/hr}$

Efficiency for a typical furnace is about 80%. Twenty percent of the heat produced rises up the flue. This heat produces draft in the stack, which is necessary for proper furnace operation. *Draft* is the movement of air across a fire and through a heat exchanger. Because modern condensing furnaces allow less heat for producing a draft, these furnaces operate with higher efficiency ratings.

Components of a furnace are the major pieces of equipment that make up the furnace. Combustion furnace components include a cabinet, blower, burner(s), heat exchanger, and filter. Electric furnace components include a cabinet, blower, resistance heating element(s) in place of the burner(s) and heat exchanger, and filter. These components are arranged differently, but all of these components are found in most furnaces. See Figure 5-4.

Cabinets

A *cabinet* is a sheet metal enclosure that completely covers and provides support for the components of a furnace. Cabinets are painted with corrosion-resistant paint to make them attractive and protect them from rust and deterioration. Cabinets for small furnaces are made of heavy-gauge sheet metal, which provides support. Larger furnaces have a frame within the

sheet metal cabinet, which supports heavier components and adds strength to the furnace.

Furnace cabinets completely enclose the other components except where the return air and supply air duct connections are made. Louvered access panels on the front of the cabinet allow combustion air flow to the burner(s) and enclose the burner vestibule, controls, and blower compartment. The *burner vestibule* is the area where the burner(s) and controls are located. See Figure 5-5.

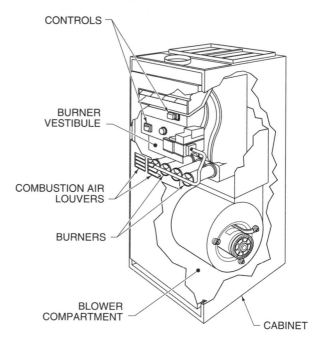

Figure 5-5. The burner vestibule is the area where the burner(s) and controls are located. Louvered access panels on the cabinet allow combustion air to flow to the burners.

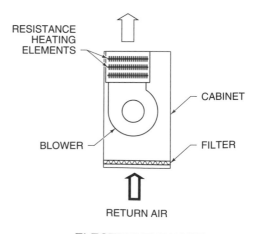

COMBUSTION FURNACE   ELECTRIC FURNACE

Figure 5-4. Combustion and electricity are used to produce heat for furnaces.

Blowers

A *blower* is a mechanical device that consists of moving blades or vanes that force air through a venturi. A *venturi* is a restriction that causes increased pressure as air moves through it. The blower in a forced-air heating system is the component that moves air through the heat exchanger or the resistance heating element and through the ductwork to building spaces. Supply and return air blowers are used to move air in large duct systems.

A blower may be a part of the furnace or a separate part within a forced-air heating system. Blowers are available in different sizes for different applications. Blowers used in forced-air systems include propeller fans, centrifugal blowers, and axial flow blowers. See Figure 5-6.

Propeller Fan. A *propeller fan* is a mechanical device that consists of blades mounted on a central hub. The hub may be mounted directly on the shaft of a motor or may be turned by a motor with a pulley and belt arrangement. The blades rotate within an opening in a plate. As the fan hub rotates, the blades move rapidly through the air. The angle at which the blades are mounted moves air through the opening. Propeller fans are used mostly in applications where there is no ductwork. These fans are most efficient when used in free air or against little or no resistance. Propeller fans are often used as blowers for outdoor condensers.

Centrifugal Blower. A *centrifugal blower* consists of a scroll, blower wheel, shaft, and inlet vanes. The *scroll* is a sheet metal enclosure that surrounds the blower wheel. The *blower wheel* is a sheet metal cylinder with curved vanes along its perimeter. The blower wheel rotates on the blower shaft. *Inlet vanes* are adjustable dampers that control the air flow to the blower. As the blower wheel rotates on its shaft, a low-pressure area is created at the center of the wheel. Return air is drawn into the air inlet and

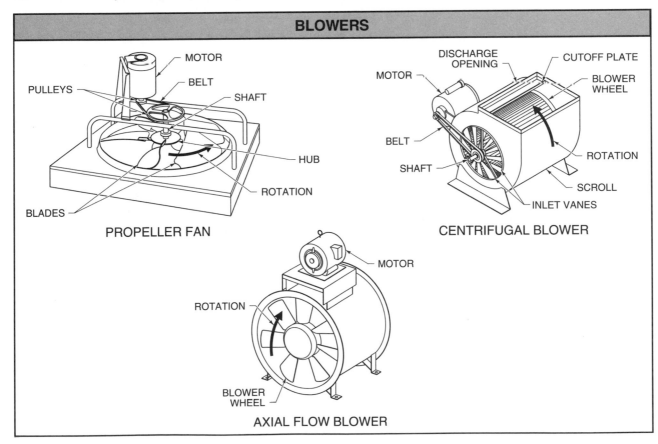

Figure 5-6. Propeller fans, centrifugal blowers, and axial flow blowers are the three types of blowers used to move air in forced-air systems.

through the inlet vanes to the center of the wheel. Air passes through the vanes of the blower wheel and is thrown off by centrifugal force through the discharge opening in the scroll. *Centrifugal force* is the force that pulls a body outward when it is spinning around a center. A cutoff plate directs the air out of the discharge opening after the air passes through the vanes of the blower wheel.

Most centrifugal blowers have forward-curved blower wheels. Forward-curved blower wheels have vanes that are inclined in the direction of the wheel rotation, which is the direction of the air flow. Blower wheels that have vanes inclined in the opposite direction of the air flow are backward-curved blower wheels. Backward-curved blower wheels produce greater pressure. See Figure 5-7.

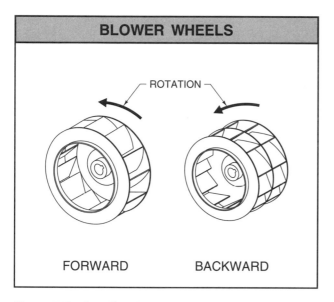

BLOWER WHEELS

ROTATION

FORWARD   BACKWARD

Figure 5-7. Centrifugal blower wheels may be forward- or backward-curved. Backward-curved blower wheels produce slightly greater pressure.

Centrifugal blowers are used in low-pressure and many medium- to high-pressure forced-air heating systems. Centrifugal blowers operate with high efficiency at pressures common in most small- to medium-size duct systems. Pressure is expressed in inches of water column. *Water column (WC)* is the pressure required to raise a column of water a given height. See Appendix. Low-pressure forced-air systems have pressures of from −.5″ WC to 2″ WC and medium- to high-pressure forced-air systems have pressures of from 3″ WC to 10″ WC.

Axial Flow Blower. An *axial flow blower* is a blower that contains a blower wheel, which works like a turbine wheel. The blower wheel is mounted on a shaft with its axis parallel to the air flow. The wheel turns at high speed. The angle of the blades moves air by compressive and centrifugal force. *Compressive force* is the force that squeezes air together. Axial flow blowers are used in medium- to high-pressure forced-air heating systems. Axial flow blowers are more efficient in high-pressure applications than other kinds of blowers. High-rise buildings often have high-pressure duct systems that use axial flow blowers.

Blower motors are electric motors that provide the mechanical power for turning blower wheels. A *blower drive* is the connection from an electric motor to a blower wheel, which is a motor-to-wheel connection. Two kinds of motor-to-wheel connections used with blowers are belt drive and direct drive systems.

A *belt drive system* is a motor-to-wheel connection that has a blower motor mounted on the scroll. The blower motor is connected to the blower wheel through a belt and sheave arrangement. A *sheave* is a pulley, which is a grooved wheel. One or more sheaves are mounted on the motor shaft and the blower wheel shaft. V belts are closed-looped belts made of rubber, nylon, polyester, and rayon. They are used to transmit power from the motor shaft to the blower wheel shaft. See Figure 5-8.

The speed of the blower wheel, in revolutions per minute (rpm), determines the volume of air that will flow through a blower. The speed of the blower wheel is indirectly proportional to the diameters of the sheaves used in the system. In a belt drive system, an adjustable blower sheave makes blower wheel speed adjustments. The outer flange of the sheave is threaded on the hub of the inner flange. With this arrangement, the distance between the two flanges can be changed. The V belt rides the outer part of the sheave if the flanges are close together and the inner part of the sheave if the flanges are far apart. The speed of a belt drive blower is found by applying the formula:

$$N_b = \frac{N_m \times PD_m}{PD_b}$$

where

N_b = speed of blower wheel (in rpm)

N_m = speed of motor (in rpm)

PD_m = diameter of motor sheave (in inches)

PD_b = diameter of blower sheave (in inches)

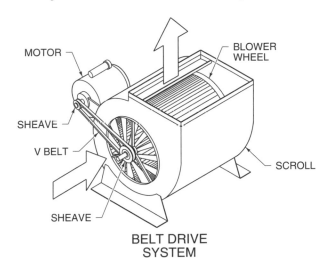

BELT DRIVE
SYSTEM

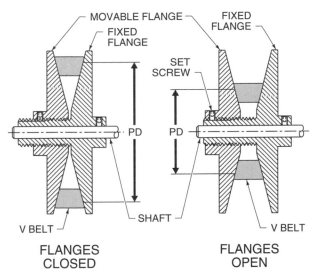

Figure 5-8. A belt drive system is a motor-to-wheel connection that has a blower motor connected to the blower wheel through a belt and sheave arrangement. The speed of the blower is changed by changing the diameter of the sheaves.

Example: Finding Belt Drive Blower Speed

A belt drive blower has a motor speed of 1725 rpm, a 3″ motor sheave, and a 7″ blower sheave. Find the speed of the blower.

$$N_b = \frac{N_m \times PD_m}{PD_b}$$

$$N_b = \frac{1725 \times 3}{7}$$

$$N_b = \frac{5175}{7}$$

$N_b =$ **739.29 rpm**

A *direct drive system* is a motor-to-wheel connection that has a blower wheel mounted directly on the motor shaft. The blower wheel turns as the motor turns the shaft. The motor is mounted in the center of the air inlet of the blower. A direct drive blower is normally connected to a multispeed motor. Changing the speed (rpm) of the blower changes the speed of the motor. Large blowers use four or five speed motors and small blowers use two or three speed motors. See Figure 5-9.

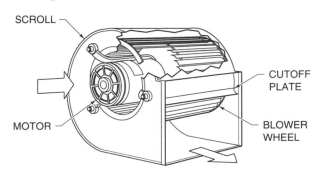

DIRECT DRIVE SYSTEM

Figure 5-9. A direct drive system is a motor-to-wheel connection that has a blower wheel mounted directly on the motor shaft. Changing the speed of the blower is accomplished by changing the speed of the motor.

The performance characteristics of a blower are horsepower (HP), speed, volume, and static pressure. These characteristics are found on blower performance charts. *Static pressure* is pressure that acts through weight only with no motion. Static pressure is expressed in inches of water column. Blower performance charts are developed for individual blowers or families of blowers (blowers with similar characteristics). Blower performance charts illustrate the performance characteristics of a particular type and size of blower in cubic feet per minute, static pressure, revolutions per minute, and horsepower. For example, a blower with a 1 HP motor has a speed of 1733 rpm and moves 1350 cfm of air to produce 2.5″ WC static pressure. See Figure 5-10.

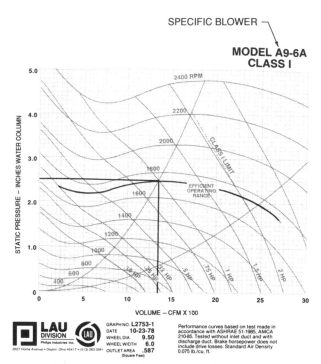

BLOWER PERFORMANCE CHART

Lau, a Division of Tomkins Industries

Figure 5-10. Blower performance charts show performance characteristics, which are horsepower, speed, volume, and static pressure.

Burners

A *burner* is the heat producing component of a combustion furnace. Some furnaces use one large burner while others use many smaller burners. A burner mixes air and fuel to provide a combustible mixture, supplies the air-fuel mixture to the burner face where combustion takes place, and meters fuel to maintain a controlled firing rate. Atmospheric burners, power burners, and low excess air burners are three types of burners.

Atmospheric Burners. *Atmospheric burners* are burners that use ambient air supplied at normal atmospheric air pressure for combustion air. *Ambient air* is unconditioned atmospheric air. Atmospheric burners burn either gas fuel or fuel oil to produce heat.

A *gas fuel-fired atmospheric burner* is a burner that mixes ambient air with a gas fuel to create a flame. The gas is directed through a manifold to burner tubes. A *manifold* is a pipe that has outlets for connecting other pipes. A *burner tube* is a tube that has an opening on one end and burner ports located along the top. A spud (fitting) that contains a small orifice is located in the outlets of the manifold. An *orifice* is a precisely sized hole through which gas fuel flows. The gas fuel flows from the manifold through the spud and orifice into the burner tubes. See Figure 5-11.

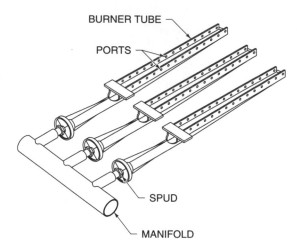

Figure 5-11. Gas fuel flows from the manifold through the spud into the burner tubes in a gas fuel-fired atmospheric burner.

A burner tube has an adjustable shutter, which is the primary air inlet. The gas fuel enters the burner tube at a high velocity, which draws air in through the adjustable shutter. See Figure 5-12.

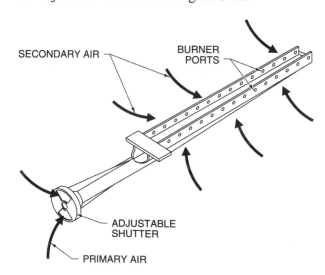

Figure 5-12. Primary air is drawn into the burner through the adjustable shutter. Secondary air is drawn into the flame at the burner ports.

Primary air mixes with the gas and produces a combustible air-fuel mixture. The air-fuel mixture passes through the burner tube and out the burner ports. The air-fuel mixture is ignited at the burner ports with a pilot light or an electric spark. Secondary air is drawn into the flame at the burner ports.

A *pilot burner* is a small burner located near the burner tubes. The pilot burner produces a pilot light, which is a small standing flame. The pilot light ignites the air-fuel mixture when the gas fuel valve opens. The pilot light may be a standing light (one that burns constantly) or may be lighted electrically on each call for heat.

An *electric spark igniter* is a device that produces an electric spark. The electric spark is used to ignite either a pilot burner or main burner. On a call for heat, an electrode is energized near the pilot burner or burner face. When the flame is established, the electrode igniter is de-energized until the next call for heat. Pilot burners and electric spark igniters are controlled by combustion safety control systems. See Figure 5-13.

The flow of fuel to a gas fuel-fired atmospheric burner is controlled by a gas fuel valve. A *gas fuel valve* is a 100% shutoff safety valve that controls the flow of fuel to the main burner and the pilot burner. If combustion does not occur on a call for heat, the valve will close and will not allow gas fuel to flow to the main burner nor to the pilot burner. A combination valve is a modern valve that has a built-in fuel pressure regulator, which regulates the pressure of the gas fuel that enters the burner. See Figure 5-14.

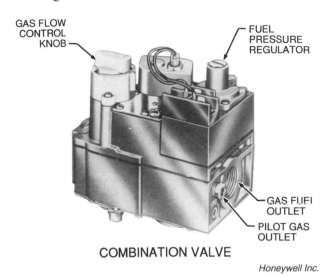

COMBINATION VALVE

Honeywell Inc.

Figure 5-14. A combination valve controls the flow of gas fuel to the main burner and the pilot burner.

The heat output of an atmospheric burner depends on the size of the burner and orifice and the pressure of the fuel. Most natural gas burners are designed to operate with pressure of 3.5″ WC. LP gas burners are designed to operate with pressure of 10.5″ WC. The holes in the orifice are sized to provide the amount of fuel that will produce the rated heat output for the particular burner.

A *fuel oil-fired atmospheric burner* is a burner that consists of an open pot into which fuel oil flows at a controlled rate. A carburetor (metering device)

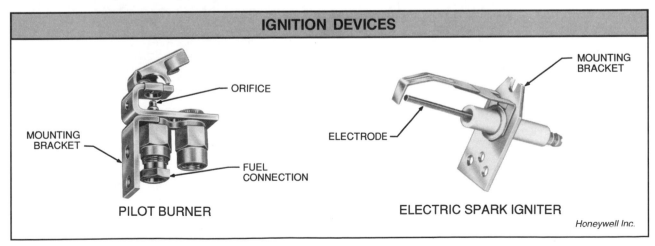

Figure 5-13. The air-fuel mixture in a gas-fuel fired atmospheric burner is ignited at the burner ports by a pilot burner or an electric spark igniter.

controls the rate of the flow of fuel oil from a reservior into a pot burner. A *float valve* is a hollow ball that floats on the liquid in a reservoir and closes the valve when the liquid reaches a certain level. The float valve is adjusted to maintain a small quantity of oil in the reservoir. The oil flows from the reservoir into the pot at a controlled rate.

Most fuel oil-fired atmospheric burners, which are also known as pot burners, must be ignited manually. After a flame is established, heat from the fire vaporizes oil in the pot and the vapors burn at the burner ring (burner face) on the pot. These burners are used in space heaters and small, manually controlled heating units.

Power Burners. *Power burners* use a fan or blower to supply and control combustion air. Air and fuel are introduced under pressure at the burner face. Two basic kinds of power burners are gas fuel and fuel oil power burners.

A *gas fuel power burner* uses natural or LP gas and contains a fan or blower, which is located on the outside of the combustion chamber. The blower provides combustion air, which is blown into the combustion chamber under pressure through a perforated bulkhead. See Figure 5-15.

through the holes in the perforated bulkhead. Combustion air and the fuel mix to provide a combustible air-fuel mixture. The proper air-fuel mixture is attained by using a properly sized orifice in the gas fuel line and by adjusting the combustion air blower. An electric spark igniter starts combustion.

Gas fuel power burners are used in commercial gas fuel furnaces with an output greater than 240,000 Btu/hr. Combustion air is easily controlled with a power burner.

A *fuel oil power burner* is a burner that atomizes fuel oil. A fuel oil power burner consists of a fuel oil pump, burner assembly, and combustion air blower. See Figure 5-16.

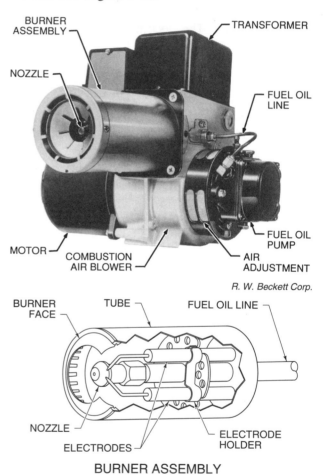

R. W. Beckett Corp.

Figure 5-16. The oil in a fuel oil power burner is atomized as it flows through the nozzle. Combustion air mixes with the oil at the burner face.

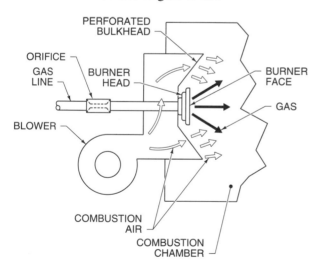

Figure 5-15. Combustion air in a gas fuel power burner is provided by a fan or blower.

Gas fuel flows from a gas fuel line through an orifice at a regulated pressure to the burner head, which is located inside the perforated bulkhead. Gas fuel flows through the burner head as air flows

The fuel oil pump draws oil from a storage tank and pumps it through the burner assembly at about 100 psi. The burner assembly consists of a tube that holds a fuel oil line, nozzle, and electrodes. Combustion air is forced through the tube by the combustion air blower. The fuel oil line carries oil from the pump to the nozzle. The nozzle releases atomized fuel oil, which mixes with combustion air at the burner face. The electrodes produce an electric spark that ignites the air-fuel mixture. The fuel oil pump and the combustion air blower are usually driven by the same electric motor.

Fuel oil power burners are available in many sizes. Different burners burn different grades of fuel oil. The smallest burners have output of 50,000 Btu/hr. The heating capacity of a fuel oil power burner depends on the fuel oil pressure and size of the nozzle. Nozzles are sized in gallons per hour (gph) of fuel oil flow.

Low Excess Air Burners. Low excess air burners are used in many new furnace designs. *Low excess air burners* are burners that use only the amount of air necessary for complete combustion. This ensures that the combustion will be as efficient as possible. Pulse burners are low excess air burners.

A *pulse burner* is a low excess air burner used in both forced-air and hydronic heating systems. A pulse burner introduces the air-fuel mixture to the burner face in small amounts or pulses. See Figure 5-17.

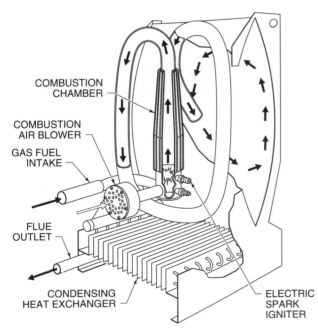

COMBUSTION
CHAMBER

COMBUSTION
AIR BLOWER

GAS FUEL
INTAKE

FLUE
OUTLET

CONDENSING
HEAT EXCHANGER

ELECTRIC
SPARK
IGNITER

Figure 5-17. A pulse burner is more efficient than conventional burners because it uses low excess air.

The pulses of the mixture are controlled by the pressure in the firing chamber. The pulses are introduced to the firing chamber through a flapper valve. A *flapper valve* is a valve that opens to let the air-fuel mixture in but closes when a firing cycle begins because of back pressure. *Back pressure* is the pressure produced by the ignition of the air-fuel mixture against the normal pressure of the gas flow. The cycles are short but continuous.

A pulse burner is more efficient than a conventional burner because no excess air is introduced. This ensures complete combustion, which produces cleaner products of combustion. Draft over the fire is produced within the burner so that no heat rises up the stack. Most pulse burners are more than 90% efficient. Conventional burners are seldom more than 80% efficient.

A *resistance heating element* is an electric heating element that consists of a grid of electrical resistance wires that are attached to a support frame with ceramic insulators. Nichrome, a nickel chromium alloy, is used for electrical resistance wire. Nichrome wire resists the flow of electricity and becomes red-hot when connected to an electrical circuit. Resistance heating elements are installed in a furnace so that air is heated as it passes through the furnace and flows over the hot elements. The electrical circuit is protected against overcurrent and overheating by fuses and temperature-sensing limit switches. See Figure 5-18.

Heat Exchangers

A heat exchanger is anything that transfers heat from one substance to another without allowing the substances to mix. The heat exchanger in a forced-air furnace transfers heat from the hot products of combustion to cool air. The products of combustion often contain toxic chemical compounds. A heat exchanger uses the heat generated by combustion but keeps the harmful products of combustion from mixing with the air used to heat a building. While the furnace is ON, metal heat exchangers expand and contract. Heat exchangers are corrugated and curved to allow movement but are strong enough to prevent deformation. Heat exchangers are often coated with corrosion-resistant ceramic glazing.

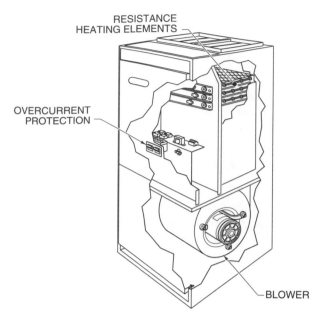

Figure 5-18. Resistance heating elements provide heat for an electric furnace.

The two types of heat exchangers used in conventional forced-air furnaces are the clam shell heat exchanger and drum heat exchanger. The clam shell heat exchanger is usually used with multiple-burner atmospheric burners. The drum heat exchanger is used with fuel oil-fired atmospheric burners and power burners. Many high-efficiency furnaces use secondary heat exchangers (condensing heat exchangers), which extract additional heat from the products of combustion.

Clam Shell Heat Exchanger. A *clam shell heat exchanger* is a heat exchanger that has multiple clam-shaped sections. The sections are made of medium-gauge sheet metal or cast iron. Each section consists of two clam-shaped pieces of metal that are placed edge-to-edge and then welded together to produce an airtight fit.

Combustion takes place at the burner openings, which are located at the bottom of each section. A burner is installed inside each burner opening. When the burner is ignited, hot products of combustion rise through the inside of the sections of the heat exchanger and out the flue openings at the top of each section. See Figure 5-19. Air is blown between the sections by a blower. Heat from the products

of combustion is transferred through the metal to the air that flows outside the sections.

Heat transfer occurs because of the temperature difference between the hot products of combustion and the cool air. The ignition temperature for natural gas is approximately 1200°F. The temperature of the flue gas when burning natural gas is about 400°F. The difference between these temperatures, which is 800°F (1200 − 400 = 800), is a result of the heat being transferred through the walls of the heat exchanger. Air temperature increases in the range of 80°F to 100°F as the air flows through the heat exchanger. Heat exchangers used with combustion burners are designed to provide a 100°F temperature rise between the initial and final temperature of the air.

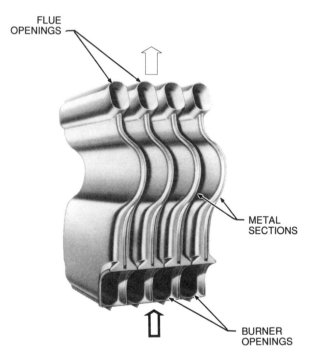

CLAM SHELL HEAT EXCHANGER

Lennox Industries Inc.

Figure 5-19. Products of combustion flow through the clam-shaped sections of a clam shell heat exchanger.

Hot flue gas that leaves the flue openings is collected in the draft diverter before the gas rises up the flue. A *draft diverter* is a box made of sheet metal that runs the width of the heat exchanger. The draft diverter is open across the bottom, which allows dilution air to mix with the flue gas as the flue gas leaves the heat exchanger. *Dilution air* is atmospheric air that mixes with, dilutes, and cools the products of combustion. A draft diverter assures a constant draft and eliminates downdrafts in the flue that would affect burner operation. See Figure 5-20.

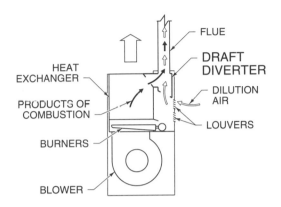

Figure 5-20. A draft diverter eliminates downdrafts, which affect burner operation.

Drum Heat Exchanger. A *drum heat exchanger* is a round drum or tube that is located on a combustion chamber to make the products of combustion flow through it. The *combustion chamber* is the area in a heating unit where combustion takes place. Combustion chambers are designed to retain heat, which helps ignite the fuel. The flame from a power burner is usually contained by a drum heat exchanger. A drum heat exchanger may be positioned horizontally or vertically depending on the unit in which it is used. A flue connection at the top of the drum carries the products of combustion away from the heat exchanger.

As the hot products of combustion flow through a drum heat exchanger, air is blown around and across the outside surface of the heat exchanger. Heat is transferred through the walls of the heat exchanger because of the temperature difference between the hot products of combustion inside and the cool air outside. Drum heat exchangers are made from heavy gauge sheet metal and may be coated with ceramic glazing to resist corrosion. Drum heat exchangers have secondary or multiple tubes for multiple-pass operation. Multiple-pass operation allows the heat exchanger to be small but have a large heating capacity. See Figure 5-21.

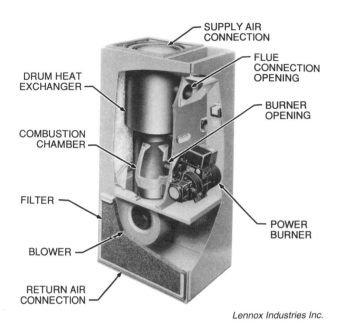

Figure 5-21. Products of combustion flow from the combustion chamber through a drum heat exchanger in a furnace with a power burner.

A drum heat exchanger used with a power burner is connected directly to the flue. A *damper* is a device that controls air flow. A barometric damper is installed in the flue above the drum heat exchanger. A *barometric damper* is a metal plate positioned in an opening in the flue so that atmospheric pressure can control the air flow through the combustion chamber and flue. The plate is balanced so atmospheric pressure can open and close the damper to allow dilution air to enter the flue to control draft. See Figure 5-22.

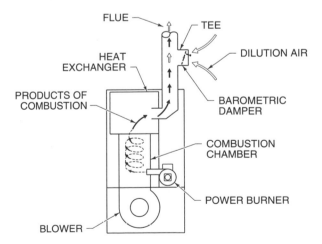

Figure 5-22. A barometric damper controls the draft of a furnace with a power burner by allowing dilution air to enter the flue.

High-efficiency furnaces use condensing heat exchangers. A *condensing heat exchanger* is a heat exchanger that reduces the temperature of the flue gas below the dew point temperature of the heat exchanger. A condensing heat exchanger removes the latent heat of vaporization from the products of combustion. Removing this heat causes the moisture produced during combustion to condense inside the heat exchanger. A high-efficiency furnace must be provided with a drain to remove the condensate. See Figure 5-23.

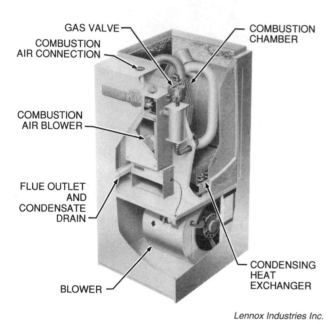

GAS VALVE
COMBUSTION AIR CONNECTION
COMBUSTION CHAMBER
COMBUSTION AIR BLOWER
FLUE OUTLET AND CONDENSATE DRAIN
BLOWER
CONDENSING HEAT EXCHANGER

Lennox Industries Inc.

Figure 5-23. A high-efficiency furnace removes the latent heat of vaporization from the products of combustion.

High-efficiency furnaces reduce the temperature of the flue gas to the dew point temperature of the condensing heat exchanger. The lower temperature provides less energy to carry the flue gas up the flue and out the stack. A combustion air blower or draft inducer is required to provide positive pressure in the flue. *Positive pressure* is pressure greater than atmospheric pressure. A *combustion air blower* is a blower used to provide combustion air at a positive pressure at the burner face. A *draft inducer* is a blower installed in the flue pipe to provide positive pressure in the flue, which carries the products of combustion up the stack. A smaller flue such as a 2½″ diameter plastic pipe can be used with a draft inducer.

The heating capacity of a heat exchanger depends on the thickness of the metal, the conduction factor of the metal, and the surface area of the heat exchanger. The conduction factor represents the amount of heat that will pass through the heat exchanger material per degree Fahrenheit temperature difference on each side of the material. Conduction (k) factors of a metal are multiplied by the surface area of the heat exchanger to find the amount of heat transferred through the walls of the heat exchanger. The number of sections in a heat exchanger determines the surface area of the heat exchanger. A heat exchanger with a given surface area produces a given heating capacity.

Filters

Filter media is any porous material that removes particles from a moving fluid. The filter media is the part of a filter that separates particles from air. Filters clean return air before the air enters a furnace. All forced-air systems should contain filters. Some furnaces have racks in the blower compartment for filters. If there is no filter rack in the furnace, a rack should be built in the return duct at the duct connection to the furnace.

Furnaces used in residential buildings are supplied with low-efficiency filters. *Low-efficiency filters* are filters that contain filter media made of fiberglass or other fibrous material. The fibers are treated with oil to help them hold dust and dirt. These filters are 1″ to 2″ thick and are mounted in a light cardboard frame. Low-efficiency filters are often called slab filters because of their shape. Low-efficiency filters remove about 40% of large airborne particles such as dust and dirt and should be disposed of when they accumulate dirt. Low-efficiency filters are used for normal residential filtering applications. See Figure 5-24.

Medium-efficiency filters are filters that contain filter media made of dense fibrous mats or filter paper. These filters are used in applications such as office buildings that require more filtering efficiency than low-efficiency filters provide. Medium-efficiency filters are usually inserted into a metal frame and disposed of when they accumulate dust and dirt. Medium-efficiency filters remove from 40% to 80% of common-size particulate matter.

High-efficiency filters are filters that contain filter media made of large bags of filter paper. The bag shape increases the surface area of the filter, which reduces the velocity of the air through the filter media. The reduced velocity increases the filtering efficiency.

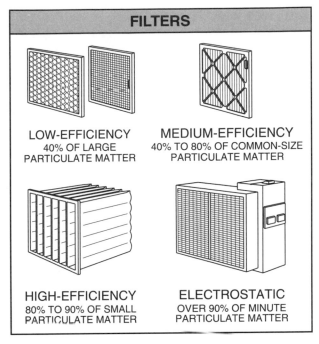

FILTERS

LOW-EFFICIENCY
40% OF LARGE
PARTICULATE MATTER

MEDIUM-EFFICIENCY
40% TO 80% OF COMMON-SIZE
PARTICULATE MATTER

HIGH-EFFICIENCY
80% TO 90% OF SMALL
PARTICULATE MATTER

ELECTROSTATIC
OVER 90% OF MINUTE
PARTICULATE MATTER

Figure 5-24. Filters remove particulate matter from the air. The kind of filter used depends on air quality requirements.

Bag filters are installed in racks or frames with prefilters. *Prefilters* are filters installed ahead of bag filters in the air stream to filter large particulate matter. Bag filters are used in hospitals, laboratories, or electronic component production facilities where a high degree of filtration is required. Filtering efficiency with bag filters is 80% to 90% of small particulate matter.

The highest degree of filtering efficiency is attained with an electrostatic filter. *Electrostatic filters* are devices that clean the air as the air passes through electrically charged plates and collector cells. Electrostatic filters remove much smaller particulate matter from the air than other kinds of filters. Filtering efficiency with electrostatic filters is about 90% to 99% of particulate matter as small as bacteria. Prefilters are usually installed ahead of electrostatic filters to remove the large particulate matter. Electrostatic filters are often used in stadiums to remove smoke and other particles from air.

DISTRIBUTION SYSTEMS

An air distribution system is the supply air ductwork and registers and return air ductwork and grills that are used to circulate air through a building. The distribution system directs heated supply air from the furnace to the building spaces that require heat and returns air from building spaces to the furnace. See Figure 5-25.

Ductwork

Air distribution systems are categorized by the layout of the ductwork. Three basic layouts are perimeter loop, radial, and trunk and branch.

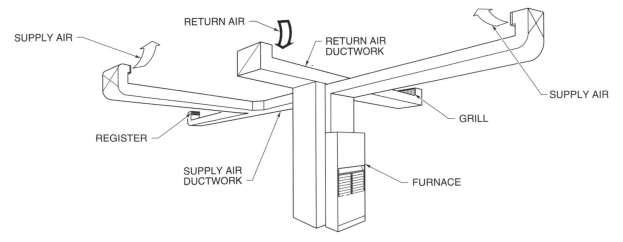

Figure 5-25. An air distribution system consists of supply air ductwork and registers and return air ductwork and grills.

Perimeter Loop System. A *perimeter loop system* consists of a single loop of ductwork with feeder branches that supply air to the loop. Perimeter loop systems are used in special limited situations. The supply plenum is located in the center and the branches extend outward from it. A *supply plenum* is a sealed sheet metal chamber that connects the furnace supply air opening to the supply ductwork. See Figure 5-26.

Radial System. A *radial system* consists of branches that run out radially from the supply plenum of a furnace. Radial systems are used where ductwork can be run in a crawlspace, attic, or duct chase. A *duct chase* is a special space provided in a building for installing ductwork.

Trunk and Branch System. A *trunk and branch system* consists of one or more trunks that run out from the supply plenum of a furnace. A *trunk* is a main supply duct that extends from the supply plenum. Branches extend from the trunks to each register. A trunk and branch system is installed with the trunks running parallel with beams or support members of a building and the branches running at

right angles to the trunks. Trunk and branch systems make it possible to keep ductwork close to building support members. In many cases the ductwork is run in joist spaces and then covered by the building finish materials.

Registers and Grills

Registers and grills are part of the air distribution system in a building. Registers are located on the supply side of the system and distribute heated air to building spaces. Grills are located on the return side of the system and return the air from building spaces back to the furnace.

A *register* is the device that covers the opening of the supply air ductwork. A register consists of a panel with vanes, frame for mounting, and a damper for controlling air flow. See Figure 5-27. The air flow pattern from the register is controlled by the vanes on the panel. Registers throw air straight out from the register and/or spread air out evenly in all directions. Registers are sized according to the quantity and velocity of air required in each building space. Registers are located for efficient distribution of air to each building space.

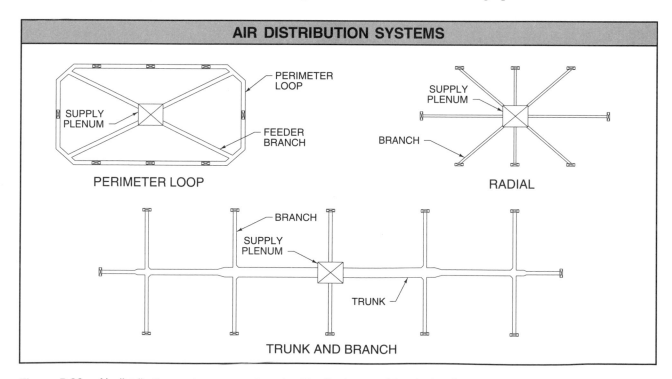

Figure 5-26. Air distribution systems are categorized by the layout of the ductwork.

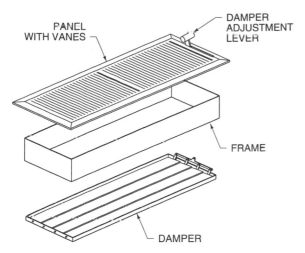

Figure 5-27. Registers consist of a panel with vanes, a frame for mounting, and damper for controlling air flow.

The air flow pattern determines the area of influence of the air from the register. The *area of influence* is the area from the front of the register to a point where the air velocity drops below 50 fpm. Air velocity above 50 fpm produces an uncomfortable draft. See Figure 5-28.

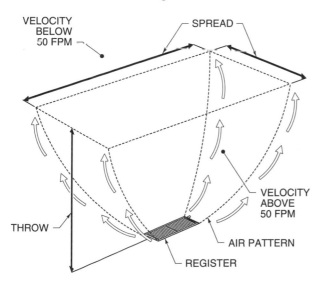

Figure 5-28. Registers throw air straight out from the front of the register and/or spread the air out evenly in all directions.

A *grill* is the device that covers the opening of the return air ductwork. A grill consists of a decorative panel with vanes and frame that holds the panel in place. The vanes are arranged to block the view to the ductwork. Grills are sized and located to return the supply air back to the furnace. Grills

are located on ceilings, walls, or floors depending on the ductwork system and the location of the ductwork in the building frame. Buildings have fewer grills than registers because grills are larger and more centrally located than registers. It is also common for grills to be located high in the spaces being conditioned, which prevents temperature stratification of air in the rooms. See Figure 5-29.

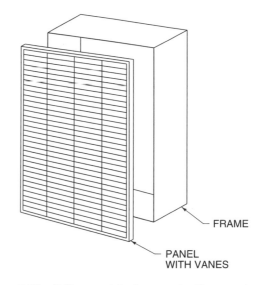

Figure 5-29. Grills consist of a panel with vanes and a frame that holds the panel in place.

CONTROLS

Controls operate the individual components of a furnace. Controls are installed on a furnace by the manufacturer or an air conditioning technician to maintain safe, efficient furnace operation. Power, operating, safety, and combustion safety controls are the different kinds of controls.

Power Controls

Power controls control the flow of electricity to a furnace. Power controls are installed in the electrical circuit between the power source and the furnace. Power controls include disconnects, fuses, and circuit breakers. Power controls should be installed by a licensed electrician per Article 424 of the National Electrical Code®.

Disconnects. A *disconnect* is a switch that, when open or closed, controls the flow of electricity to a furnace. When a disconnect is open, no electric current flows. When the disconnect is closed, electric current flows. At least two disconnects are installed in the electrical power circuit to a furnace. One disconnect is located in the electrical service panel where the circuit originates. The other disconnect is located on or near the furnace. See Figure 5-30. Most heating systems use manual disconnects. A *manual disconnect* is a protective metal box that contains fuses or circuit breakers and the disconnect. A manual disconnect has a handle that extends outside the box. The handle allows manual opening or closing of the disconnect without opening the box.

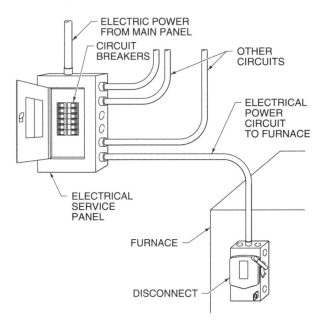

Figure 5-30. Disconnects in the electrical circuit to a furnace are located at the electrical service panel and on or near the furnace.

Conductors are the electrical wiring in an electrical power circuit between the power supply and the equipment. Conductors in an electrical power circuit are sized to carry the electricity required to operate the equipment. If excessive current flows through a conductor, the conductor will overheat. Fuses or circuit breakers are placed in the electrical power circuit to protect the conductors from excessive current flow.

Fuses. A *fuse* is an electric overcurrent protection device located in an electrical power circuit. A fuse

will blow (burn out) if an overcurrent condition occurs, which breaks the circuit and shuts OFF the current flow before the circuit is damaged. See Figure 5-31. Large heating systems use cartridge fuses to protect electrical circuits. A *cartridge fuse* is a fibrous or plastic tube that contains a fuse wire that carries a specific amount of current. If an overcurrent condition occurs, the fuse wire will melt and the circuit will open.

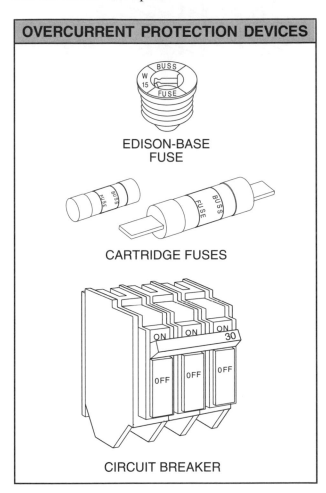

Figure 5-31. Fuses and circuit breakers protect the wiring and components in an electrical circuit from excessive current flow.

Circuit Breakers. A *circuit breaker* is a current-sensing device that is designed to open a circuit automatically if an overcurrent condition occurs. Circuit breakers that are used as disconnects have a switch that can be used to manually open and close the circuit. Circuit breakers are used for combination disconnects and fusing in small heating systems.

Operating Controls

Operating controls are controls that cycle equipment ON or OFF. Operating controls include transformers, thermostats, blower controls, relays, contactors, magnetic starters, and solenoids.

Transformers. A *transformer* is an electric device that changes the voltage in an electrical circuit. A transformer consists of a primary coil and a secondary coil. Each coil is a wire that is wound on a metal core. See Figure 5-32. The primary side of the transformer is connected to an electrical power source. The secondary side of the transformer is connected to an electrical load.

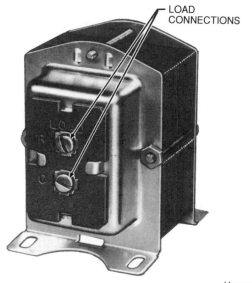

Honeywell Inc.

Figure 5-32. A transformer is used to change voltage from line voltage (120 V) to a lower voltage (24 V) for control systems.

When current flows through the primary coil, an electromagnetic field is produced. The electromagnetic field causes an electric current in the secondary coil. If the coil on the primary side has more windings than the coil on the secondary side, the voltage on the secondary side is less than the voltage on the primary side. This is a step-down transformer. If the coil on the secondary side has more windings than the coil on the primary side, the voltage on the secondary side is greater than the voltage on the primary side. This is a step-up transformer. Most control circuit transformers are step-down trans-

formers with a secondary voltage of 24 V. These step-down transformers are installed at the factory as part of the furnace control package.

Thermostats. A *thermostat* is a temperature-actuated electric switch. A thermostat operates and controls the burner(s) or heating elements in a heating unit. When the temperature at the thermostat falls below a setpoint, the thermostat closes a switch, which completes the electrical circuit. When the temperature at the thermostat rises above the setpoint, the thermostat opens the switch, which opens the electrical circuit. See Figure 5-33.

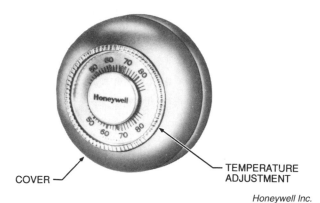

Honeywell Inc.

Figure 5-33. A thermostat is a temperature-actuated electric switch that controls and operates the burner(s) or heating elements in a heating unit.

Thermostats may be low-voltage or line voltage. Low-voltage and line voltage thermostats carry different voltages. Low-voltage thermostats carry 24 V and are more sensitive to temperature change than line voltage thermostats. Line voltage thermostats carry 120 V or 240 V. The contacts and the temperature-sensing element in line voltage thermostats must be large enough to handle the high voltage. A problem with line voltage thermostats is that the high voltage heats the temperature-sensing element and opens the contacts too soon.

The operating parts of a thermostat are the sensor, switch, and setpoint adjustor. The sensor in a thermostat may be a bimetal element, remote bulb, or electronic circuit. A *bimetal element* is a sensor that consists of two different kinds of metal that are bonded together into a strip or a coil. The metals expand at different rates when heated. When a bimetal element is heated, the element bends away from the metal with the greater rate of expansion. When the bimetal element is cooled, the element bends toward the metal with the greater rate of expansion. The sensitivity and temperature range of the bimetal element increases when the element is coiled. The movement of the element actuates a switch that controls the flow of electric current. See Figure 5-34.

A *remote bulb* is a sensor that consists of a small, refrigerant-filled metal bulb connected to the thermostat by a thin tube. The refrigerant in the metal bulb is the sensor in a remote bulb thermostat. The refrigerant in the metal bulb vaporizes or condenses in response to temperature changes. Pressure exerted by the refrigerant vapor is transmitted to a bellows element in the thermostat. The pressure change expands or contracts the bellows

element, which actuates the switch that controls the flow of electric current.

Electronic sensors are electronic devices that sense temperature changes. A *thermistor* is an electronic device that changes resistance in response to a temperature change. Thermistors are direct- or reverse-acting. The electrical resistance of a direct-acting thermistor changes directly with a temperature change. The electrical resistance of a reverse-acting thermistor changes indirectly with a temperature change.

Open contact and mercury bulb are the two types of switches used in thermostats. Open contact switches may be sealed in glass to protect the contacts from dirt and oxide buildup. The switch mechanism opens or closes the contacts quickly with positive action to prevent electrical arcing (sparks) and burning of the contacts, which causes oxide buildup.

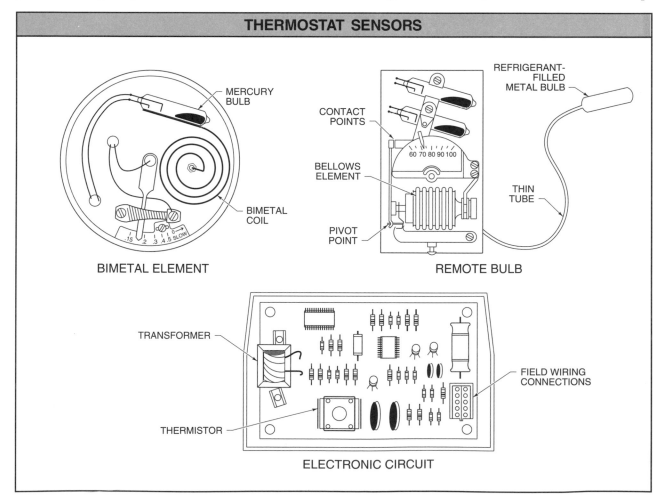

Figure 5-34. The sensor in a thermostat may be a bimetal element, remote bulb, or electronic circuit.

Open contact switches are closed to complete the circuit and opened to break the circuit. The contacts consist of a material that readily conducts electricity. Open contact switches are often actuated by a bimetal element but may be actuated by any mechanical motion. See Figure 5-35.

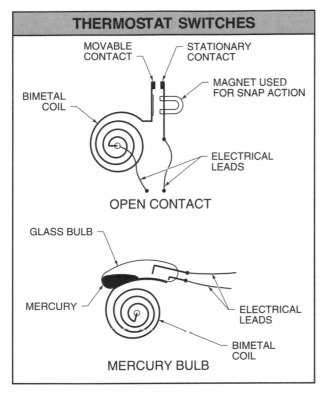

Figure 5-35. Open contact and mercury bulb switches are used in thermostats.

Mercury bulb switches are closed glass bulbs that contain a drop of mercury. The bulb is mounted on a bimetal coil. Exposed electrical leads are embedded in the bulb. When the bimetal coil expands or contracts, the glass bulb tips one way or the other. When the mercury drop moves to the end that holds the electrical leads, contact is made and the circuit is closed. When the drop moves to the other end, contact is broken and the circuit is open.

Setpoint temperature is the temperature at which the switch in a thermostat will open and close. Most thermostats are built so that the setpoint can be changed manually. The *setpoint adjustor* is a lever or dial that indicates the temperature on an exposed scale. See Figure 5-36. The setpoint adjustor for a bimetal element thermostat is located inside the thermostat assembly on the mounting base that holds

the sensor. When the adjustor is moved, it rotates the mounting on a pivot, which increases or decreases the temperature at which the switch opens or closes.

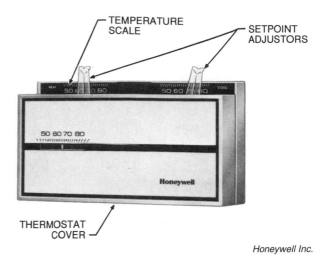

Honeywell Inc.

Figure 5-36. The setpoint adjustor indicates the temperature at which the switch opens or closes.

Setpoint temperature on a remote bulb thermostat is adjusted on the linkage between the remote bulb and the switch. Moving the adjustor shortens or extends the linkage. The movement increases or decreases the temperature at which the switch opens or closes. Adjusting the setpoint on an electronic thermostat is done by adjusting resistance devices within the thermostat.

When a thermostat senses the setpoint temperature and shuts OFF the burner(s), the heat that remains in the furnace will raise the temperature in a building space above the setpoint. The switch turns the burner(s) ON when the temperature drops approximately two degrees below the setpoint and shuts the burner(s) OFF when the temperature rises approximately two degrees above the setpoint. The *differential* is the difference between the temperature at which the switch in the thermostat turns the burner(s) ON and the temperature at which the thermostat turns the burner(s) OFF. The differential is necessary to prevent rapid cycling of the burner(s). Because of the differential, the actual temperature in a building space usually rises slightly above the setpoint during an ON cycle and slightly below the setpoint during an OFF cycle. A heat anticipator is used to prevent the temperature from rising above the setpoint.

A *heat anticipator* is a small heating element that is located inside a thermostat. The heat anticipator is wired with the thermostat contacts. When the thermostat calls for heat, the heat anticipator produces heat. A heat anticipator improves temperature control in a building space by providing enough heat to turn the thermostat OFF before the room temperature increases above the setpoint. When enough heat is produced, the heat anticipator is disconnected from the circuit and does not provide any false heat. See Figure 5-37.

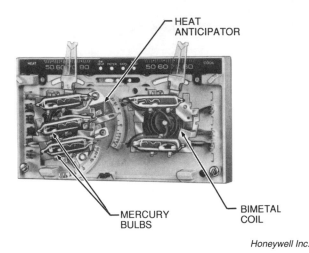

Honeywell Inc.

Figure 5-37. A heat anticipator improves temperature control in a building space.

Blower Controls. A *blower control* is a temperature-actuated switch that controls the blower motor of a furnace. A blower control consists of a bimetal element that operates an electric switch. The electric switch closes when the temperature increases and opens when the temperature decreases.

A blower control is installed in a furnace so the bimetal element is located in the air stream of the furnace near the heat exchanger. When the furnace burner ignites, the air near the heat exchanger is heated. The temperature change closes the electric switch, which turns the blower motor ON. When the furnace burner shuts OFF, the air around the heat exchanger cools. This temperature change opens the electric switch, which turns the blower motor OFF. See Figure 5-38.

Relays. A *relay* is an electric device that controls the flow of electric current in one circuit with

another circuit. Relays are often used in control circuits where a low-voltage circuit controls a line voltage circuit. The different types of relays used in heating system controls are electromechanical relays, contactors, and magnetic starters.

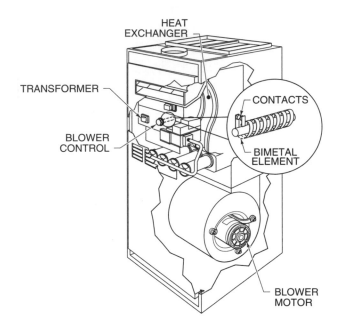

Figure 5-38. A blower control is located so that the bimetal element is near the heat exchanger.

An *electromechanical relay* is an electric device that uses a magnetic coil to open or close one or more sets of contacts. One contact is fixed, and the other contact is located on a movable arm that is controlled by the coil. The moving contact is held in the initial position by a spring. When the coil is energized, the contacts open or close depending on the action desired. See Figure 5-39.

Relays are used for different applications. They are identified by poles and by the position of the contacts when the control circuit is de-energized. *Poles* are the number of load circuits that contacts control at one time. SP stands for single-pole circuit. DP stands for double-pole circuit. The position of the contacts are identified as normally open (NO) or normally closed (NC). *Throws* are the number of different closed contact positions per pole, which is the number of circuits that each individual pole controls. Circuits are identified as single-throw (ST) circuits or double-throw (DT) circuits. Relays are available for many combinations of poles, but the

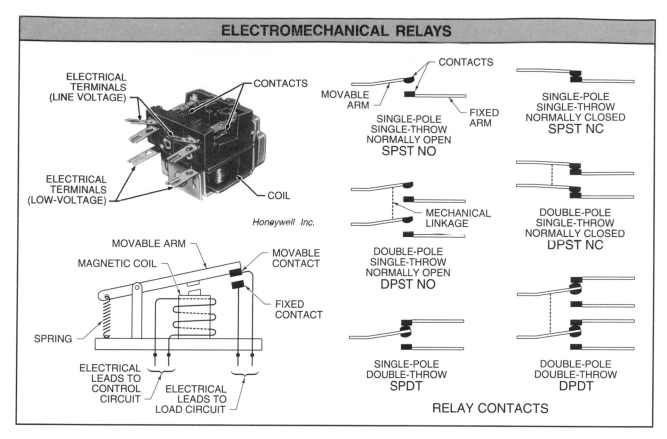

Figure 5-39. An electromechanical relay uses a magnetic coil to open or close a set of contacts.

contact positions are generally limited to NO, NC, or a combination of the two.

A *contactor* is a heavy-duty relay. A contactor has a coil and contacts that are designed to operate with high electric current, which is required to run large electric motors. Contactors may be used for controlling compressors or large motors in a system.

A magnetic starter is a contactor with overload relays added to it. *Overload relays* are electric switches that protect a motor against overheating and mechanical overloading. Overload relays open the circuit to a motor if excessive electric current or heat is present. Magnetic starters are used in motor circuits when the motors do not have internal overloads. See Figure 5-40.

When a motor starts, it draws a tremendous momentary inrush of current that can be six to eight times normal running current. Fuses or circuit breakers must handle the momentary inrush of current without opening the circuit when the motor starts. If the current increases while a motor is running,

the motor can be damaged without blowing the fuses or tripping the circuit breakers.

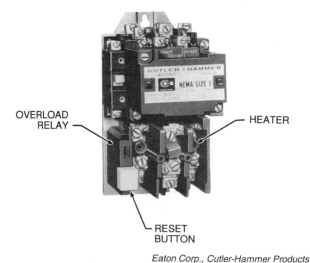

Eaton Corp., Cutler-Hammer Products

Figure 5-40. A magnetic starter is a contactor that has overload relays added to it.

Motors overheat when they are overloaded. *Overload* occurs when a motor is connected to an excessive load. For example, a $\frac{1}{2}$ HP motor is overloaded when connected to a $\frac{3}{4}$ HP load. When a motor is overloaded, it draws more electric current than it is designed to carry. Overload causes motors to overheat and breaks down wiring insulation in the motor.

A *bimetal overload relay* contains a set of contacts that are actuated by a bimetal element. If the temperature around the bimetal element rises because of excessive current flow, the element opens the contacts, which shuts the motor OFF. When the temperature drops, the element closes the contacts, which turns the motor ON. See Figure 5-41.

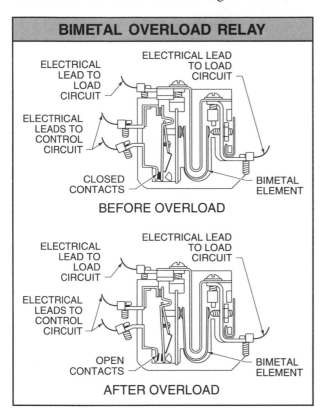

Figure 5-41. Overload relays are electric switches that shut a motor OFF if it overheats.

Solenoids. A solenoid is similar to a relay in that it controls one electrical circuit with another. A *solenoid* is an electric switch consisting of a hollow coil that surrounds a metal core. The core moves back and forth inside the coil. When the coil is energized, the core closes a set of contacts. When the coil is de-energized, the core opens the set of contacts. The core is spring-loaded so it returns to the open position when the coil is de-energized.

Safety Controls

Safety controls are controls that monitor the operation of a furnace. If a furnace causes a hazard to personnel or equipment, safety controls will shut the furnace OFF.

Limit Switches. A *limit switch* is an electric switch that shuts a furnace OFF if the furnace overheats. If the furnace temperature rises above a safe temperature, the limit switch shuts the burner(s) or electric heating element OFF. A bimetal element senses the temperature of the air around the switch and opens the electric switch if the temperature rises above a setpoint. A faulty fuel valve, broken blower belt, or faulty blower motor could cause a furnace to overheat. Most limit switches are automatic electric switches, which reset automatically when the temperature drops below the setpoint. Manual-reset limit switches are used for some applications. See Figure 5-42.

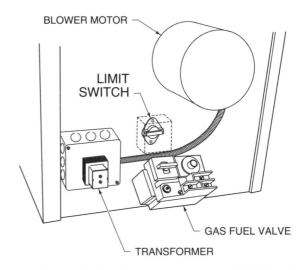

Figure 5-42. Limit switches shut a furnace OFF if the furnace overheats.

Combustion Safety Controls

Combustion safety controls are safety controls that shut down the burner(s) if a malfunction occurs. Combustion safety controls monitor firing to make sure that ignition occurs and that the flame remains ON during a call for heat. Some combustion safety controls reset automatically after a shutdown, but others must be reset manually. Manual-reset safety controls require that the burner(s) be checked before the furnace is re-ignited. Combustion safety controls include stack switches, pilot safety controls, flame rods, and flame surveillance controls.

Stack Switches. A *stack switch* is a mechanical combustion safety control device that contains a bimetal element that senses flue-gas temperature and converts it to mechanical motion. A stack switch is installed in the flue near the furnace. On a call for heat, current flows through a safety switch heater in the burner control circuit. The safety switch heater is wired to a set of cold contacts (normally closed contacts). The current flow through the cold contacts allows the fuel valve to remain open.

If the flue-gas temperature does not rise in approximately 90 seconds, the safety switch heater opens the cold contacts. If ignition does occur, the bimetal element expands. This moves a metal rod that closes a set of hot contacts (normally open contacts) in the burner control circuit. This allows the fuel valve to remain open. See Figure 5-43. A mechanical linkage connects the bimetal element and the contacts in the control box. The contacts open or close as the bimetal element moves.

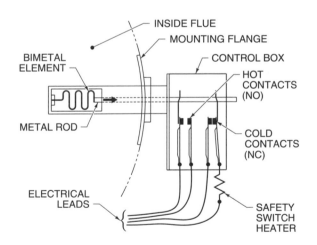

Figure 5-43. A stack switch is a combustion safety control device that converts temperature to mechanical motion.

Pilot Safety Controls. A *pilot safety control* is a safety control that determines if the pilot light is burning. Burners that use a pilot light for initial combustion have pilot safety controls. A pilot safety

control is an electric combustion safety control that contains a thermocouple. A *thermocouple* is a pair of electrical wires (usually constantan and copper) that have different current-carrying characteristics welded together at one end (hot junction). The thermocouple is installed near the pilot flame, which heats the hot junction. When the hot junction is heated, a low-voltage electric signal is produced. The free ends of the wires (cold junction) are connected to an electromagnetic coil in a safety valve ahead of the gas fuel valve or in a special valve within the gas fuel valve.

When the pilot light is burning, the low-voltage electric signal generated by the thermocouple produces a magnetic field in the coil. The magnetic field holds the safety valve open. See Figure 5-44. If the pilot light goes out, there is no magnetic field and the safety valve closes. A pilot safety control monitors the pilot light in a burner and does not allow the gas fuel valve to open or fuel oil burner to ignite unless the pilot flame is established. Pilot safety controls are used mainly on gas fuel-burning equipment that has a standing pilot.

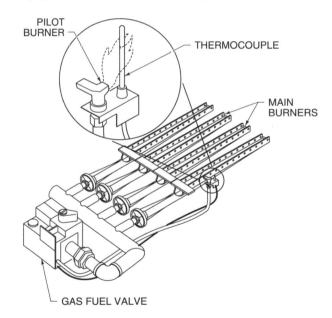

Figure 5-44. A pilot safety control monitors the pilot light in a burner and does not allow the gas fuel valve to open unless the pilot flame is established.

Flame Rods. A flame rod is an electronic combustion safety control used on large commercial furnaces. A *flame rod* is an electric device that uses a flame to conduct electricity. As a furnace is firing, a control device sends out a low-voltage electric signal to a metal rod located in the flame. If a flame is established on a call for heat, the safety circuit is closed by the flame. If a flame is not established within a reasonable length of time, the furnace shuts OFF. See Figure 5-45. Flame rods are smaller and more durable than stack switches or thermocouples. Flame rods are used in systems that have intermittent pilots. Intermittent pilots establish the pilot flame when the thermostat calls for heat. Intermittent pilots extinguish the pilot flame when the burner flame is extinguished.

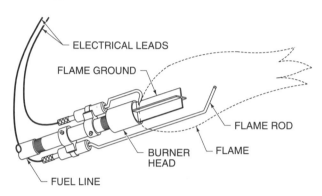

Figure 5-45. A flame rod uses the burner flame as an electrical conductor.

Flame Surveillance Controls. A *flame surveillance control* is an electronic combustion safety control that consists of a light-sensitive device that detects flame. Cadmium sulfide is a light-sensitive substance used in flame surveillance controls. The resistance of cadmium sulfide to electric current depends on the intensity of the light that strikes the material. When a cadmium sulfide cell (cad cell) is exposed to light, the resistance to the flow of electricity through the cell is low. When the cad cell is in darkness, the resistance through the cell is high. Electrical leads are connected to each side of the cell. Current flow through the cell is monitored to determine if the cell detects light.

The cad cell is mounted in a burner in such a way that is in direct line-of-sight with the flame from the burner. If the burner ignites and a flame is established on a call for heat, the resistance through the cell is low. Low resistance allows the electric signal to pass to the control center, which actuates the furnace. If the burner does not ignite, the resistance through the cell is high, which pre-

vents the electric signal from reaching the control center. After a reasonable time, the control center shuts the furnace OFF. Cad cell combustion safety control devices are used on furnaces that contain fuel oil burners. See Figure 5-46.

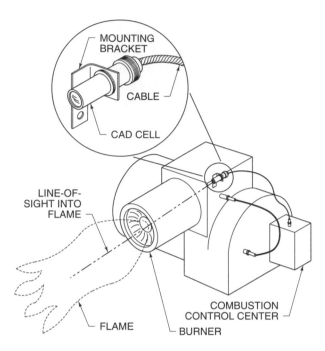

Figure 5-46. A cad cell is a combustion safety control device that detects flame.

SPECIAL PURPOSE HEATING UNITS

Special purpose heating units are furnaces built for special applications. These applications include large open spaces, spaces with a common air supply but not a central heating unit, spaces that use 100% makeup air, and spaces that do not require heat but have people that need heat.

Unit Heaters

A *unit heater* is a self-contained heating unit that is not connected to ductwork. Unit heaters burn gas fuel, use resistance heating elements, or have hot water or steam coils. Unit heaters are used as space heaters. See Figure 5-47.

A blower in the unit heater draws air in through the back of the unit and heated air is blown out through louvers in the front. A unit heater is normally suspended from the ceiling or from the roof frame of a building. A fuel line, control circuit, and electric conductors are run to the heater.

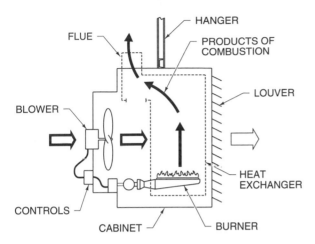

Figure 5-47. A unit heater is a self-contained heating unit that is not connected to ductwork.

Duct Heaters

A *duct heater* is a unit heater that is installed in a duct and supplied with air from a remote blower. See Figure 5-48. The blower for a duct heater is usually controlled by a central control system. Each duct heater has an air-proving limit switch that shuts the unit OFF if no air is moving in the duct. Duct heaters are usually resistance heating elements but may be heated by hot water or steam.

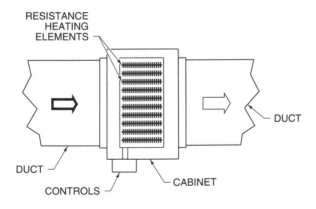

Figure 5-48. A duct heater is located inside ductwork and uses air from a remote blower.

Duct heaters are used in large systems that have a central blower for reheat on heating and ventilation systems. *Reheat* is heat that is supplied at the point of use while an air supply comes from a central location.

Direct-fired Heaters

A *direct-fired heater* is a unit heater that does not have a heat exchanger. The blower blows air directly through the combustion chamber and out the supply end of the unit. Direct-fired heaters are used in applications where 100% makeup air is required for ventilation and 100% of the air is constantly exhausted. See Figure 5-49.

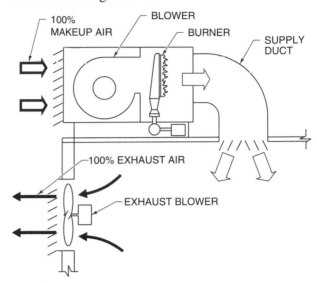

Figure 5-49. A direct-fired heater is a unit heater that does not have a heat exchanger.

Infrared Radiant Heaters

An *infrared radiant heater* is a heating unit that heats by radiation only. An infrared radiant heater has a heat source such as a gas fuel flame, resistance heating elements, or hot water coil that heats a surface to a temperature high enough to radiate energy. See Figure 5-50. Infrared radiant heaters do not heat the air. Radiant energy is converted to heat when the radiant energy waves strike an opaque object.

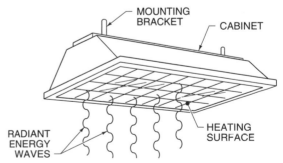

Figure 5-50. The surface of an infrared radiant heater is at a temperature high enough to radiate energy waves.

Review Questions

1. List and describe the two main forced-air heating systems.

2. List the three different furnaces as categorized by the direction of heat flow.

3. What is the relationship between ductwork and air flow in a furnace?

4. List and describe the basic parts of a furnace.

5. Why are louvers located on the access panels of a furnace cabinet?

6. List and describe the three different blowers used in forced-air heating systems.

7. Why are centrifugal blowers used most commonly in low- to medium-pressure forced-air heating systems?

8. List and describe the two drives used on blowers in forced-air systems.

9. Describe the process of changing the speed of the blower wheel of a belt drive blower.

10. How is the speed of the blower wheel of a direct drive blower changed?

11. List and explain the three functions of a burner.

12. What is the main difference between an atmospheric burner and a power burner?

13. List and describe the two most common heat exchangers.

14. What is the main function of a heat exchanger in a furnace?

15. List the four general categories of filters and the filtering efficiency of each.

16. Describe input and output ratings for a conventional furnace.

17. What are the components of an air distribution system?

18. List and describe the three duct system layouts used in forced-air systems.

19. When the temperature rises in a forced-air heating system, does the switch in a thermostat open or close? Explain.

20. What kind of sensor is used most often in thermostats?

21. How would a blower control operate in the furnace of a forced-air heating system on a call for heat?

22. Describe the magnetic action of a relay or a solenoid.

23. How do overload relays protect against overcurrent conditions? Explain.

24. List and describe four combustion safety controls.

25. List and describe four special-purpose heating units.

Hydronic Heating Systems

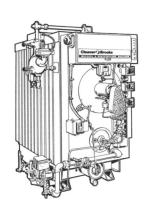

Hydronic heating systems use water or steam to carry heat. Water has a relatively high specific weight and specific heat, which make water an excellent medium for carrying heat. Hydronic heating systems heat water in a boiler at a central location. The heated water is distributed by a circulating pump through supply piping to building spaces. In the building spaces where the heat is required, the hot water flows through terminal devices. Terminal devices transfer heat from the hot water to the air. The water returns to the boiler from the terminal devices through return piping.

HYDRONIC HEATING SYSTEMS

A *hydronic heating system* is a heating system that uses water, steam, or other fluid to carry heat from the point of generation to the point of use. Hydronic heating systems are used where heating equipment is located far from building spaces that require heat. Hydronic heating systems are also used where several buildings are heated by a central heating plant. Hydronic heating systems consist of a boiler, fit-

tings, piping system, circulating pump, terminal devices, and controls. See Figure 6-1.

A boiler heats and stores water or steam. The fittings maintain safe and efficient operation of a hydronic heating system. The piping system distributes hot water or steam to building spaces. Terminal devices transfer heat from the hot water or steam to the air in the building spaces that require heat. The circulating pump moves the water or steam from

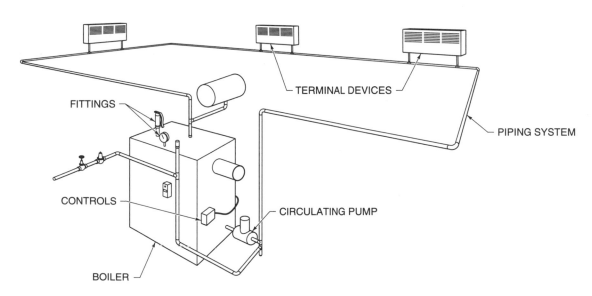

Figure 6-1. Hydronic heating systems consist of a boiler, fittings, piping system, circulating pump, terminal devices, and controls.

the boiler through the piping system to the terminal devices and back to the boiler. Controls regulate the temperature of the water and steam in a hydronic heating system and the temperature of the air in the building spaces. Controls operate the boiler and circulating pump, which produce and distribute hot water or steam to the building spaces.

BOILERS

The boiler is the central element of a hydronic heating system. A *boiler* is a pressure vessel that safely and efficiently transfers heat to water. A *pressure vessel* is a tank or container that operates at a pressure that is greater than atmospheric pressure. Boilers may use hot water or steam to carry heat. Hot water boilers heat water. The heat is then transferred to the air in building spaces. Hot water boilers and piping systems are completely filled with water. Steam boilers heat water to the boiling point, which changes the water to steam. Steam boilers and piping systems are not completely filled with water. Boilers are classified based on working pressure and temperature, method of heat production, and materials of construction.

Low-pressure Boilers

Boilers are designed to operate at different working pressures and temperatures. Low-pressure boilers operate at working pressures up to 15 psi steam or up to 160 psi water. Low-pressure boilers must meet ASME *Boiler and Pressure Vessel Code*, Section 4, Rules for Construction of Heating Boilers. *ASME (American Society of Mechanical Engineers)* is an educational association of engineers that establishes codes and standard practices for boiler operation. ASME has established standards for boilers since 1911. Low-pressure steam boilers are used primarily for heating warehouses, factories, schools, and apartments. Low-pressure hot water boilers operate with water temperatures up to 250°F and are used for residential heating.

High-pressure Boilers

High-pressure boilers operate at working pressures above 15 psi steam or above 160 psi water. High-pressure boilers are constructed to meet ASME *Boiler and Pressure Vessel Code*, Section 1, Rules for Construction of Power Boilers. High-pressure steam boilers are used for processing operations in industry and for generating electricity. High-pressure hot water boilers operate with water temperatures of 250°F or greater and are used for heating.

Heat Production

Boilers burn fuel or use electricity to produce heat, which heats water. Combustion boilers burn fuel such as natural gas, fuel oil, and coal to produce heat. Natural gas- and fuel oil-fired boilers are used in many buildings. Coal boilers are used in large installations such as commercial or industrial plants. Electric boilers use resistance heating elements to produce heat. Electric boilers are used mainly in residential or small commercial buildings.

Combustion Boilers. Boilers that produce heat by combustion are made of steel or cast iron. Steel boilers are welded together to form a single unit that has a specific output rating. Steel boilers have removable manhole covers for maintenance and service. Cast iron (sectional) boilers consist of hollow sections that are bolted together. Openings between the sections are provided for water and flue-gas connections. Hot water flows inside the sections and hot products of combustion flow outside the sections. Heat is transferred from the products of combustion through the cast iron to the water. The number of sections in a cast iron boiler determines the output rating of the boiler. A required output rating is achieved by adding or removing sections of the boiler. See Figure 6-2.

The fire side of a boiler contains hot products of combustion and the water side contains water. The heat exchange area is as large as possible so that a large surface area of water is exposed to the hot products of combustion. A large surface area increases the rate of heat transfer from the hot products of combustion to the water and increases boiler efficiency. The heat exchange area of steel boilers is composed of horizontal, vertical, or slanted tubes. These tubes may be fire tubes or water tubes.

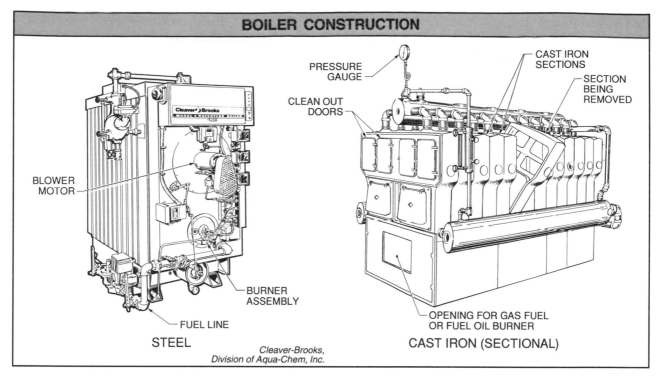

BOILER CONSTRUCTION

Figure 6-2. Steel boilers are usually welded together to form a single unit with a specific output rating. Cast iron (sectional) boilers consist of hollow sections that are bolted together.

Fire tubes hold hot products of combustion. A firetube boiler heats water as the hot products of combustion pass through the fire tubes, which are located on the water side of the boiler. The combustion chamber is located below or at one end of the boiler. A breeching connected to the stack collects the hot products of combustion. See Figure 6-3.

Water tubes hold water that is heated by the hot products of combustion. In a watertube boiler, water is heated as hot products of combustion pass over the outside surfaces of the water tubes. The water tubes run from the ends of the boiler through the fire side. Water fills the ends of the boiler and the water tubes. The combustion chamber is located below or at one end of the boiler. A breeching connected to the stack collects the hot products of combustion.

The combustion chamber (furnace) of a boiler is the area where combustion takes place. By increasing the heating surface of a boiler, more heat can be transferred to the water. *Heating surface* is the

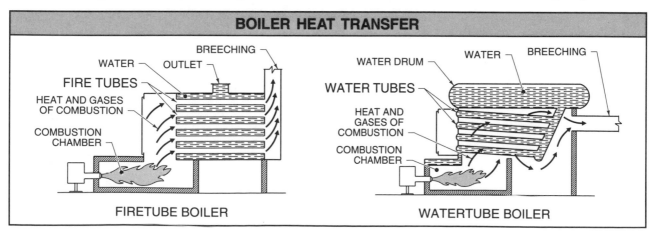

BOILER HEAT TRANSFER

Figure 6-3. Fire tubes hold the hot products of combustion, which heat the water that surrounds the fire tubes. Water tubes hold water heated by the hot products of combustion, which surround the water tubes.

boiler metal that has heat from the combustion chamber on one side and water on the other side. The relationship between the water in a boiler and the combustion chamber determines whether a boiler is a dry base boiler, wet leg boiler, or wet base boiler. A *dry base boiler* has no water below the combustion chamber. A *wet leg boiler* has water along the sides of the combustion chamber. A *wet base boiler* has water on all sides of the combustion chamber. See Figure 6-4.

Boiler burners are similar to burners used in forced-air furnaces. One large burner or multiple smaller burners are used to heat water in a boiler. The burner(s) may be atmospheric or power burners. Atmospheric burners use ambient air supplied at normal atmospheric pressure for combustion air. Primary air enters at the point where the fuel enters the burner. Primary air mixes with the fuel to provide a combustible air-fuel mixture. Secondary air enters the flame at the burner face. Power burners use combustion air provided by a combustion air blower. Air is forced into the combustion area with the fuel. Draft across the flame is maintained by the pressure of the air.

Electric Boilers. Electric boilers use heat produced by resistance heating elements to heat water and are therefore classified separately from combustion boilers. The resistance heating elements used in boilers are immersion heaters. *Immersion heaters* are copper rods enclosed in an insulated waterproof tube that are installed so that water surrounds the heater. No heat exchanger is required because heat

is transferred directly from the immersion heater to the water. See Figure 6-5.

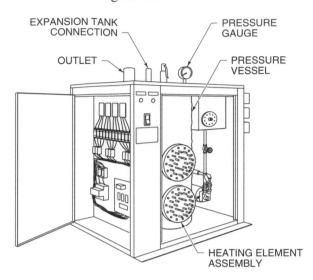

Figure 6-5. Electric boilers use heat produced by resistance heating elements that heat water directly.

Input rating is the heat produced by a boiler in Btu per hour per unit of fuel burned. Output rating is the actual heat output of the boiler in Btu per hour after heat losses from draft. Output is normally 75% to 80% of input for combustion boilers. *Gross unit output* is the heat output of a boiler when it is fired continuously. *Net unit output* is gross boiler output multiplied by a percentage of loss because of pickup. *Pickup* is additional heat that is needed to warm the water in a hydronic heating system after a period of off-time such as overnight. The pickup factor can be as high as 30% to 50%. Electric boilers have output ratings of 100% because

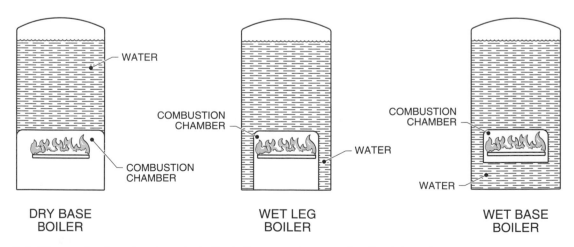

Figure 6-4. The combustion chamber of a boiler is the area where combustion takes place.

all the energy used to produce heat is transferred to the water.

The amount of heat produced by a combustion boiler depends on the size of the boiler and the kind of fuel used. Gas fuel- and fuel oil-fired boilers have output ratings from 100,000 Btu/hr for residential applications to hundreds of thousands of Btu per hour for commercial applications. Coal-fired boilers are used for large industrial applications and may have output ratings of millions of Btu per hour. The amount of heat produced by an electric boiler depends on the size of the boiler and the voltage used.

FITTINGS

Fittings are pieces of equipment on a boiler or in the boiler piping that improve operation and efficiency of a hydronic heating system. Fittings include pressure-temperature gauges, expansion tanks, air controls, aquastats, and control and safety valves. All fittings must be manufactured and installed per the ASME code that corresponds to the operating pressures and temperatures of the boiler. See Figure 6-6.

Pressure-Temperature Gauge

A pressure-temperature gauge ensures safe and efficient boiler operation. A *pressure-temperature gauge* is a gauge that measures the temperature and

pressure of the water at the point where the gauge is located. A pressure-temperature gauge is installed at the top and front of the boiler so that it can be seen easily. See Figure 6-7.

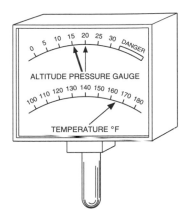

Figure 6-7. A pressure-temperature gauge measures the pressure and temperature of boiler water.

Expansion Tank

All components, piping, and terminal devices of hydronic heating systems operate while full of water. Water expands as it is heated. An *expansion tank* is a vessel that allows the water in a hydronic heating system to expand without raising the water pressure to dangerous levels. An expansion tank is normally filled with air. As water in the system expands, the water flows into the tank. The air in the tank acts as a cushion. As the water expands, the air in the tank compresses. See Figure 6-8.

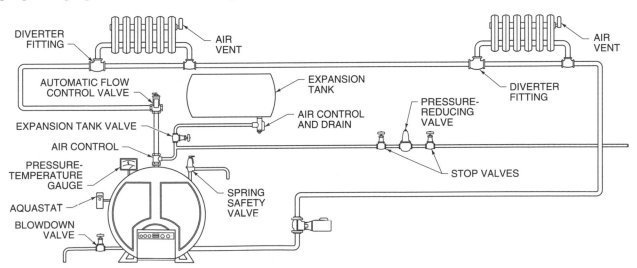

Figure 6-6. Fittings are pieces of equipment on a boiler or in boiler piping that improve operation and efficiency of a hydronic heating system.

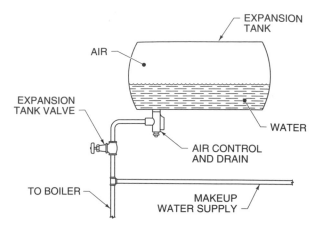

Figure 6-8. An expansion tank takes in hot water as the boiler water expands.

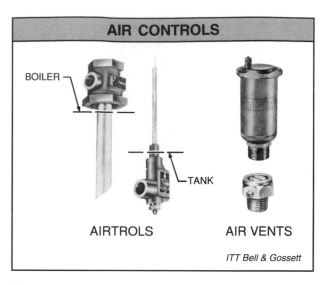

Figure 6-9. Air controls vent air from the water in a hydronic heating system.

Air Controls

Air controls in a hydronic heating system prevent air from building up inside a boiler, boiler piping, or terminal devices. Air in a hydronic heating system can cause many problems such as water hammer and corrosion. *Water hammer* is the condition that occurs when water pounds against the inside of a pipe. The pounding can cause damage to valves and other boiler fittings. *Corrosion* is a condition that occurs when oxygen in air reacts and breaks down metal. Corrosion causes early failure of metal parts such as the boiler, circulating pump, and piping. Air is carried into a system with makeup water or it may be drawn in through leaks on the inlet side of the circulating pump. *Makeup water* is water added to a boiler to replace water lost by leaks in a hydronic heating system. The air enters a boiler mixed in solution with the makeup water. As the water is heated and circulated through the hydronic heating system, the air separates from the water. Air controls pick up air from the water as the air-water solution circulates through the control. Air controls include Airtrols® and air vents.

Airtrols®. An *Airtrol®* is a mechanical device that vents air from the water in the boiler and expansion tank. Airtrols® are installed on the outlet line and on the expansion tank to remove air from the boiler and expansion tank. Airtrol® is a registered trademark of ITT Bell & Gossett. See Figure 6-9.

Air Vents. Air vents are automatic or manual devices that vent air from a hydronic heating system without allowing water to escape from or air to enter the system. Air vents are installed in the highest parts of the piping system where the air rises as it comes out of solution with the water. As an air-water solution is heated, the air separates from the solution and rises because air is lighter than water. This air is then vented from the hydronic heating system through air vents.

Valves

Valves regulate the flow of boiler water and maintain safe boiler operation in hydronic heating systems. Valves used on hydronic heating systems include feedwater, check, stop, pressure-reducing, flow control, bottom blowdown, and spring safety valves.

The boiler and piping system of a hydronic heating system must be completely filled with water at all times. The water level in a hydronic heating system is important. The water level of a boiler must be maintained at a point where the heat exchange surfaces exposed to the hottest temperatures on the fire side are covered with water on the water side. The water level changes because of water leaks and water that leaves the boiler with air that is vented from the system.

Feedwater Valves. Feedwater valves control the flow of makeup water into a boiler to make up for losses. The makeup water piping has a check valve and stop valve. *Check valves* are valves that allow flow in only one direction. Check valves control the direction of water flow. Many different kinds of check valves are available. The disk in a check valve opens when a fluid pushes against it in one direction but closes when the direction of flow reverses. Check valves are used in the makeup water piping to a boiler where water should flow in only one direction. See Figure 6-10.

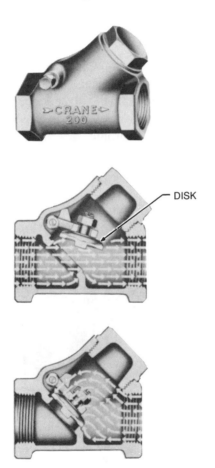

Crane Valves

Figure 6-10. Check valves allow boiler water to flow in only one direction.

Gate valves and globe valves are used as stop valves. A *gate valve* is a valve that has an internal gate that slides over the opening through which water flows. A gate valve is a two position valve that is open or closed and is not designed to throttle or regulate the flow of water. When the gate is

open, the valve allows full flow. When the gate is closed, no water flows. Gate valves have little resistance to water flow when open and are sealed tightly when closed.

A *globe valve* is a valve that has a gasket that rises or lowers over a port through which water flows. The port is at right angles to the flow of water. The water must make two 90° turns to exit the valve. A globe valve is an infinite position valve that is used to regulate water flow. Resistance and pressure drop through the valve is controlled by the amount the valve is opened. See Figure 6-11.

A *stop valve* is a valve that stops water flow. Stop valves are located anywhere in a hydronic heating system where water may have to be completely shut OFF. Gate valves are often used as stop valves because gate valves are either open or closed. Globe valves may also be used as stop valves.

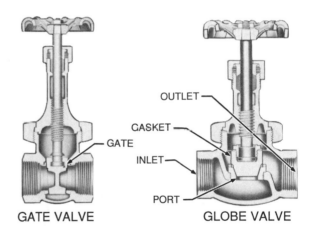

Crane Valves

Figure 6-11. Gate valves completely stop boiler water flow. Globe valves regulate boiler water flow but are sometimes used to stop boiler water flow.

Pressure-reducing Valves. A *pressure-reducing valve* reduces pressure of makeup water to approximately 12 psi to 18 psi so that the water can be used in a boiler. A pressure-reducing valve is located in the makeup water piping that leads to a boiler. See Figure 6-12.

Flow Control Valves. A *flow control valve* is a valve that regulates the flow of water in a hydronic heating system. A flow control valve may be manual or automatic.

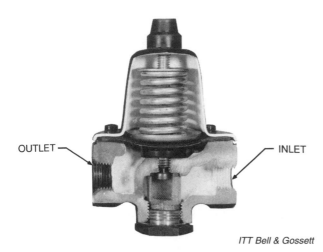

OUTLET — — INLET

ITT Bell & Gossett

Figure 6-12. A pressure-reducing valve reduces the pressure of makeup water so that the water can be used in a boiler.

A *manual flow control valve* is a globe valve that manually controls the flow of water in a hydronic heating system. An *automatic flow control valve* is a check valve that opens or closes automatically. When a circulating pump is ON, the flow control valve opens to allow water flow. When the circulating pump is OFF, the flow control valve closes and prevents water from circulating by natural circulation. The flow control valve is located on the outlet pipe of a boiler. See Figure 6-13.

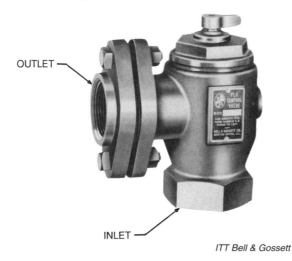

OUTLET —

INLET —

ITT Bell & Gossett

Figure 6-13. Flow control valves regulate the flow of water in a hydronic heating system.

Makeup water contains minerals and scale-forming salts. As makeup water fills hydronic heating systems, the minerals and salts in the water build up in the system. Chemicals are often introduced into the hydronic heating system with the makeup water to prevent the buildup of scale on the boiler walls and piping. *Scale* is a hard, brittle substance that forms when minerals and salts such as calcium carbonate and magnesium carbonate deposit on heating surfaces. Scale buildup reduces heat transfer through heating surfaces. The chemicals keep the minerals in suspension. The minerals settle from the boiler water to the bottom of the boiler and form sludge, which is a mud-like material. Blowdown removes boiler water and sludge from the bottom of a boiler. *Blowdown* is the rapid removal of the lower portions of boiler water.

Blowdown Valves. A *blowdown valve* is a quick-opening manual valve located at the lowest part of a boiler. A blowdown valve drains the boiler water, which removes dirt and sludge that accumulate in a hydronic heating system. See Figure 6-14.

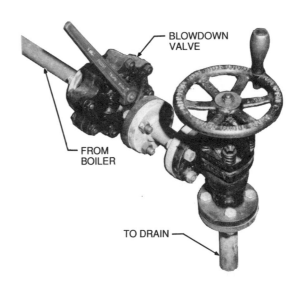

BLOWDOWN VALVE

FROM BOILER

TO DRAIN —

Figure 6-14. Blowdown valves drain boiler water and sludge from the bottom of a boiler.

Safety Valves. A *safety valve* is a valve that prevents excessive pressure from building up in a boiler and hydronic heating system in the event of a boiler malfunction. The safety valve is the most important valve on a boiler. A safety valve is mounted in the system piping on the top of a boiler with no other valves between it and the boiler.

If the pressure in the system exceeds normal working pressure, the safety valve opens automatically and exhausts hot water and/or steam. ASME code requires that every boiler have at least one safety valve. Spring safety valves are used often.

A *spring safety valve* is a valve that has a valve disk held closed against a valve seat by a compression spring. The spring compression is set to match the pressure desired in the system. If the pressure in the system exceeds the pressure exerted by the compression spring, the valve disk lifts off the seat. The spring is held down on the valve seat by the compression in the spring. A test lever is used to test the valve. Safety valves should be tested regularly. See Figure 6-15.

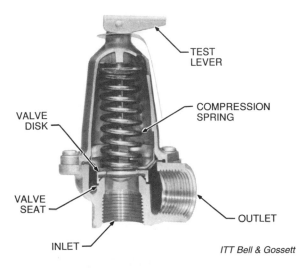

ITT Bell & Gossett

Figure 6-15. A spring safety valve opens automatically and exhausts hot water and/or steam if the pressure in a hydronic heating system exceeds normal working pressure.

PIPING SYSTEMS

Piping systems in hydronic heating systems are used to distribute heated water from a boiler and return the water to the boiler. Piping systems vary depending on the degree of temperature control desired and the specific application. In most systems the pipe is made of steel or cast iron. In small systems the pipe may be copper. Piping systems used for hydronic heating systems include one-, two-, three-, and four-pipe systems. Because of variations in buildings, each piping system is designed for a specific kind of building.

One-pipe

One-pipe systems are used in small- to medium-size buildings. The two kinds of one-pipe systems are series and primary-secondary systems.

Series. A *one-pipe series system* circulates water through one pipe and through each terminal device in turn. As the water flows through the system, a temperature drop occurs in each terminal device.

The inlet temperature at each terminal device is lower than the temperature in the preceding terminal device. The inlet temperature of the last terminal device in a system is lower than the inlet temperature of the first terminal device. The difference is the sum of the temperature drops of all of the terminal devices. A one-pipe series piping system is an economical system to install and gives good temperature control when used in small buildings. See Figure 6-16.

Primary-Secondary. A *one-pipe primary-secondary system* is a piping system that circulates boiler water through a primary loop. The terminal devices are connected to the primary loop in parallel. This connection forms a secondary loop through each terminal device. The secondary loop is connected to the main supply pipe by a diverter fitting.

A *diverter fitting* is a tee that meters the water flow from the primary loop. Water flow can be adjusted for gallons per minute required in the secondary loop through the terminal device. The return water from each terminal device returns to the primary loop so the temperature drop of each device affects others in the system, but the flow rate in each terminal device can be controlled within its own secondary loop. Flow control gives better control over the heat output of the terminal device.

Two-pipe

Two-pipe systems are used in medium- to large-size residential and commercial buildings. Two-pipe systems consist of a supply pipe that carries hot water from the boiler to terminal devices and a return pipe that carries water back to the boiler.

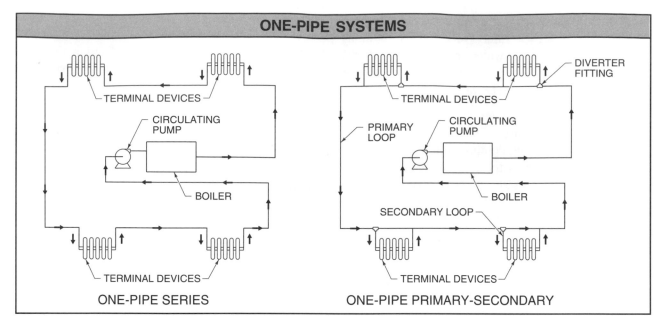

Figure 6-16. One pipe is used to supply and return boiler water in one-pipe hydronic piping systems.

Two-pipe systems include direct-return and reverse-return systems. See Figure 6-17.

Direct-return. A *two-pipe direct-return system* is a piping system that circulates the supply water in the opposite direction of the circulation of the return water. One side of a terminal device is connected to the supply piping and the other side is connected to the return piping. Water in the return piping flows from each terminal device back to the boiler through the return piping. A direct-return piping system is difficult to balance because the water that flows through the final terminal device must be pumped back through the whole piping system to the boiler.

Balancing is the adjustment of the resistance to the flow of water through the piping loops in a

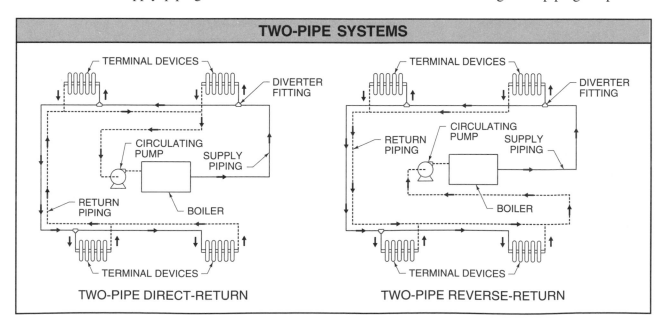

Figure 6-17. Two separate pipes are used to supply and return boiler water in two-pipe hydronic piping systems.

system. Balancing assures the proper amount of water flow through each terminal device for the amount of heat required at each terminal device.

Reverse-return. A *two-pipe reverse-return system* is a piping system that circulates return water in the same direction as supply water. The same flow direction makes the system easy to balance because the distance around the piping system is the same for the water flowing through each terminal device.

Three-pipe

A three-pipe system has two supply pipes and one return pipe. A three-pipe system is used when different parts of a system require heating and cooling simultaneously. One supply pipe is connected to a boiler and the other supply pipe is connected to a chiller. See Figure 6-18.

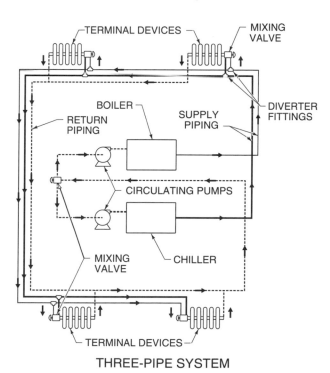

THREE-PIPE SYSTEM

Figure 6-18. A three-pipe hydronic piping system uses two supply pipes and one return pipe.

Both supply pipes are connected to each terminal device. The return piping is a common line that runs from each terminal device back to the boiler

and chiller. Mixing valves regulate the flow of water from the two supply pipes to the terminal devices in the system. On a call for heat, water from the heating supply pipe flows into a terminal device. On a call for cooling, water from the chiller supply pipe flows into a terminal device.

Three-pipe systems provide good control of air temperature but are more expensive to install than two-pipe systems. Three-pipe systems are also more expensive to operate because the hot and cold return water are mixed.

Four-pipe

A four-pipe system separates heating and cooling. The terminal devices are connected to both heating pipes and cooling pipes. Water flow to the terminal devices is controlled by mixing valves. Heating and cooling can be achieved at any terminal device when desired. Four-pipe piping systems are expensive to install but provide excellent control of air temperature and are more economical to operate than three-pipe systems. See Figure 6-19.

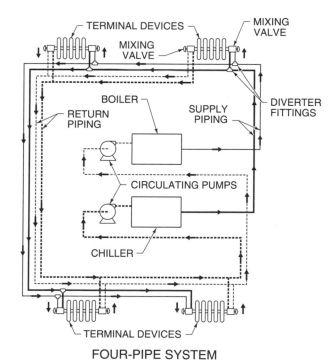

FOUR-PIPE SYSTEM

Figure 6-19. A four-pipe hydronic piping system uses supply and return heating piping and supply and return cooling piping.

CIRCULATING PUMPS

Circulating pumps are pumps that move water from a boiler and through the piping system and terminal devices of a hydronic heating system. A circulating pump may be connected on either the supply or return piping but is usually connected close to the boiler on the return piping. The number and size of circulating pumps are determined by the volume of water to be moved and resistance to flow in the system. Hydronic heating systems that have many piping loops that feed different zones in a building have a pump located on each loop.

Most circulating pumps that are used on hydronic heating systems are centrifugal pumps. *Centrifugal pumps* are pumps that have a rotating impeller wheel inside a cast iron or steel housing. An *impeller wheel* is a disk with blades that radiate from a central hub. The water inlet is an opening in the pump housing that is parallel with the impeller wheel shaft. The water outlet is an opening in the pump housing that is located on the outer perimeter of the pump housing at right angles to the impeller shaft. See Figure 6-20.

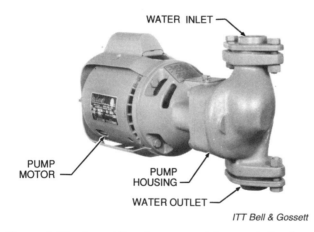

ITT Bell & Gossett

Figure 6-20. A centrifugal pump contains a rotating impeller wheel inside a cast iron or steel housing.

Return piping is connected to a centrifugal pump at the water inlet. Supply piping is connected to a centrifugal pump at the water outlet. The impeller of a circulating pump rotates at speeds from 1150 rpm to 3550 rpm. As the impeller rotates, a low-pressure area is created at the center of the impeller. The impeller throws water out of the water outlet by centrifugal force.

Circulating pumps are driven by electric motors. The motor is often part of the pump housing on a circulating pump that has a pumping rate from 0 gpm to 100 gpm. The motor on a circulating pump that has a pumping rate above 100 gpm is usually connected indirectly by a belt and sheave arrangement, or by an in-line connection on the shaft of the motor. Pipe-mounted circulating pumps are small pumps that are installed in the piping of a hydronic system. Floor-mounted circulating pumps are large pumps that must have piping run to them. See Figure 6-21.

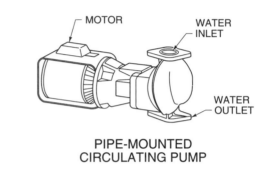

**PIPE-MOUNTED
CIRCULATING PUMP**

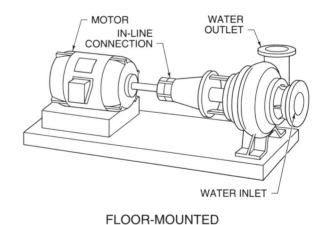

**FLOOR-MOUNTED
CIRCULATING PUMP**

Figure 6-21. Small pumps are installed in the piping of a system while large pumps are floor-mounted and piping must be run to them.

TERMINAL DEVICES

Terminal devices transfer heat from the hot water in a hydronic heating system to air in building spaces. A supply pipe runs from the boiler to each terminal device and a return pipe leaves each terminal device and runs back to the boiler. Radiant heating and convection heating terminal devices are the two types of terminal devices.

Radiant Heating

Thermal radiation is heat transfer that occurs as radiant energy waves transport energy from one object to another. Radiant heating systems are based on thermal radiation. Radiant heating occurs when a radiant surface is heated and the radiant surface gives off heat in the form of radiant energy waves.

The *radiant surface* is the heated surface from which radiant energy waves are generated. The radiant energy waves are absorbed only by opaque objects in direct line-of-sight of the radiant surface and are not absorbed by the air. Opaque objects absorb the radiant energy waves and become heated. As the objects are heated, they give up some of the heat to the air around them, which raises the temperature of the air.

A *hydronic radiant heater* is a heater that has a radiant surface that is heated by hot water to a temperature high enough to radiate energy. Three hydronic radiant heaters are surface radiation systems, radiant panels, and radiators. See Figure 6-22.

Surface Radiation Systems. A *surface radiation system* is a heating system that uses the interior surfaces of a room as radiant surfaces. A piping coil or grid is built into the construction material of the ceiling, walls, or floor of a room. Insulation is installed behind the piping and a thin layer of surfacing material is installed over it.

The radiant surface is heated by hot water that flows through the pipes. When the temperature of the radiant surface is higher than the temperature of objects in the room, heat flows from the radiant surface to the objects. The amount of heat produced is regulated by controlling the surface temperature.

Radiant Panels. A *radiant panel* is a factory-built panel with a radiant surface that is heated by a piping coil or grid built into it. Radiant panels are individual heaters that are not an integral part of a building. Radiant panels are enclosed in an insulated cabinet. Radiant panels are used for spot heating. *Spot heating* provides heat at a particular area. The majority of space in a building, much of which may be unused, is not heated with spot heating from radiant panels.

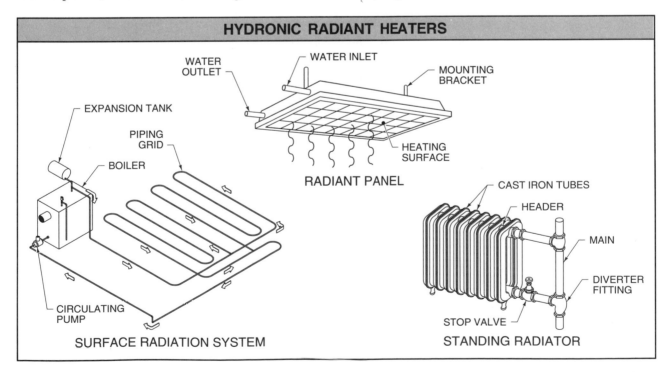

Figure 6-22. Radiant heating devices use hot water to heat a surface to a temperature high enough for heat to flow from the surface.

Radiators. *Radiators* are heat-distributing devices that consist of hollow metal coils or tubes. Hot water passes through the coils or tubes. Heat from the hot water is transferred to the tubes, which emit heat. The air in a building space is heated by radiation and convection.

A *standing radiator* is a section of vertical or horizontal tubes that are connected with headers at each end. Hot water from a boiler flows through the tubes. Heat is transferred to the air by radiation and convection when air flows over the tubes. Standing radiators are usually made of cast iron. Because the units are unattractive and usually hard to conceal, they are seldom used except in warehouses or factory buildings.

Convection Heating

Convection heating is based on heat transfer by convection. Convection is heat transfer that occurs when currents circulate between warm and cool regions of a fluid. The two kinds of convection heaters are cabinet convectors and baseboard convectors. See Figure 6-23.

Cabinet Convectors. A *cabinet convector* is a convection heater that is enclosed in a cabinet. The hot water coil or tubes and the front of the heater are covered by a sheet metal cabinet. The cabinet has a grill in the bottom for cool air to enter and a grill on the top for warm air to exit. Cabinet convectors can be assembled in sections to produce the amount of heat required to heat a building space. Cabinet convectors are available in different sizes between 24″ to 32″ high, 7″ deep and vary in length according to the heat output required. Cabinet convectors are installed under windows along the outside walls of a room.

Baseboard Convectors. A *baseboard convector* is a convection heater that is enclosed in a low cabinet that fits along a baseboard. Baseboard convectors are smaller than cabinet convectors and use a section of finned tube to transfer heat. A *finned tube* is a copper pipe with aluminum fins. The fins provide a larger heat transfer surface area. Baseboard

convectors are normally 12″ high and are installed at floor level along outside walls.

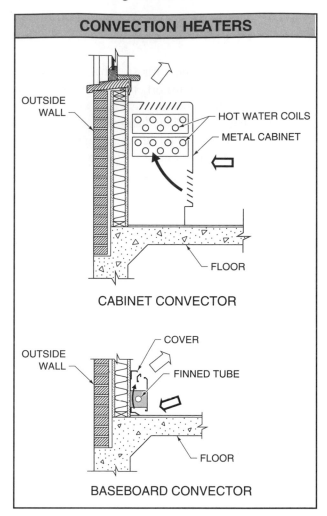

Figure 6-23. Convection heating devices contain hot water coils that heat air as the air flows over the hot water coils.

Forced Convection Heating

Forced convection heating creates convective currents by mechanically moving air past a hot water coil. Forced convection heaters use a blower or fan to move air. The blower draws air in the bottom or back of the heater, moves it past the hot water coil, and distributes the air to the building space. All forced convection heaters have a hot water coil, blower, controls, and a sheet metal cabinet. The three kinds of forced convection heaters used in hydronic heating systems are unit heaters, cabinet heaters, and unit ventilators. See Figure 6-24.

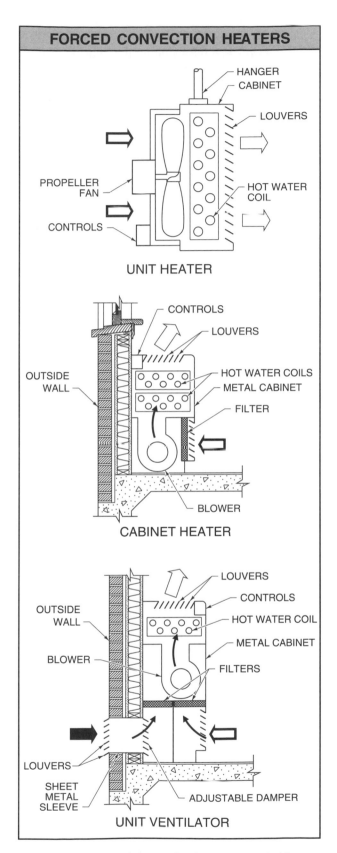

FORCED CONVECTION HEATERS

HANGER
CABINET
LOUVERS
PROPELLER FAN
HOT WATER COIL
CONTROLS

UNIT HEATER

CONTROLS
LOUVERS
OUTSIDE WALL
HOT WATER COILS
METAL CABINET
FILTER
BLOWER

CABINET HEATER

LOUVERS
CONTROLS
OUTSIDE WALL
HOT WATER COIL
METAL CABINET
BLOWER
FILTERS
LOUVERS
SHEET METAL SLEEVE
ADJUSTABLE DAMPER

UNIT VENTILATOR

Figure 6-24. Forced convection heaters contain blowers that move air.

Unit Heaters. A *unit heater* is a self-contained forced convection heater that contains a propeller fan, hot water coil, and controls in one cabinet. Unit heaters are normally used as space heaters. Unit heaters are suspended from the ceiling framework of a building space. Return air is drawn in the back of the unit by the propeller fan. The air is blown across the hot water coil and out the front of the unit through louvers.

Unit heaters are made in different sizes and have different heating capacities. The heat output of a unit is controlled by regulating the flow of water and the speed of the propeller fan.

Cabinet Heaters. A *cabinet heater* is a forced convection heater that has a blower, hot water coils, filter, and controls in one cabinet. The blower blows heated air from the hot water coil. Cabinet heaters differ from cabinet convectors in that a blower moves the air and a filter cleans the air. The blower adds energy to the air so that directional louvers can be used on the top or front of the heater to distribute the air more efficiently.

The size of a cabinet heater depends on the heat output required. Most cabinet heaters are from 30″ to 36″ in height, 6″ to 7″ deep, and vary in length. Cabinet heaters are installed along the outside wall of a room under a window.

Unit Ventilators. A *unit ventilator* is a forced convection heater that has a blower, hot water coil, filters, controls, and a cabinet that has an opening for outdoor ventilation. The blower blows heated air from the hot water coil. Unit ventilators differ from cabinet heaters in that they have an opening to bring in air from outdoors for ventilation. The cabinet fits on an outside wall so that the connection to the outdoors can be made with an opening in the back of the cabinet for an outdoor air inlet. A sheet metal sleeve extends from the inlet through the wall. A louver covers the opening to the outdoors. The outdoor air inlet has an adjustable damper on the inside for regulating the amount of air admitted to the unit ventilator. Unit ventilators are often used in buildings that have a large number of people and a need for constant ventilation.

CONTROLS

Controls operate the components of a hydronic heating system. Hydronic heating system controls include power, operating, boiler, and terminal device controls. Most power and operating controls are similar to the power and operating controls in forced-air heating systems. Boiler controls maintain safe and efficient boiler operation. Terminal device controls control the heat given off by terminal devices by controlling the flow of water and air through the terminal device.

Power Controls

Power controls control the flow of electricity to a boiler. Power controls are installed in the electrical circuit between the power source and the boiler. Power controls for hydronic heating systems include disconnects, fuses, and circuit breakers. All of these controls are also used in forced-air heating systems. Power controls should be installed by a licensed electrician per article 424 of the National Electric Code®. See Figure 6-25.

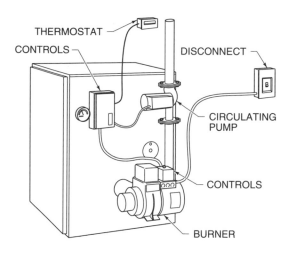

Figure 6-25. Power controls control the flow of electricity to a boiler. Operating controls cycle equipment ON or OFF as required.

Operating Controls

Operating controls cycle equipment ON or OFF as required. Operating controls for hydronic heating systems include transformers, thermostats, relays, contactors, magnetic starters, and solenoids. All of these controls are the same as the controls used in forced-air heating systems.

Boiler Controls

Boiler controls ensure that a boiler operates safely and that there is no danger to personnel or equipment. The three categories of boiler controls are operating, safety, and combustion safety controls.

Boiler Operating Controls. *Boiler operating controls* automatically and safely energize boiler burner(s) or resistance heating elements to maintain the temperature of the water at the setpoint temperature. Boiler operating controls include the aquastat and Pressuretrol®.

An *aquastat* is a temperature-actuated electric switch used to limit the temperature of boiler water. The electric switch is actuated by a bellows. The bellows is connected to a remote bulb sensor. The remote bulb sensor is charged with a small amount of refrigerant and is located either in a well in the water tank on the boiler or immersed directly in the water. See Figure 6-26.

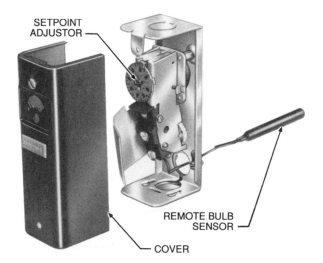

Honeywell Inc.

Figure 6-26. An aquastat senses the temperature of boiler water and cycles the burners as necessary to maintain the boiler water temperature at the setpoint.

If the temperature of the water increases, the refrigerant in the bulb will boil off and increase the pressure in the bulb. The increase in pressure pushes the bellows out. If the temperature of the water decreases, the refrigerant in the bulb will condense and decrease the pressure in the bulb. The decrease in pressure moves the bellows in. The movement of the bellows opens or closes the contacts in the circuit. If the temperature of the water in the boiler rises above the setpoint of the aquastat, the switch will open. If the temperature of the water falls below the setpoint of the aquastat, the switch will close.

An aquastat has an adjustor for the setpoint and differential. The setpoint is the temperature at which the water is maintained. The differential is the difference between the temperature at which the aquastat turns the burner ON and the temperature at which the aquastat turns the burner OFF. An aquastat has a differential to prevent the control from shutting the burner ON or OFF rapidly.

As water is pumped from a boiler and circulated through terminal devices in a system, the water temperature drops because heat is taken from the water in each terminal device. The amount of the temperature drop depends on the load on the system at the time. Since the heating load varies with outside and inside load conditions, the temperature drop of the water flowing through the system also varies. The aquastat on a boiler senses the temperature drop and cycles the burners as necessary to maintain the boiler water temperature at a setpoint.

A *Pressuretrol*® is a pressure-actuated mercury switch. A mercury switch is a switch that uses the movement of mercury in a glass bulb to control the flow of electricity in a circuit. A Pressuretrol® controls the burner on a boiler by starting or stopping the boiler burner(s) based on the pressure inside the boiler. A Pressuretrol® controls water pressure and an aquastat controls water temperature. See Figure 6-27.

A Pressuretrol® is a Honeywell Incorporated trademark, but it is widely used to refer to all basic automatic pressure controls for boilers. Other names for these pressure controls are operating limit pressure control and pressurestat.

Boiler Safety Controls. *Boiler safety controls* are controls that monitor boiler operation to prevent excessive temperature and pressure. If a malfunction that might cause unsafe operation occurs, the safety

controls will either correct the problem or shut the boiler down. Boiler safety controls include high-temperature limit controls and spring safety valves.

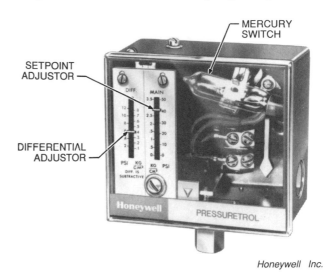

Honeywell Inc.

Figure 6-27. A Pressuretrol® controls the burner on high-pressure boilers.

A *high-temperature limit control* is a temperature-actuated electric switch that senses boiler water temperature. High-temperature limit controls on a boiler serve the same purpose as the high-temperature limit control on a forced-air furnace. A high-temperature limit control will shut a burner OFF if a malfunction occurs. If a gas valve is stuck in the open position during normal boiler operation, the temperature of the water will rise above a safe setpoint. A high-temperature limit control shuts the burner OFF. High-temperature limit control is usually a function of the aquastat.

A *spring safety valve* controls the pressure in a boiler by exhausting hot water and steam to the atmosphere when the pressure in the boiler rises above the normal working pressure of the boiler. A spring safety valve is the most important safety device on a boiler. ASME code requires that every boiler have at least one safety valve. See Figure 6-28.

Boiler Combustion Safety Controls. *Boiler combustion safety controls* are controls that are similar to the combustion safety controls used on forced-air furnaces. The type of burner used determines which boiler combustion safety control should be used. Boiler combustion safety controls detect flame in different ways. The three types of combustion safety controls are pilot safety, flame rod, and flame surveillance.

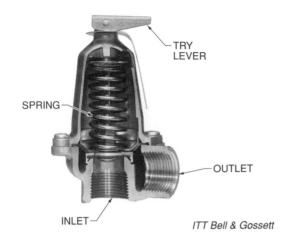

Figure 6-28. A spring safety valve is the most important safety device on a boiler.

Terminal Device Controls

Terminal device controls control the temperature of the air in a building space by regulating the air and water flow through the terminal devices. Circulating pumps, zone valves, and blowers in the terminal devices regulate air and water flow.

Circulating pumps are switched ON or OFF by a signal from a thermostat, which controls the flow of hot water in a hydronic piping system. A circulating pump control provides hot water for a complete hydronic heating system or secondary piping loop in a system. A *secondary piping loop* is a loop of pipe off a main supply and return pipe.

Zone Valves. Zone valves regulate the flow of water in a control zone or terminal device of a building. A *control zone* is any part of a building that is controlled by one controlling device. Zone valves may be manual or automatic. Manual zone valves are set by hand to regulate the flow of water in a piping loop. Automatic zone valves are opened or closed automatically by a valve motor. Automatic zone valve motors are driven by an electric or electronic control system or by air in a pneumatic control system and are controlled by a zone thermostat. See Figure 6-29.

Honeywell Inc.

Figure 6-29. Zone valves regulate the flow of water in a control zone or terminal device of a building.

Valves used for controlling water flow are either digital or modulating valves. A *digital valve* is a two position (ON/OFF) valve. The valve is either completely open or completely closed. A *modulating valve* is an infinite position valve. The valve may be completely open, completely closed, or at any intermediate position in response to the control signal the valve receives.

Low-temperature Limit Controls. A low-temperature limit control is used on terminal devices that use outdoor air for ventilation. A *low-temperature limit control* is a temperature-actuated electric switch that will energize the damper motor and shut the damper if the ventilation air temperature drops below a setpoint, usually 32°F. If the outdoor air temperature drops below 32°F in the heating coil of a terminal unit, the water coil may freeze and must have freeze-protection controls. The damper motor also opens the water valve to ensure water continues to circulate through the coil while the temperature is low. See Figure 6-30.

Thermostats. Thermostats used with terminal devices regulate water and air flow by controlling the circulating pump(s), zone valves, or terminal device blowers. A thermostat is used to control a digital valve. On a call for heat, the thermostat opens the valve. When the temperature reaches the thermostat setpoint, the thermostat closes the valve.

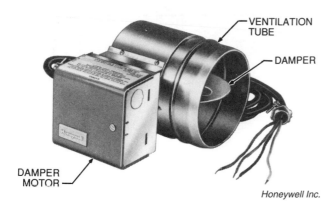

Honeywell Inc.

Figure 6-30. A low temperature limit control energizes a damper motor if the ventilation air temperature drops below the setpoint.

A proportional thermostat is used to control a modulating valve. A *proportional thermostat* contains a potentiometer that sends out an electric signal that varies as the temperature varies. A *potentiometer* is a variable-resistance electric device that divides voltage proportionally between two circuits. The potentiometer receives the signal and converts it to mechanical action in the valve motor.

Blowers. The blowers in terminal devices are controlled by relays, which cycle the blower in the terminal device ON or OFF. The relays are the same as those used in forced-air heating systems. A signal from the thermostat closes the relay contacts, which supply electricity to the blower motor. Thermostats control the blower in a unit heater or unit ventilator. Thermostats used to control terminal devices may be located inside the cabinet of the unit that they control or located remotely in the space being heated. See Figure 6-31.

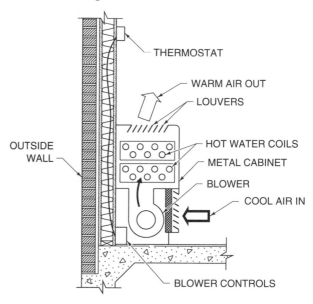

Figure 6-31. The blowers in terminal devices are controlled by relays.

Review Questions

6

1. List three ways that hydronic heating systems carry heat.

2. List the six main parts of a hydronic heating system.

3. What is the main purpose of a boiler?

4. List the four ways boilers are classified.

5. Where is the water in relation to the hot products of combustion in a firetube boiler?

6. Where is the water in relation to the hot products of combustion in a watertube boiler?

7. What does input relate to when finding the amount of heat produced by a boiler?

8. What is the difference between gross output and net unit output when finding the amount of heat produced by a boiler?

9. An automatic feedwater valve provides makeup water to a hydronic heating system. What is makeup water?

10. Why does a safety valve open? What happens when a safety valve opens?

11. What is the primary purpose of a check valve?

12. What does an expansion tank do for a hot water heating system?

13. List the two types of one-pipe hydronic piping systems. Explain how they are different.

14. List the two types of two-pipe hydronic piping systems. Explain how they are different.

15. What part of a circulating pump actually moves the water?

16. What is the function of the terminal devices used in hydronic heating systems?

17. Describe how a radiant heating terminal device heats the air in a building space.

18. Describe the basic design of a convection heater.

19. What must be added to a convection heater to change it to a forced convection heater?

20. What is the distinguishing feature of a unit heater?

21. What is the distinguishing feature of a unit ventilator?

22. List and describe the function of three categories of controls used in hydronic heating systems.

23. What is the main function of an aquastat?

24. What is the main function of combustion safety controls?

25. What terminal device requires some kind of freeze-protection control? Why?

Refrigeration Principles

Chapter 7

Refrigeration is the process of moving heat from an area where it is undesirable to an area where it is not objectionable. Refrigeration is based on a law of physics that states that matter gains or loses heat when it changes state. The two main types of refrigeration processes are mechanical compression and absorption. Mechanical compression refrigeration uses mechanical equipment to produce a refrigeration effect. Absorption refrigeration uses the absorption of one chemical by another chemical and heat transfer to produce a refrigeration effect.

REFRIGERATION PRINCIPLES

Refrigeration is the process of moving heat from an area where it is undesirable to an area where the heat is not objectionable. According to the second law of thermodynamics, heat always flows from a material at a high temperature to a material at a low temperature.

A *refrigeration system* is a closed system that controls the pressure and temperature of a refrigerant to regulate the absorption and rejection of heat by the refrigerant. One side of a refrigeration system decreases the pressure on and temperature of the refrigerant, which causes the refrigerant to absorb heat from air or water in the system. The air or water is cooled when the heat is absorbed out of the air or water. The air or water is then used for cooling. The other side of a refrigeration system increases the pressure on and temperature of the refrigerant, which causes the refrigerant to reject heat to the air or water in the system. The air or water is heated and used for heating or exhausted to the atmosphere.

Refrigeration applications include commercial and industrial refrigeration and air conditioning. A commercial or industrial refrigeration system uses mechanical equipment to produce a refrigeration effect for applications other than human comfort.

An air conditioning system produces a refrigeration effect to maintain comfort within a building space.

MECHANICAL COMPRESSION REFRIGERATION

Mechanical compression refrigeration is a refrigeration process that produces a refrigeration effect with mechanical equipment. A mechanical compression refrigeration system consists of a compressor, refrigerant, condenser, expansion device, evaporator, refrigerant lines, and accessories. See Figure 7-1.

A *compressor* is a mechanical device that compresses refrigerant or other fluid. A compressor increases the temperature of and pressure on refrigerant vapor and produces the high pressure in the high-pressure side of the system. A *refrigerant* is a fluid that is used for transferring heat in a refrigeration system. Refrigerants have a low boiling (vaporization) point, which makes refrigerants boil and vaporize at room temperature.

A *condenser* is a heat exchanger that removes heat from high-pressure refrigerant vapor. High-pressure refrigerant vapor flows through the condenser and a condensing medium passes across the outside of the condenser. Heat flows from the hot refrigerant vapor to the cold condensing medium.

A *condensing medium* is a fluid (air or water) that has a lower temperature than the refrigerant, which causes heat to flow to the medium. A condensing medium removes heat from a refrigerant because it has a lower temperature than the refrigerant. Air and water are condensing mediums used in refrigeration systems. As the refrigerant vapor gives up heat to the condensing medium in a condenser, the vapor condenses to a liquid.

An *expansion device* is a valve or mechanical device that reduces the pressure on a liquid refrigerant by allowing the refrigerant to expand. As the pressure on the liquid refrigerant decreases, some of the liquid refrigerant vaporizes because it has a low boiling point. The refrigerant absorbs heat as it vaporizes, which cools the remainder of the refrigerant. The refrigerant then flows as a liquid-vapor mixture to the evaporator.

An *evaporator* is a heat exchanger that adds heat to low-pressure refrigerant liquid. Low-pressure refrigerant liquid flows through the evaporator and an evaporating medium passes across the outside of the evaporator. Heat flows from the hot evaporating medium to the cold refrigerant. An *evaporating medium* is a fluid (air or water) that is cooled when heat is transferred from the medium to the cold refrigerant. An evaporating medium adds heat to a refrigerant because it has a higher temperature than the refrigerant. As the liquid refrigerant absorbs heat from the evaporating medium, the refrigerant boils and vaporizes.

Refrigerant lines carry refrigerant and connect the components of a mechanical compression refrigeration system. Accessories monitor the system to ensure proper operation.

Pressure-Temperature Relationships

Pressure is the force per unit area that is exerted by a fluid. Pressure is expressed in pounds per square inch (psi). *Atmospheric pressure* is the force exerted by the weight of the atmosphere on the earth's surface. Atmospheric pressure at sea level is 14.7 psi. Atmospheric pressure is expressed in pounds per square inch and measured with a mercury barometer. A *mercury barometer* is an instrument used to measure atmospheric pressure. A mercury barometer is commonly calibrated in inches of mercury (in. Hg). A mercury barometer consists of a glass tube that is closed on one end

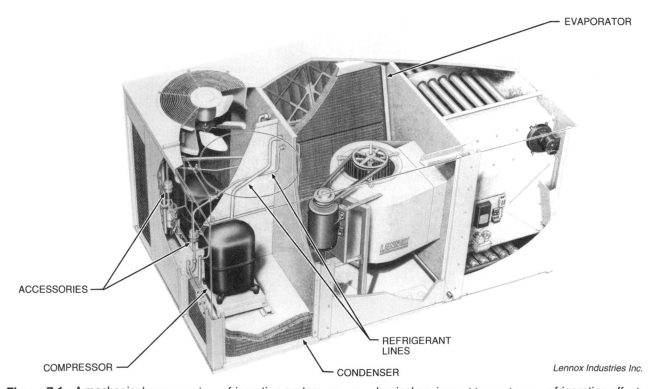

EVAPORATOR

ACCESSORIES

COMPRESSOR

REFRIGERANT LINES

CONDENSER

Lennox Industries Inc.

Figure 7-1. A mechanical compression refrigeration system uses mechanical equipment to produce a refrigeration effect.

and filled completely with mercury. The tube is inverted in a dish of mercury. A vacuum is created at the top of the tube as the mercury tries to run out of the tube. *Vacuum* is a pressure lower than atmospheric pressure. Vacuum is expressed in inches of mercury. The pressure of the atmosphere on the mercury in the open dish prevents the mercury in the tube from running out of the tube. The height of the mercury in the tube corresponds to the pressure of the atmosphere on the mercury in the open dish. Minute pressure changes are expressed in inches of water column (in. WC). See Figure 7-2. See Appendix.

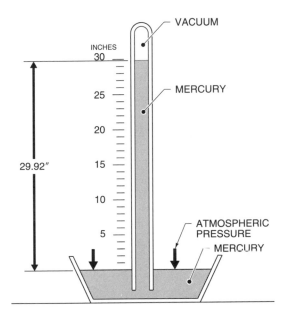

PSIA	PSIG or PSI	in. Hg	in. WC
45	30	61.1	—
30	14.7	29.92	408
.07	1	2.03	27.68
15.2	.49	1	—
15.13	.43	.88	12
.003	.04	.07	1
14.7	0	0	0
10	−5	10	—
5	−10	20	—
0	−14.7	29.92	−408

Figure 7-2. A mercury barometer is used to measure atmospheric pressure.

Gauge pressure is pressure above atmospheric pressure that is used to express pressures inside a closed system. Gauge pressure is expressed in pounds per square inch gauge (psig). Gauge pressure assumes that atmospheric pressure is zero (0 psi). *Absolute pressure* is pressure above a perfect vacuum. Absolute pressure is the sum of gauge pressure plus atmospheric pressure. Absolute pressure is expressed in pounds per square inch absolute (psia).

Pressure outside a closed system such as normal air pressure is expressed in pounds per square inch absolute. The difference between gauge pressure and absolute pressure is the pressure of the atmosphere at sea level at standard conditions (14.7 psi). A pressure gauge reads 0 psig at normal atmospheric pressure. To find absolute pressure when gauge pressure is known, add the atmospheric pressure of 14.7 psi to the gauge pressure. Absolute pressure is found by applying the formula:

$$psia = psig + 14.7$$

where

$psia$ = pounds per square inch absolute

$psig$ = pounds per square inch gauge

14.7 = constant

Example: Finding Absolute Pressure

A gauge reads 68 psig on the low-pressure side of an operating refrigeration system. Find the absolute pressure.

$$psia = psig + 14.7$$

$$psia = 68 + 14.7$$

$$psia = \textbf{82.7 psia}$$

Boiling point is the temperature at which a liquid vaporizes. Boiling point is directly related to the pressure on the liquid. If the pressure on a liquid increases, the boiling point will be higher. If the pressure on a liquid decreases, the boiling point of the liquid will be lower. For example, at 14.7 psia the boiling point of water is 212°F. If the pressure on the water increases, the boiling point of the water will be higher. At 29.8 psia, the boiling point of water is 250°F. If the pressure on the water decreases, the boiling point of water will be lower. At 19.03″ Hg, which is in the vacuum range, the boiling point of water is 190°F. See Figure 7-3.

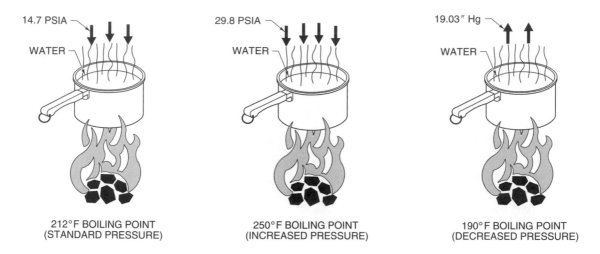

Figure 7-3. The temperature at which a liquid boils or condenses is directly related to pressure.

Condensing point is the temperature at which a vapor condenses to a liquid. If the pressure on a vapor decreases, the temperature at which the liquid condenses will decrease. If the pressure on a vapor increases, the temperature at which the vapor condenses will increase. All substances follow this pressure-temperature relationship.

Heat Transfer

In a mechanical compression refrigeration system, heat transfer occurs in the condenser and the evaporator. In the condenser, heat is transferred from the refrigerant that flows through the condenser to the condensing medium that passes across the outside of the condenser. The refrigerant condenses as it rejects heat to the condensing medium. This process heats the condensing medium. See Figure 7-4.

In the evaporator, heat is transferred from the evaporating medium that passes across the outside of the evaporator to the refrigerant that flows through the evaporator. The refrigerant vaporizes as it absorbs heat from the evaporating medium. This process cools the evaporating medium.

Pressure Control

Mechanical compression refrigeration systems have a high-pressure side and a low-pressure side. Pressure is controlled in the low-pressure side by reduced refrigerant flow, reduced pressure, and a relatively low suction pressure. See Figure 7-5.

The expansion device meters the flow of refrigerant in a refrigeration system. An expansion device is located just ahead of the evaporator in the liquid line. The *liquid line* is the refrigerant line that connects the condenser and the expansion device.

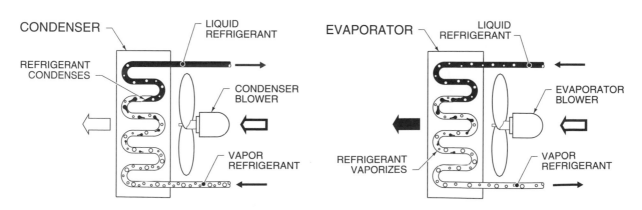

Figure 7-4. In a mechanical compression refrigeration system, heat transfer occurs in the condenser and the evaporator.

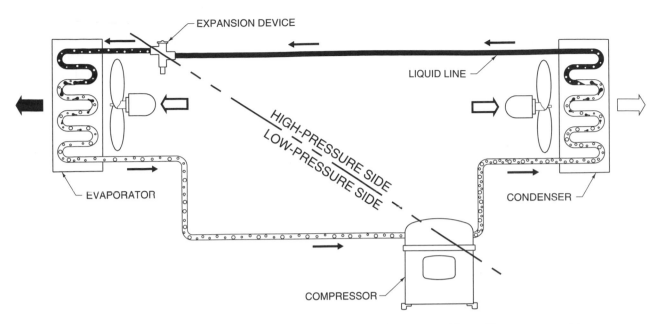

Figure 7-5. Mechanical compression refrigeration systems have a high-pressure side and low-pressure side.

As liquid refrigerant flows through the expansion device, the pressure on the refrigerant decreases. The decreased pressure causes some of the refrigerant to vaporize. This vaporization takes heat from and decreases the temperature of the rest of the liquid refrigerant. The liquid-vapor mixture flows from the expansion device directly to the evaporator. The pressure of the refrigerant remains the same through the low-pressure side of the system except for decreases in pressure caused by the evaporator, lines, or fittings. See Figure 7-6.

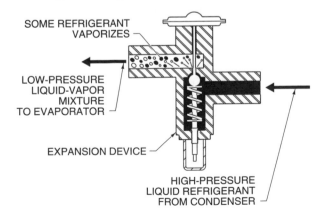

Figure 7-6. As liquid refrigerant flows through an expansion device, the decrease in pressure causes some of the refrigerant to vaporize.

Compressor suction pressure in the compressor maintains the low pressure in a mechanical refrig-

eration system. *Compressor suction pressure* is the pressure produced by the compressor when refrigerant is drawn into the compressor. Compressor suction pressure maintains low pressure in the low-pressure side of the system. Because refrigerant is drawn out of the evaporator to the suction inlet on the compressor as fast as it is introduced through the expansion device, the pressure on the low-pressure side of the system remains constant.

Compressor discharge pressure maintains high pressure in the high-pressure side of a refrigeration system. *Compressor discharge pressure* is the pressure produced by the compressor when refrigerant is discharged from the compressor. The compressor compresses and circulates the refrigerant. Refrigerant leaves the evaporator as a vapor and flows to the compressor. The refrigerant vapor is compressed in the compressor and leaves the compressor at a higher pressure. See Figure 7-7. Because the refrigerant absorbs heat from the compression process and the compressor motor windings, the refrigerant that leaves the compressor is hotter than the refrigerant in the rest of the refrigeration system. This hot refrigerant vapor flows to the condenser. The pressure through the high-pressure side is constant except for some pressure decrease from friction in the fittings and refrigerant lines.

Because of the low pressure in the evaporator and compressor suction pressure, the temperature of the refrigerant can be decreased to a temperature lower than the temperature of the evaporating medium. The temperature difference causes heat to flow from the evaporating medium to the refrigerant. Because of the high pressure in the condenser and the compressor discharge pressure, the temperature of the refrigerant can be increased to a temperature higher than the temperature of the condensing medium. This temperature difference causes heat to flow from the refrigerant to the condensing medium.

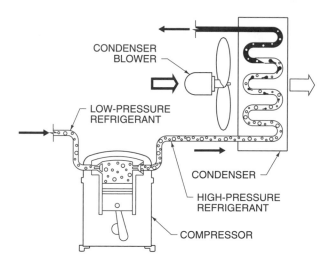

Figure 7-7. Refrigerant vapor is compressed in the compressor and leaves the compressor at a higher pressure and temperature.

An example of a mechanical compression refrigeration system is a household refrigerator. Inside a refrigerator, a refrigerant circulates through the evaporator and absorbs heat from food. This heat vaporizes the refrigerant. The refrigerant vapor is drawn out of the evaporator by the compressor suction pressure. The compressor compresses the refrigerant, which raises the pressure and temperature of the refrigerant. The refrigerant circulates to a condenser, which is located on the sides of the refrigerator. The heat absorbed from the food inside the refrigerator is released to the air outside the refrigerator. See Figure 7-8.

Refrigerants

The first substance to be used as a refrigerant was probably ice, which was used for cooling food and drinks. As ice melts, it changes state from a solid to liquid. The ice absorbs heat from the air or liquid that surrounds it. The removal of heat cools the air or liquid. Water in the liquid state is also used as a refrigerant. In dry climates, water-soaked canvas bags are used to cool water. As the water evaporates, it absorbs heat from the water inside the bag. The evaporation provides a cooling effect on the water inside the bag.

In the early 1900's, mechanical refrigeration systems that used air or chemicals as refrigerants became common. Some of the chemicals used were ammonia, carbon dioxide, sulfur dioxide, methyl chloride, and hydrocarbons. Most of these chemicals have properties that make using them impractical. Some of the chemicals require very high pressures, require heavy equipment, and consume large amounts of power. Some of these refrigerants are toxic, flammable, or corrosive.

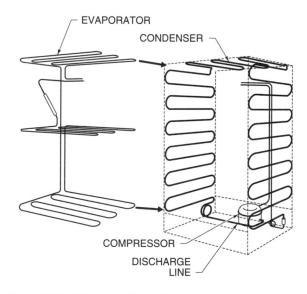

Figure 7-8. A household refrigerator removes heat from food and releases the heat to the air outside the refrigerator through the condenser.

Refrigerants used in modern refrigeration systems are derived from methane (CH_4) or ethane (C_2H_6). Fluorine (F) or chlorine (Cl) atoms are used to replace the hydrogen atoms in methane or ethane to give the chemicals different properties. These refrigerants are known as halocarbon refrigerants. Halocarbon refrigerants operate at pressures easily attained in a refrigeration system, are nontoxic, nonflammable, and relatively safe.

Halocarbon refrigerants were first developed by the General Motors Corporation. The refrigerants were later made jointly with Du Pont Co. Du Pont now produces refrigerants in their Freon® division. Freon® is a registered trademark of the Du Pont Co. and has been accepted as the common name for all halocarbon refrigerants.

Refrigerants are identified by numbers. The numbers are assigned according to the physical and chemical composition of a refrigerant. Numbers less than 170 are assigned to refrigerants based on the chemical composition of the refrigerant. Numbers above 170 are assigned arbitrarily.

Refrigerants derived from methane are assigned two-digit numbers. The first digit in two-digit numbers equals one more than the number of hydrogen atoms in each molecule. The second digit is the number of fluorine atoms in each molecule. Refrigerants derived from ethane are assigned three-digit numbers under 170. The first digit in three-digit numbers is arbitrarily assigned. The second digit in three-digit numbers equals one more than the number of hydrogen atoms in each molecule. The third digit is the number of fluorine atoms in each molecule. Atoms that are not accounted for that are not carbon are chlorine. See Appendix. For example, the chemical formula for refrigerant 12 (R-12) is CCl_2F_2 and the chemical name is dichlorodifluoromethane. R-12 is a methane-base refrigerant that contains two atoms of fluorine and no atoms of hydrogen, which leaves room for two atoms of chlorine. The carbon atom is from the original ethane. The number of a refrigerant identifies a particular refrigerant regardless of the manufacturer. See Figure 7-9.

There is concern about the effect of refrigerants on the earth's atmosphere. Refrigerants that use chlorofluorocarbons (CFCs) and fully halogenated chlorofluorocarbons (HCFCs) such as R-11, R-12, and R-22 are believed to be responsible for the depletion of the earth's ozone layer, which may cause global warming. As a result, venting CFCs from stationary equipment was banned in July, 1992 under provisions of the Clean Air Act. Contractors must be certified and document refrigeration system maintenance as required by federal, state, and local regulations. When a refrigeration system is serviced, the refrigerant has to be recovered and may be reclaimed or recycled.

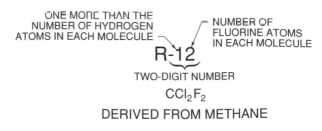

ONE MORE THAN THE NUMBER OF HYDROGEN ATOMS IN EACH MOLECULE

NUMBER OF FLUORINE ATOMS IN EACH MOLECULE

R-12

TWO-DIGIT NUMBER

CCl_2F_2

DERIVED FROM METHANE

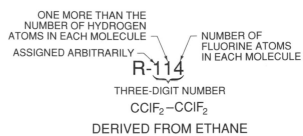

ONE MORE THAN THE NUMBER OF HYDROGEN ATOMS IN EACH MOLECULE

ASSIGNED ARBITRARILY

NUMBER OF FLUORINE ATOMS IN EACH MOLECULE

R-114

THREE-DIGIT NUMBER

$CCIF_2-CCIF_2$

DERIVED FROM ETHANE

Figure 7-9. Refrigerants derived from methane are assigned two-digit numbers. Refrigerants derived from ethane are assigned three-digit numbers under 170.

Reclaiming a refrigerant will have better results than recycling the refrigerant on-site. When used refrigerants are recycled, they are recovered, cleaned, and replaced into the refrigeration system. Recycling a refrigerant increases the risks of poor equipment performance. When used refrigerants are reclaimed, they are recovered and shipped to the manufacturer for purification. The refrigeration system is then refilled with new refrigerant, which guarantees proper system performance.

The production of CFCs will decrease until the year 2000 when the production of CFCs will be phased out completely. In addition, excise taxes on CFCs will increase over 400% by the year 2000. The increased cost of the refrigerants will make conservation and leak detection imperative.

By the year 2000, companies will have to convert to alternative refrigerants that have improved environmental characteristics. Du Pont Co. is currently producing new SUVA® alternative refrigerants. SUVA® is the Du Pont Co. registered trademark for the new alternative refrigerants. SUVA® refrigerants are currently available for new and retrofitted systems. *Retrofitting* is the process of furnishing a system with new parts that were not available at the time the system was manufactured. Refrigeration systems that currently use CFCs can be retrofitted with new equipment to make the system compatible with the new, safer refrigerants.

SUVA® refrigerants provide similar performance to CFCs with reduced potential for ozone depletion and global warming. Du Pont Co. and refrigeration system manufacturers have successfully retrofitted chillers with these alternative refrigerants.

Properties. Thermodynamic and physical properties are the most important properties of a refrigerant. Properties of refrigerants determine efficiency, cooling, and flow rates. Properties are the amount of heat involved in a change of state, the temperature at which a change of state occurs, and the volume of the refrigerant in liquid and vapor states. Refrigerant property tables contain information about thermodynamic and physical properties of refrigerants. See Appendix.

A *refrigerant property table* is a table that contains values for and information about the properties of a refrigerant at saturation conditions or at other pressures and temperatures. *Saturation conditions* are the temperature and pressure at which a refrigerant changes state. The properties listed on a refrigerant property table are volume, density, enthalpy, and entropy.

Determining the efficiency of the heat transfer in the evaporator and condenser is done by comparing the temperatures of the refrigerant in the evaporator and condenser to the temperatures of the evaporating and condensing medium. The volume and quantity of refrigerant needed to provide a certain amount of cooling in a system determines the size of the components needed to hold the refrigerant. Refrigerants are selected for a specific applications based on refrigerant properties.

The following are desirable refrigerant properties:
   low boiling point
   low freezing point
   nontoxic
   high critical point
   low specific volume
   high density
   high latent heat
   low compression ratio
   nonflammable
   noncorrosive
   stable
   miscible with oil

A refrigerant must have a boiling point that is lower than the desired temperature of the air that leaves the evaporator. The actual boiling point is controlled by the pressure in the low-pressure side of the system. The boiling point should be low enough so that the temperature can be attained without special equipment. The pressure in the low-pressure side should be positive pressure, which is pressure above atmospheric pressure. Positive pressure in the system prevents air from leaking in if a leak occurs.

The refrigerant in an air conditioning system must have a boiling point that is about 20°F lower than the temperature of the air that leaves the evaporator. Because the temperature of the air that leaves the evaporator is about 60°F, the temperature of the refrigerant must be about 40°F or lower. R-12 or R-22 are often used in air conditioning systems. The boiling point of R-12 is about –20°F at atmospheric pressure and the boiling point of R-22 is about –40°F at atmospheric pressure.

Refrigerants with lower boiling points are required for lower-temperature refrigeration systems. R-502 is used for lower-temperature refrigeration systems. R-502 has a boiling point of –49.8°F at normal atmospheric pressure and is used in medium-temperature freezer applications. R-13, R-14, or R-503 may be used for lower-temperature applications as low as –130°F. These refrigerants are used in two-stage compression or cascade systems. A *two-stage compression system* is a compression system that uses more than one compressor to raise the pressure of a refrigerant. The system raises the pressure above the pressure that can be achieved with a single compressor. A *cascade system* is a compression system that uses one refrigeration system to cool the refrigerant in another system.

A refrigerant should have a low freezing point. A refrigerant should be in the liquid state or the vapor state at all times and should never be in the solid state in a refrigeration system. If a refrigerant changes to a solid at normal operating pressures, the equipment will not operate. A refrigerant with a low freezing point does not change from a liquid to a solid at the temperatures in a typical refrigeration system.

A refrigerant should be nontoxic. Toxic materials can cause illness or death if inhaled or absorbed through the skin. A refrigerant should be nontoxic because refrigerant could escape when a system is charged or serviced and because refrigerant leaks could occur during operation. Ammonia, sulfur dioxide, and other refrigerants used in early refrigeration systems are toxic and have been replaced with halocarbons.

A refrigerant should have a high critical point. *Critical point* is the pressure and temperature above which a material does not change state regardless of the absorption or rejection of heat. All refrigerants have critical points. If a system operates with a high-pressure side close to the critical point of a refrigerant, the cost of operation will be high because the compressor must circulate the refrigerant against the higher pressure. For efficiency, a system should operate with the high-pressure side far below the critical pressure of the refrigerant.

Specific volume is the volume of a substance per unit of the substance. The specific volume of a refrigerant is the volume (in cubic feet) that 1 lb of the refrigerant occupies at a given pressure. In a system at operating pressures, a refrigerant that has a high specific volume occupies more space than a refrigerant that has a low specific volume. Because circulating a larger volume of refrigerant requires more energy, operating a system that uses a refrigerant with a high specific volume costs more than operating a system that uses a refrigerant with a low specific volume.

The volume of refrigerant that a compressor moves at required pressures determines the size of the compressor. A refrigeration system that has a refrigerant with a high specific volume requires a larger compressor than a refrigeration system that has a refrigerant with a low specific volume.

Density is the weight of a substance per unit of volume. The density of a refrigerant is the weight (in pounds) of 1 cu ft of the refrigerant. Density is the reciprocal of the specific volume at a given pressure. If a refrigerant has a low specific volume, it will have a high density. A system that circulates a high-density refrigerant operates more efficiently than a system that circulates a low-density refrigerant. A centrifugal compressor, however, is designed to move large volumes of refrigerant at relatively low pressures. If a refrigerant with a low density is used in a system with a centrifugal compressor, the system will operate inefficiently.

A refrigerant should have high latent heat content. Latent heat is heat identified by a change of state and no temperature change. Heat transfer in a refrigeration system causes a refrigerant to change state from a liquid to a vapor in the evaporator and from a vapor to a liquid in the condenser. The refrigeration effect of a system is the amount of latent heat involved in the change in the evaporator.

A refrigerant should be nonflammable, noncorrosive, and nontoxic. A flammable refrigerant could ignite when work such as brazing and soldering is done during installation or repair of a refrigeration system. Some early refrigerants were highly flammable and therefore dangerous.

A refrigerant should be noncorrosive. A corrosive substance reacts chemically with materials and causes corrosion of surfaces and components. A refrigerant must not react chemically with any of the materials in a refrigeration system. A corrosive refrigerant breaks down parts of the system such as the refrigerant lines, the compressor, the motor in a hermetic compressor, or materials that may be used for brazing, soldering, or repairing a leak. A *hermetic compressor* is a compressor in which the motor and compressor are sealed in the same housing. If acid in a system causes severe corrosion of the refrigerant lines, the compressor could form a copper plating. Corrosion or plating of mechanical parts cause system failure.

A refrigerant should be stable at the pressures and temperatures in a refrigeration system. When exposed to temperatures and/or pressures above the critical point, a nonstable refrigerant can break down chemically into base compounds or recombine into new compounds. If the refrigerants are exposed to high temperatures and pressures, the properties of the refrigerant will change. A refrigerant must be stable at all pressures and temperatures within a system including high pressures and temperatures.

A refrigerant should be miscible with oil. *Miscibility* is the ability of a substance to mix with other substances. The oil that lubricates a compressor circulates with the refrigerant in a refrigeration system. A refrigerant should be able to mix with oil so that the refrigerant can carry the oil through the compressor. If the oil separates from the refrigerant and remains at any other point in the system, the oil will block the flow of refrigerant and cause high pressures and reduced refrigeration effect.

Handling. Refrigerants are stored under pressure in steel cylinders. Most refrigerants boil at temperatures that are lower than normal room temperatures at atmospheric pressure. Refrigerants must be kept under pressure when stored and when transferred from a storage cylinder to a system. The pressure prevents vaporization.

Control valves on the outlet of the cylinder control the flow of refrigerant that leaves the cylinder. The valve outlet is threaded so that a hose can be connected. The hose is connected to a gauge set. A *gauge set* is a combination gauge that measures the pressure on each side of the refrigeration system. A gauge set is used for charging a system. Storage cylinders for refrigerants are available in a variety of sizes. Large cylinders are used for storage and shop use. Small cylinders can be carried to a job site by a service technician. See Figure 7-10.

REFRIGERANT CODES	
Freon® Type	**Color**
R–11	Orange
R–12	White
R–22	Light Green
R–500	Yellow
R–502	Light Purple

CONTROL VALVE

Du Pont Co.

Figure 7-10. Storage cylinders for refrigerants are available in a variety of sizes and are color-coded for identification.

Refrigerant cylinders are color-coded for safety and identification. Refrigerants should never be mixed. When changing the refrigerant in a system, the original refrigerant must be removed completely and the system must be cleaned thoroughly before the new refrigerant is added.

Hazards. Most of the hazards of working with refrigerants are caused by the high pressure and low boiling points of refrigerants. Refrigerants are stored and used at high pressures, which make them hazardous. The high temperatures in parts of a refrigeration system can be higher than the critical point of the refrigerant. The chemical breakdown that occurs could be hazardous.

Some refrigerants are toxic, corrosive, or flammable. Refrigerant cylinders should never be overheated or overfilled. Safety precautions must be taken when working with or around refrigerants to protect against hazards. See Figure 7-11.

REFRIGERANT SAFETY PRECAUTIONS
• Know which lines and tanks contain high pressure.
• Tag or put signs on hot pipes.
• Know which refrigerant is inside a system.
• Do not fill a refrigerant cylinder more than 85% full of liquid refrigerant.
• Reduce the gauge pressure to 0 psig before opening part of a refrigeration system.
• Never weld or solder on a sealed line.
• Never use an open flame to heat a refrigerant container.
• Wear goggles and protective clothing when working with caustic materials.
• If a leak occurs in a system that contains a toxic refrigerant, evacuate the area immediately.

Figure 7-11. Safety precautions protect against the hazards of refrigerants.

Many refrigerants are nontoxic, nonflammable, and nonpoisonous, but become acidic or toxic when they are exposed to pressures or temperatures above the critical point of the refrigerant. For example, if the motor of a hermetic compressor burns out, the refrigerant will be exposed to high temperatures in the system and will become acidic.

If a system runs at extremely high pressures for a length of time, the refrigerant and oil in the system can react chemically to form acids or sludge. Refrigerant from a burned-out system should be removed carefully to protect against acid burns.

When the refrigerant from a burned-out system is purged, the acid normally stays in the oil used in the compressor. The oil is acidic and should be carefully removed and discarded.

The pressure exerted by a refrigerant in a cylinder is the pressure that corresponds to the saturation temperature of the refrigerant. This means that the refrigerant is a liquid inside the cylinder. The temperature of the refrigerant is the same as the temperature of the air around the cylinder. If a refrigerant cylinder is heated, the temperature and pressure of the refrigerant will increase. If a refrigerant cylinder is left out in the sun or near a heat source, the pressure in the cylinder will rise and the cylinder may burst. Refrigerant cylinders must never be filled more than 85% full.

Connecting a service hose to a cylinder valve must be done carefully. Most refrigerants flash to a vapor when released to the atmosphere because of the lower pressure. The temperature of the refrigerant falls rapidly and may cause frostbite if it contacts the skin.

Operation

1. R-22 enters the evaporator of a refrigeration system with a pressure of 68.5 psig, a temperature of 40°F, and a heat content of approximately 34.4 Btu/lb of refrigerant. See Figure 7-12. The boiling point of R-22 at 68.5 psig is 40°F. The temperature of the refrigerant remains at about 40°F as it moves through the evaporator. Air at about 80°F that passes across the outside of the evaporator is warmer than the refrigerant in the evaporator. The refrigerant absorbs the heat and vaporizes.

The evaporator is large enough to allow the refrigerant to completely vaporize before it leaves the evaporator. The refrigerant absorbs superheat in the last few rows of coils in the evaporator. *Superheat* is heat added to a material after it has changed state. The refrigerant leaves the evaporator at a temperature higher than the saturated temperature for its pressure. The refrigerant has more heat than if it were saturated because of the superheat absorbed by the refrigerant in the evaporator.

2. The refrigerant vapor leaves the evaporator with a pressure of 68.5 psig, a temperature of 52°F, and a heat content of approximately 109.1 Btu/lb. The refrigerant absorbs heat from the evaporating medium, which raises the temperature and heat content of the refrigerant. The refrigerant leaves

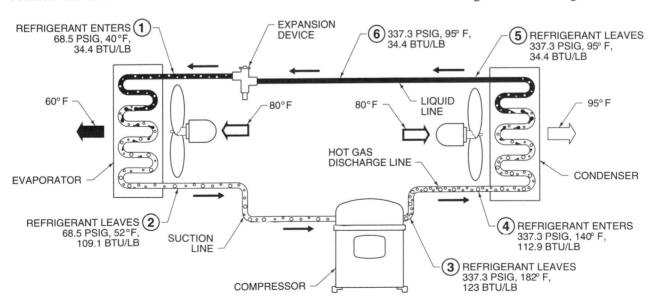

Figure 7-12. Mechanical compression refrigeration produces a refrigeration effect with mechanical equipment.

the evaporator through the suction line. The *suction line* is the line that connects the evaporator and the compressor.

In the compressor the pressure of the refrigerant rises to 337.3 psig. The temperature of the refrigerant rises because of the heat added by compression and, in certain compressors, from cooling the compressor motor windings.

3. The refrigerant vapor leaves the compressor with a pressure of 337.3 psig, a temperature of 182°F and a heat content of 123 Btu/lb of refrigerant. The saturated temperature of the refrigerant at 337.3 psig is 140°F, but the actual temperature of refrigerant as it leaves the compressor is about 182°F because the refrigerant is superheated. The refrigerant leaves the compressor through the hot gas discharge line and moves to the condenser.

The *hot gas discharge line* is the line that connects the compressor to the condenser. The hot gas discharge line contains the hot gas (refrigerant vapor) that is cooled in the condenser. While the refrigerant moves through the hot gas discharge line, it loses some of the superheat it absorbed in the compressor. By the time the refrigerant reaches the condenser, it is close to saturated temperature.

4. The refrigerant enters the condenser from the hot gas discharge line with a pressure of 337.3 psig, a temperature of 140°F, and a heat content of 112.9 Btu/lb. Heat flows from the refrigerant to the condensing medium because the temperature of the condensing medium is lower than the temperature of the refrigerant in the condenser. The amount of heat rejected by the refrigerant in the condenser is the same as the amount of heat absorbed by the refrigerant in the evaporator and compressor. As the refrigerant rejects heat in the condenser, the refrigerant changes state from a vapor back to a liquid.

5. The refrigerant leaves the condenser at the same pressure, 337.3 psig, but it is now a liquid. Most condensers have extra capacity so the refrigerant completely condenses to a liquid and extra cooling (subcooling) takes place in the liquid state. *Subcooling* is the cooling of a material such as a refrigerant to a temperature that is lower

than the saturated temperature of the material for a particular pressure. Because of subcooling, the refrigerant leaves the condenser with a temperature of 95°F and a heat content of approximately 34.4 Btu/lb.

The refrigerant leaves the condenser through the liquid line and moves either directly to the expansion device or to a receiver tank. In either case the refrigerant enters the expansion device at the same pressure and temperature that it had when it left the condenser.

6. The refrigerant flows through the expansion device. The restriction in the expansion device causes a pressure decrease. The pressure decrease is the difference between the high-pressure side and low-pressure side of the system. The decreased pressure allows 15% of the refrigerant to vaporize, which causes a temperature decrease from 95°F to 40°F. The boiling point of the refrigerant on the low-pressure side is 40°F.

From the expansion device, the refrigerant enters the evaporator at 68.5 psig as a liquid-vapor mixture and the cycle begins again.

ABSORPTION REFRIGERATION

An absorption refrigeration system produces a refrigeration effect when a refrigerant absorbs heat as it vaporizes and rejects the heat as it condenses. The components of the system control the flow of the refrigerant, which produces the refrigeration effect. See Figure 7-13.

An absorption refrigeration system works because a liquid under pressure expands rapidly, vaporizes, and produces a cooling effect when the pressure on the liquid decreases. An example of this cooling effect is a carbon dioxide cartridge that is pierced by a pin. Frost forms on the end of the cartridge as the carbon dioxide escapes because the carbon dioxide evaporates and absorbs heat from the end of the cartridge and the air that surrounds it. In an absorption refrigeration system, a refrigerant is raised to a relatively high pressure by adding heat to it. The high-pressure liquid refrigerant vaporizes into a low-pressure area where heat is absorbed from an evaporating medium.

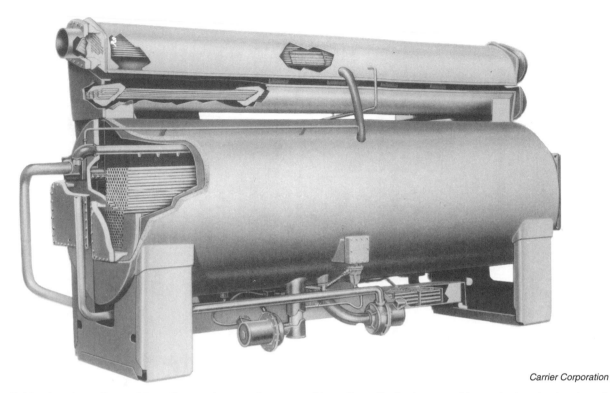

Figure 7-13. An absorption refrigeration system produces a refrigeration effect when a refrigerant absorbs heat as it vaporizes and rejects the heat as it condenses.

Absorption refrigeration is used in refrigeration systems in large buildings. Large absorption refrigeration systems may have a capacity of several hundred tons of cooling. A ton of cooling is the amount of heat required to melt a ton of ice. Some medium-size absorption air conditioning systems have been built for cooling the air in residential buildings, but they are more common in small and large sizes.

The four components in an absorption refrigeration system are the absorber, generator, condenser, and evaporator. An absorption refrigeration system includes a circulating pump and an orifice, which produce a refrigeration effect. See Figure 7-14.

The *absorber* is the component in which a refrigerant is absorbed by the absorbent. An *absorbent* is a fluid that has a strong attraction for another fluid. As absorption occurs, the combined volume of the refrigerant-absorbent solution decreases, which causes lower pressure in the absorber. Different combinations of refrigerants and absorbents may be used. Ammonia and water or water and lithium bromide are the two most common combinations for refrigerant-absorbent solutions. In both cases the absorbent has a strong attraction for the refrigerant and absorbs it in large proportions.

In the absorber, refrigerant vapor from the evaporator mixes with the absorbent. Most of the refrigerant vapor is absorbed by the absorbent and the solution is pumped to the generator by a circulating pump.

In the generator, heat is added to the refrigerant-absorbent solution to vaporize the refrigerant, raise the pressure of the refrigerant, and separate the refrigerant from the absorbent. Once the refrigerant and absorbent are separated, the refrigerant vapor moves to the condenser and the absorbent returns to the absorber.

In the condenser, the high-pressure refrigerant vapor condenses to a liquid when heat is removed from it with a heat exchanger. The condensing medium that runs through the condenser carries the heat away where it can be exhausted or used.

In the evaporator, the low-pressure, low-temperature refrigerant evaporates as it absorbs heat from the evaporating medium. The evaporating medium flows through a heat exchanger where it contacts the refrigerant. The evaporating medium is then used for cooling.

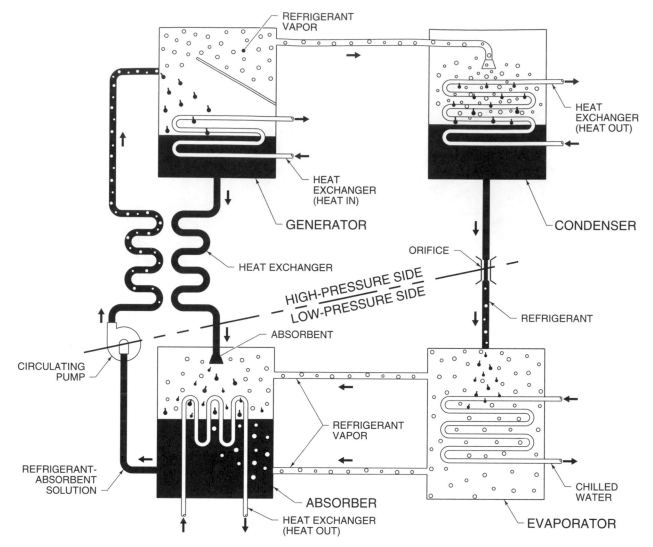

Figure 7-14. The four major components in an absorption refrigeration system are the absorber, generator, condenser, and evaporator.

A circulating pump is located on the refrigerant-absorbent line between the absorber and the generator. A circulating pump moves the refrigerant-absorbent solution from the absorber to the generator. The pump helps to maintain low pressure in the low-pressure side of the system by removing the refrigerant-absorbent solution from the low-pressure side as rapidly as it enters through the orifice. Not all absorption refrigeration systems have circulating pumps.

The orifice is a restriction in the refrigerant line that leads from the condenser to the evaporator. The orifice causes a pressure decrease in the refrigerant as it flows through the line. The refrigerant before the orifice is at high pressure, the refrigerant after the orifice is at low pressure.

High- and Low-pressure Sides

An absorption refrigeration system has a low-pressure side and a high-pressure side. The low-pressure side includes the orifice between the condenser and the evaporator, the evaporator, absorber, and the refrigerant-absorbent solution line that leads to the inlet of the circulating pump. The high-pressure side includes the circulating pump discharge, generator, condenser, and the refrigerant line that leads to the inlet of the orifice.

On the low-pressure side of an absorption system, the refrigerant expands as it absorbs heat from the evaporating medium in the refrigeration system. The pressure decreases as the refrigerant recombines with the absorbent in the absorber.

The high-pressure side of an absorption system contains the components that remove heat from the refrigerant. The pressure on the refrigerant-absorbent solution increases at the circulating pump discharge. The pressure increases more when the temperature increases because of heat added in the generator. As the pressure increases, the temperature of the refrigerant increases. Because the temperature of the refrigerant that enters the condenser is higher than the temperature of the condensing medium, heat flows from the refrigerant to the condensing medium. The pressure of the refrigerant controls the temperature of the refrigerant.

The pressure in the high-pressure side of an absorption refrigeration system is maintained by introducing liquid refrigerant-absorbent solution through the circulating pump as rapidly as the vaporized refrigerant and absorbent flow through the orifice and bypass back to the absorber. The absorbent is separated from the refrigerant in the generator and returns to the absorber for reuse.

Heat is added to the refrigerant-absorbent solution in the generator. The heat separates the refrigerant from the absorbent and increases the pressure in that part of the system. In small systems that do not have a circulating pump, the generator provides the power that circulates the refrigerant through the system and increases the pressure of the refrigerant.

Refrigeration effect is produced in the evaporator. The refrigerant vaporizes in the evaporator because the pressure in the evaporator is low. The pressure is low because of the low pressure in the absorber. The low pressure in the absorber is caused by the orifice, and because refrigerant absorbs heat from the evaporating medium.

Heat Transfer

In an absorption system, heat is transferred from the evaporating medium to the refrigerant in the evaporator. This heat transfer cools the evaporating medium, which is used for cooling. Heat is transferred from the refrigerant to the condensing me-

dium in the condenser. This heat is transferred to a place where it is exhausted or used as auxiliary heat. Heat is added to the refrigerant-absorbent solution in the generator.

Heat transfer occurs in the evaporator of an absorption system when liquid refrigerant flows from the condenser to the evaporator. The liquid refrigerant passes through the orifice, which causes a pressure decrease. The refrigerant vaporizes because the pressure in the evaporator is lower than the pressure ahead of the orifice. The evaporating medium flows through the heat exchanger, which is located in the evaporator. As the refrigerant vaporizes, it absorbs heat from the evaporating medium. The added heat cools the evaporating medium. The evaporating medium is used for cooling. See Figure 7-15.

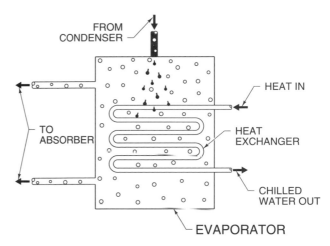

Figure 7-15. In an absorption refrigeration system, heat is transferred in the evaporator when liquid refrigerant flows from the condenser and is vaporized in the evaporator.

The *refrigeration effect* is the amount of heat in Btu/lb that the refrigerant absorbs from the evaporating medium. The amount of heat absorbed is equal to the amount of cooling the refrigeration system produces. The refrigeration effect of the system is found by applying the formula:

$$RE = h_\text{l} - h_\text{e}$$

where

RE = refrigeration effect (in Btu/lb)

h_l = enthalpy of refrigerant leaving evaporator (in Btu/lb)

h_e = enthalpy of refrigerant entering evaporator (in Btu/lb)

Example: Finding Refrigeration Effect — Absorption Refrigeration System

A refrigerant enters an evaporator of an absorption refrigeration system with an enthalpy of 77.9 Btu/lb and leaves the evaporator with an enthalpy of 1080.0 Btu/lb. Find the refrigeration effect.

$RE = h_l - h_e$

$RE = 1080.0 - 77.9$

$RE = 1002.1$ Btu/lb

In the condenser of an absorption refrigeration system, heat is rejected from the high-pressure refrigerant vapor to the condensing medium. The *heat of rejection* is the amount of heat in Btu/lb rejected by the refrigerant in the condenser. As the refrigerant flows through the condenser, heat is rejected to the condensing medium that flows through a heat exchanger in the condenser. As heat is rejected from the refrigerant vapor, the vapor condenses back to a liquid. See Figure 7-16.

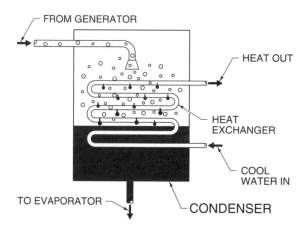

Figure 7-16. In an absorption refrigeration system, heat is rejected in the condenser as high-pressure refrigerant vapor from the generator condenses.

The heat of rejection in the condenser is found by applying the formula:

$HR = h_e - h_l$

where

HR = heat of rejection (in Btu/lb)

h_e = enthalpy of refrigerant entering condenser (in Btu/lb)

h_l = enthalpy of refrigerant leaving condenser (in Btu/lb)

Example: Finding Heat of Rejection — Absorption Refrigeration System

A refrigerant enters the condenser in an absorption refrigeration system with an enthalpy of 1080.0 Btu/lb. The refrigerant leaves the condenser as a liquid with an enthalpy of 77.9 Btu/lb. Find the heat of rejection.

$HR = h_e - h_l$

$HR = 1080.0 - 77.9$

$HR = 1002.1$ Btu/lb

Because the refrigeration effect equals the heat of rejection, it appears that the system operates with no energy cost. However, heating and cooling the refrigerant and absorbent in the heat exchanger must be considered in a complete analysis of operation.

Heat is added to the refrigerant-absorbent solution in the generator of an absorption refrigeration system. The heat vaporizes the refrigerant and separates it from the absorbent. The source of the heat may be a flame, an electric heater, or a hot water or steam coil. The absorbent that remains after the separation returns to the absorber and is used again. See Figure 7-17.

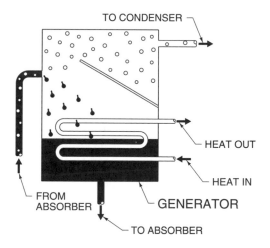

Figure 7-17. Heat vaporizes the refrigerant-absorbent solution in the generator. Vaporization separates the refrigerant from the absorbent.

Pressure Control

The temperature of the refrigerant in an absorption refrigeration system is controlled by the pressure of the refrigerant in the evaporator and the condenser. The pressure on the refrigerant decreases as the refrigerant passes through the orifice in the refrigerant line to the evaporator. The decreased pressure vaporizes some of the refrigerant at the orifice. The vaporization draws heat out of the remaining refrigerant, which decreases the temperature of the remaining refrigerant. The rest of the refrigerant vaporizes as it absorbs heat from the evaporating medium.

This added heat gives the refrigerant in the condenser a higher pressure and temperature than the refrigerant in the rest of the system. The refrigerant in the condenser is warmer than the condensing medium. Heat flows from the refrigerant to the condensing medium in the condenser heat exchanger because of the temperature difference.

Refrigerants

The refrigerants used in absorption refrigeration systems are solutions that consist of two chemical compounds. One of the two compounds must have a strong attraction for the other compound. Two commonly used chemical compounds are water and lithium bromide or ammonia and water. A water and lithium bromide solution uses water as the refrigerant and lithium bromide as the absorbent. An ammonia and water solution uses ammonia as the refrigerant and water as the absorbent. The percentage of each compound used in the solution is determined by the system in which the solution is used.

The following requirements should be met for a compound in a refrigerant-absorbent solution to be suitable for use in an absorption refrigerant system:

Absence of solid state at operating conditions. Each compound should be in the liquid state or the gas (vapor) state at operating conditions. When a refrigerant changes to the solid state, the flow is interrupted and the system malfunctions.

Volatile. The refrigerant should be much more volatile than the absorbent so the two can be easily separated in the generator. A *volatile* substance vaporizes readily at relatively low temperatures.

Strong attraction. The absorbent must have a strong attraction for the refrigerant under the temperatures and pressures attained in the absorber.

Moderate operating pressures. Operating pressures are attained by physical changes in the refrigerant. High pressures require heavy equipment and low pressures require large amounts of refrigerants.

Stable. The compounds must be stable under all operating conditions. The system should operate over a long period of time and the materials must remain stable. Changing pressures and temperatures affect the stability of refrigerants.

Noncorrosive. When used with common tubing, valves, and pumps, the compounds must not cause corrosion or undue wear.

Nontoxic and nonflammable. The compounds must be nontoxic and nonflammable for safety during operation and service.

Low viscosity. The compounds should have a low viscosity to help in mass transfer. Viscosity is the ability of a liquid or a semiliquid to resist flow.

High latent heat. The latent heat of the refrigerant should be as high as possible at the operating pressures in the system in order to reduce the amount of refrigerant required.

Although no known refrigerant-absorbent solution satisfies all of these requirements, ammonia and water or water and lithium bromide satisfy most of them.

Operation

Water and lithium bromide are often used as refrigerant and absorbent in small- to medium-size absorption refrigeration systems. The water and lithium bromide are brought together in the absorber where they readily combine. See Figure 7-18. As the substances combine, the lithium bromide absorbs the water and forms a strong-bonding refrigerant-absorbent solution. Some heat is produced during absorption. The absorber may contain a heat exchanger with cool water flowing through it to help control the temperature of the solution.

1. In an absorption refrigeration system, the refrigerant-absorbent solution leaves the absorber with a temperature of 100°F and a heat content of 47.2 Btu/lb of refrigerant.

2. Most systems have a heat exchanger between the absorber and the generator. Heat flows from the absorbent going back to the absorber to the refrigerant-absorbent solution going to the generator. The temperature of the refrigerant-absorbent solution rises to 170°F with a heat content of 79 Btu/lb.

In the generator, heat added by a heat exchanger separates the refrigerant from the refrigerant-absorbent solution.

3. The refrigerant leaves the generator as a vapor at a temperature of 200°F and a heat content of 1150 Btu/lb. The refrigerant moves to the condenser.

4. The absorbent that has separated from the refrigerant leaves the generator at a temperature of 210°F and a heat content of 107 Btu/lb. The absorbent returns to the absorber through the heat exchanger between the generator and absorber.

The refrigerant vapor in the condenser condenses when cooled as heat is removed by a heat exchanger.

5. The refrigerant flows from the condenser to the evaporator through the orifice. The pressure decrease as the refrigerant flows through the orifice decreases the temperature of the refrigerant to 110°F and heat content to 77.9 Btu/lb.

As the refrigerant flows through the evaporator, it vaporizes and absorbs heat from the evaporating medium. Most of the refrigerant leaves the evaporator as a vapor.

6. When the refrigerant leaves the evaporator, it is at a temperature of 41°F and heat content of 1080 Btu/lb. The refrigerant returns to the absorber where the cycle begins again.

An absorption refrigeration system always works with a coefficient of performance of less than one.

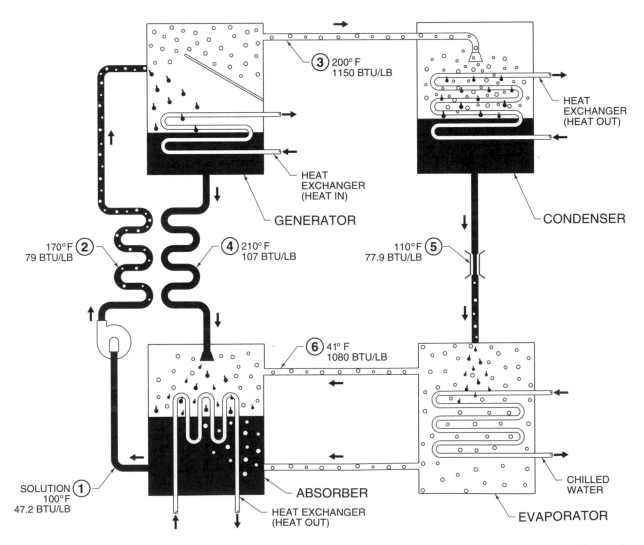

Figure 7-18. An absorption refrigeration system produces a refrigeration effect when one chemical is absorbed by another and heat is added.

Coefficient of performance (COP) is the cooling capacity produced from the heat input. COP is found by dividing the cooling capacity produced from a system by the energy used in the system. COP is found by applying the formula:

$$COP = \frac{CC}{EU}$$

where

COP = coefficient of performance

CC = cooling capacity (in Btu/hr)

EU = equivalent of energy used (in Btu/hr)

Example: Finding Coefficient of Performance

A refrigeration system produces 48,500 Btu/hr of cooling while using the equivalent of 70,000 Btu/hr of energy. Find the COP of the system.

$$COP = \frac{CC}{EU}$$

$$COP = \frac{48,500}{70,000}$$

$$COP = .69$$

Review Questions

7

1. Describe refrigeration effect.

2. List and describe the function of the main parts of a mechanical compression refrigeration system.

3. What is the source of the heat that increases the temperature of the refrigerant as it moves through a compressor?

4. What causes the temperature of the refrigerant to decrease as it leaves the expansion device and enters the evaporator?

5. Explain the relationship between the boiling point of a liquid and the pressure on the liquid.

6. Define *superheat*. Explain the function of superheat in a refrigeration system.

7. List two functions of an expansion device.

8. What two devices in a mechanical compression refrigeration system separate the low-pressure side from the high-pressure side?

9. Describe how water provides a refrigeration effect when it vaporizes.

10. Describe how the number of a refrigerant identifies the chemical composition of the refrigerant.

11. List the desirable refrigerant properties.

12. Why is boiling point an important property of a refrigerant?

13. Is a refrigerant with a low specific volume better than a refrigerant with a high specific volume?

14. Define *critical point*. Why is it important?

15. Why is the compression ratio of an absorption refrigeration system important?

16. What happens if a mechanical compression refrigeration system operates above the critical point of the refrigerant?

17. Define *miscibility*. Why is it important that a refrigerant be miscible?

18. List five hazards of working with refrigerants.

19. In what two ways are mechanical compression and absorption refrigeration systems alike?

20. List and describe the function of each component of an absorption refrigeration system.

21. What happens to the refrigerant-absorbent solution in the generator of an absorption refrigeration system?

22. How is heat removed from a refrigerant in the condenser of an absorption refrigeration system?

23. How does the evaporator of an absorption refrigeration system cool the evaporating medium?

24. What device is located before the evaporator in an absorption refrigeration system? What does it do?

Pressure-Enthalpy Diagrams

Pressure-enthalpy diagrams define the relationships between the thermodynamic properties of a refrigerant as the refrigerant flows through a refrigeration system. The thermodynamic properties found on a pressure-enthalpy diagram are the pressure, enthalpy, constant temperature, constant quality, constant volume, and constant entropy of a refrigerant. The diagrams are graphic representations of the properties of a refrigerant as it flows through a refrigeration system. The graphic representation is helpful for visualizing and evaluating the performance of a refrigeration system.

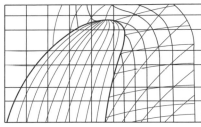

PRESSURE-ENTHALPY DIAGRAMS

A *pressure-enthalpy diagram* is a graphic representation of the thermodynamic properties of a refrigerant. A pressure-enthalpy diagram is also referred to as a Mollier diagram. On a pressure-enthalpy diagram, pressure is plotted against enthalpy. The curves on a pressure-enthalpy diagram represent the pressure (psia) and the enthalpy (Btu/lb) of a refrigerant at different properties. All of the properties are plotted in relation to pressure and enthalpy.

If any two properties of a refrigerant are known, a pressure-enthalpy diagram can be used to find the other properties of the refrigerant. The properties that can be found on a pressure-enthalpy diagram are pressure, enthalpy, temperature, constant quality, constant volume, and constant entropy. Each refrigerant has a specific pressure-enthalpy diagram because the properties of each refrigerant are different. See Figure 8-1.

By drawing a graph on a pressure-enthalpy diagram according to the pressure and enthalpy of a refrigerant, the other properties of the refrigerant at any point in the refrigeration system can be found. The graph on the diagram represents the properties of a refrigerant as it flows through a mechanical compression refrigeration system.

Pressure

System pressure is the pressure of the refrigerant in a refrigeration system. The system pressure of an operating refrigeration system depends on the type of refrigerant used and the temperatures required in the system. The pressure scales on a pressure-enthalpy diagram are located on the left and right sides of the diagram. On the diagram, system pressure is expressed in pounds per square inch absolute (psia). System pressure at different points is found on the horizontal lines that run across a pressure-enthalpy diagram.

System pressure is measured with a gauge and is expressed in gauge pressure. Gauge pressure is the pressure inside a closed system that is measured with a pressure gauge such as a Bourdon tube. Atmospheric pressure has no effect on a closed system. Gauge pressure is always used to measure and express pressure in a closed system.

A *Bourdon tube* is a tube inside a mechanical pressure gauge. It is a circular stainless steel or bronze tube that is flattened to make it more flexible. As pressure enters the gauge, it causes the tube to try to straighten out. The amount of movement in the tube is proportional to the amount of pressure that enters the tube.

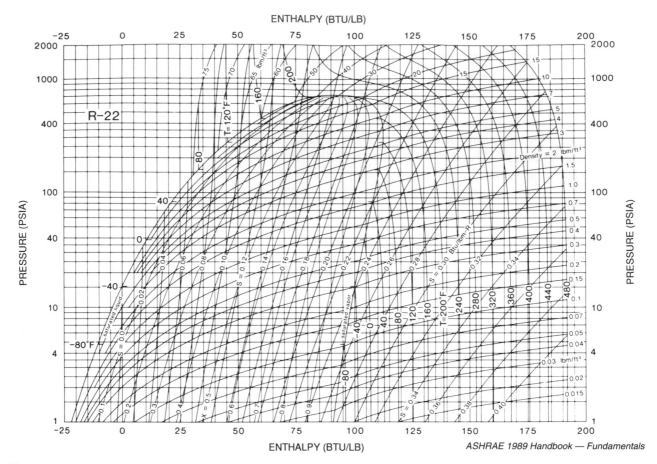

Figure 8-1. Pressure is plotted against enthalpy on a pressure-enthalpy diagram.

The tip of a Bourdon tube is connected to small gears and springs. The gears and springs change the movement of the tube to a movement of a pointer on the face of the pressure gauge. See Figure 8-2.

Saturation conditions are the temperature and pressure of a refrigerant at which the refrigerant changes state. Both liquid and vapor are present when a refrigerant changes state (vaporizes or condenses). *Saturation pressure* is the pressure at which a substance such as a refrigerant changes state. At saturation pressure, both liquid and vapor are present. Saturation pressure varies with the temperature of the substance.

The saturated pressure and saturated temperature of a refrigerant are found on a table of the properties of saturated liquid and saturated vapor. See Appendix. A refrigerant has a specific saturation pressure for each temperature. As a refrigerant changes state, the temperature remains constant as long as the pressure remains constant.

Enthalpy

Enthalpy is the total heat contained in a substance. Enthalpy includes sensible and latent heat. When calculating the difference in enthalpy between two temperatures or pressures, a base of 0 Btu at −40°F is used. The enthalpy scale on a pressure-enthalpy diagram does not begin at 0 Btu because the enthalpy for refrigerants is considered to be 0 Btu/lb at −40°F. This is not technically correct, but because nearly all calculations are related to differences during a change of state, the results are satisfactory. Enthalpy is found on the vertical lines that run from the bottom to the top of a pressure-enthalpy diagram. Enthalpy values found on a pressure-enthalpy diagram are used to calculate the performance of a refrigeration system.

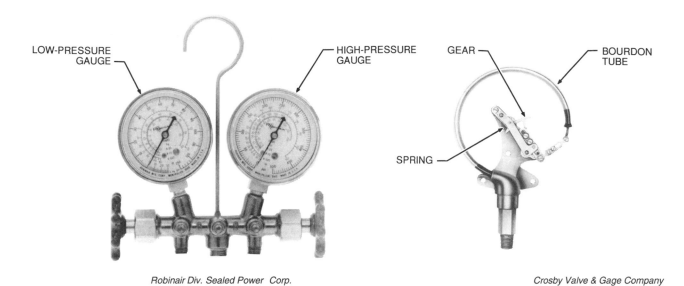

LOW-PRESSURE GAUGE

HIGH-PRESSURE GAUGE

GEAR

BOURDON TUBE

SPRING

Robinair Div. Sealed Power Corp.

Crosby Valve & Gage Company

Figure 8-2. Gauge pressure is expressed in pounds per square inch gauge and is measured with a Bourdon tube or other pressure gauge.

Temperature

Temperature is the intensity of heat. Temperature is measured with a thermometer. A change in sensible heat causes a temperature change. *Saturation temperature* is the temperature at which a substance such as a refrigerant changes state. Saturation temperature varies with the pressure on the substance. At saturation temperature, both liquid and vapor are present. The saturation temperature of a refrigerant is found on a table of the properties of saturated liquid and saturated vapor for the particular refrigerant. In an evaporator, a refrigerant absorbs heat and vaporizes. While the refrigerant vaporizes, the liquid-vapor mixture is at the saturation temperature for the pressure in the evaporator.

Saturated liquid is liquid at a certain temperature and pressure that will vaporize if the pressure and temperature increase. *Saturated vapor* is vapor at a certain temperature and pressure that will condense if the pressure or temperature decreases.

Superheat is sensible heat that is added to a substance after the substance has changed state. The amount of superheat is the amount of heat added above the saturation temperature. The temperature of a liquid remains constant during vaporization. After vaporization, heat can be added to the vapor to raise the temperature of the vapor. See Figure 8-3. For example, water absorbs heat as the water

flows through a boiler. When the water has absorbed enough heat, it vaporizes and becomes steam. If additional heat is added to the steam, the temperature of the steam will increase. See Appendix.

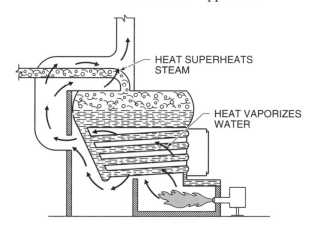

HEAT SUPERHEATS STEAM

HEAT VAPORIZES WATER

Figure 8-3. Superheat is sensible heat added to a substance after the substance has completely vaporized.

Superheat is added to the refrigerant in the evaporator. When the refrigerant leaves the evaporator, it flows to the compressor through the suction line. As the refrigerant flows through the compressor, heat of compression is also added. *Heat of compression* is the thermal energy equivalent of the mechanical energy expended by the motor that turns the compressor. The heat of compression adds additional superheat to the refrigerant. See Figure 8-4. The amount of superheat is found by using the temperature difference between the temperature of the refrigerant and the saturation temperature at a given pressure.

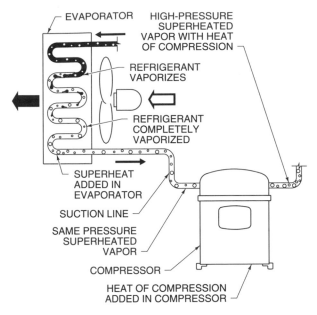

Figure 8-4. Superheat is added to a refrigerant in the evaporator and in the compressor.

Subcooling is the process of cooling of a substance such as a refrigerant to a temperature that is lower than the saturated temperature of the substance at a given pressure. When a vapor condenses to a liquid, the temperature of the liquid-vapor mixture remains constant. If additional heat is rejected after a refrigerant has completely condensed, the temperature of the liquid refrigerant will decrease.

As liquid refrigerant flows through a condenser, the refrigerant is at the saturation temperature for the specific pressure. Heat is rejected by the refrigerant as it flows through the condenser. When the latent heat is rejected, the refrigerant changes state from a vapor to a liquid. After the refrigerant has completely condensed to a liquid, additional sensible heat is rejected in the last rows of the condenser coil. When the refrigerant leaves the condenser, the refrigerant is subcooled because the refrigerant rejects additional heat. See Figure 8-5. For example, a refrigerant rejects heat as it flows through a condenser coil. The rejection of heat causes the refrigerant to condense back to the liquid state. The refrigerant condenses to a liquid in the first 90% of the condenser coil. The liquid refrigerant that

flows through the last 10% of the condenser coil is subcooled as the liquid rejects heat.

On a pressure-enthalpy diagram, the temperature of a saturated liquid refrigerant is found on the saturated liquid line, which is a curved line that begins near the lower left corner of the diagram and curves up to the right. The saturated liquid line curves right sharply as it approaches the critical pressure of the refrigerant and continues to the critical point for the refrigerant. The critical point of a refrigerant is the point at which refrigerant does not change state regardless of absorption or rejection of heat. The temperature of a saturated refrigerant vapor is found on the saturated vapor line, which begins at a point on the bottom of the diagram and curves left to meet the saturated liquid line at the critical point.

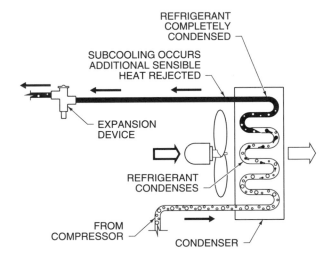

Figure 8-5. Subcooling occurs in the last rows of the condenser coil.

Temperature scales that indicate the temperature of the saturated liquid and saturated vapor are found on the saturated liquid and saturated vapor lines. The temperature scales start with a negative temperature at the bottom and increase as the lines curve up. Temperature lines start on the saturated vapor line and curve down to the right. The temperature lines indicate the temperature of the refrigerant at any point on the lines. See Figure 8-6.

Constant Quality

Constant quality is a percentage that expresses the ratio of vapor to liquid as a refrigerant changes state.

Constant quality is 0% at the saturated liquid line and 100% at the saturated vapor line. The area between the saturated liquid and saturated vapor lines is divided into ten segments. The two saturated liquid and saturated vapor lines curve up from the bottom of a pressure-enthalpy diagram and meet at the critical point for the refrigerant.

Constant Volume

Specific volume is the volume of a substance per unit of the substance. Specific volume is used to express the volume of a refrigerant at a given pressure in a refrigeration system. Specific volume is expressed in cubic feet per pound, which is the

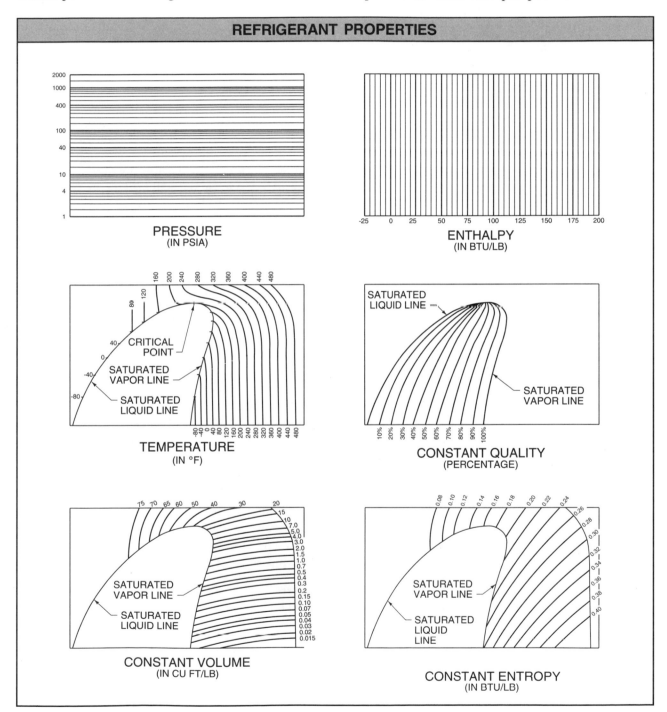

Figure 8-6. Pressure-enthalpy diagrams show the properties of a refrigerant as it flows through a refrigeration system.

number of cubic feet that 1 lb of the refrigerant occupies. Specific volume of a refrigerant is found on refrigerant property tables or pressure-enthalpy diagrams. A refrigerant property table includes the volume of the refrigerant in both the liquid state and the vapor state for different temperatures. In an operating refrigeration system, the volume of a refrigerant changes as the pressure, state, or temperature of the refrigerant changes.

Constant volume is the volume of the refrigerant vapor that remains constant because of the relationship between the pressure and enthalpy of the refrigerant. Constant volume is expressed in cubic feet per pound. Constant volume is found on a pressure-enthalpy diagram on the lines that begin on the saturated vapor line and run up to the right. The constant volume scale is found on these lines. The volume decreases as pressure increases.

Constant Entropy

Constant entropy is a calculated value that indicates energy lost to the disorganization of the molecular structure of a substance when heat is transferred. Constant entropy is the ratio of the amount of heat added to a substance to the absolute temperature of the substance at the time the heat is added.

Constant entropy lines begin on the saturated vapor line and run steeply up to the right. Constant entropy is only of value on a pressure-enthalpy diagram because it defines the properties of a refrigerant as it moves through an ideal compression cycle.

System Graphs

A *system graph* is a small graphic representation of the operation of a refrigeration system on a pressure-enthalpy diagram. System graphs are constructed by drawing lines that correspond to the system pressure on the low-pressure side and high-pressure side of the refrigeration system, the change in pressure and enthalpy of the refrigerant in the compressor, and the change in pressure of the refrigerant in the expansion device. All of the other properties of the refrigerant in a refrigeration system can be found on a system graph.

A system graph represents the performance of an operating refrigeration system. The properties of

the refrigerant at different points are plotted on a pressure-enthalpy diagram and the resulting figure is the system graph, which is used to identify the other thermodynamic properties of a refrigerant as it flows through a refrigeration system. System graphs are a valuable tool for visualizing refrigeration system performance and obtaining data to calculate system performance.

IDEAL CYCLE

In an ideal refrigeration cycle, superheating and subcooling do not occur. The system graph of an ideal refrigeration system is plotted on a pressure-enthalpy diagram by drawing lines that correspond to the system pressure on the low-pressure side and high-pressure side of the refrigeration system, the change in pressure and enthalpy of the refrigerant in the compressor, and the change in pressure of the refrigerant in the expansion device.

If the system pressure on the low-pressure and high-pressure side of a refrigeration system and the type of refrigerant are known, the system graph can be plotted by applying the procedure:

1. To plot the system pressure for each side of a refrigeration system on a pressure-enthalpy diagram, the system pressure must be converted from gauge pressure to absolute pressure. Pressure inside closed systems is expressed in gauge pressure. Gauge pressure plus atmospheric pressure (14.7 psi) equals absolute pressure. Absolute pressure is used to express system pressure on pressure-enthalpy diagrams. When the gauge pressures that correspond to the system pressure on the low-pressure and high-pressure side of a refrigeration system are converted to absolute pressure, the lines for the low-pressure side and the high-pressure side can be drawn on a pressure-enthalpy diagram.

The pressure gauge on a refrigeration system that contains R-22 reads 90 psig on the low-pressure side and 200 psig on the high-pressure side. The line that represents the low-pressure side of the refrigeration system is located on a horizontal line at 104.7 psia (90 psig + 14.7 = 104.7 psia). The low-pressure line is drawn from the left of the saturated liquid line to the right of the saturated vapor line. See Figure 8-7.

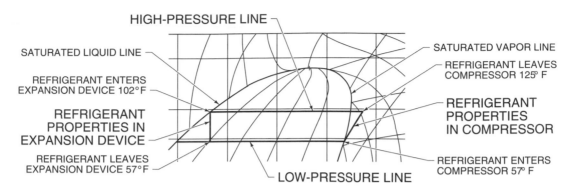

Figure 8-7. A system graph consists of four lines that correspond to the properties of the refrigerant as it flows through the parts of a refrigeration system.

2. The line that represents the high-pressure side of the refrigeration system is located on a horizontal line at 214.7 psia (200 psig + 14.7 = 214.7 psia). The high-pressure line is drawn from the left of the saturated liquid line to the right of the saturated vapor line. These two lines are the top and bottom of the system graph.

3. The temperature and pressure changes that occur in the compressor are also part of a system graph. A line begins at the point where the low-pressure line intersects the saturated vapor line and runs parallel with the closest constant entropy line until it intersects the high-pressure line. This line represents the refrigerant properties, which are the temperature and pressure of the refrigerant in the compressor. Points on the line represent the properties of the refrigerant as it flows through the compressor.

As a compressor compresses refrigerant, the mechanical energy produced by the compressor is converted to thermal energy. The refrigerant absorbs this heat as the refrigerant flows through the compressor. The additional heat from the compressor raises the temperature of the refrigerant vapor. This heat increases the temperature of the refrigerant as it flows through the compressor. The temperature of the refrigerant that leaves the compressor is found where the temperature line curves down across the high-pressure side line.

4. Pressure change in the compressor is the difference between the low-pressure and high-pressure lines on the system graph. Pressure change in the compressor is equal to the distance between the low-pressure and high-pressure lines on the system graph.

5. In the expansion device, a restriction produces a pressure decrease in the refrigeration system. The pressure decrease causes the refrigerant to vaporize, which causes the temperature of the refrigerant to change in the expansion device. As the refrigerant vaporizes, it absorbs sensible heat from the remaining liquid refrigerant. Removing this sensible heat causes the temperature of the refrigerant to decrease.

The refrigerant properties in the expansion device are found on a line that begins at the intersection of the high-pressure line and the saturated liquid line. The line is drawn vertically down to the low-pressure line.

Pressure decrease and temperature decrease through the expansion device are found on a system graph. The pressure decrease through the expansion device is equal to the difference between the low-pressure and high-pressure lines. The temperature decrease through the expansion device is equal to the difference in the saturated temperatures of the refrigerant at the low-pressure side and the high-pressure side.

Ideally, no heat from outside a refrigeration system is involved in the process of vaporizing the refrigerant as it flows through the expansion device. An *adiabatic change* is a change in the pressure and temperature of a substance in a closed system that occurs without heat transfer from outside the system.

Example: Plotting a System Graph — Ideal Refrigeration Cycle

A mechanical compression refrigeration system contains R-22. The low-pressure side is 75 psig and the high-pressure side is 185 psig. Plot the system graph of an ideal refrigeration cycle on a pressure-enthalpy diagram for R-22.

1. The low-pressure line is located on the pressure-enthalpy diagram along the 89.7 psia line (75 psig + 14.7 = 89.7 psia). The low-pressure line is drawn from the left of the saturated liquid line to the right of the saturated vapor line. See Figure 8-8.

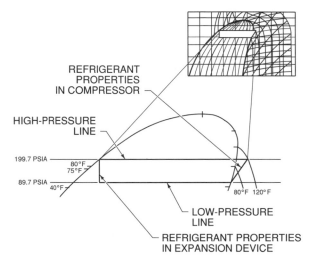

Figure 8-8. If the gauge pressure on the low-pressure and high-pressure side of the system and the type of refrigerant are known, the system graph for the refrigerant can be plotted.

2. The high-pressure line is located on the 199.7 psia line (185 psig + 14.7 = 199.7 psia). The high-pressure line is drawn from the left of the saturated liquid line to the right of the saturated vapor line.

3. The line that represents the refrigerant properties in the compressor begins at the point where the low-pressure line intersects the saturated vapor line and is drawn parallel with the closest constant entropy line until it intersects the high-pressure line.

4. Pressure change through the compressor is the difference between the low-pressure and high-pressure lines on the system graph (199.7 psia − 89.7 psia = 110 psia)

5. The refrigerant properties in the expansion device are found on the line that begins at the intersection of the high-pressure line and the saturated liquid line. The line runs vertically down to the low-pressure line. The figure that

results from drawing the lines is the system graph, which represents the performance of an ideal refrigeration cycle.

Heat Transfer

The amount of heat transfer in the different parts of a refrigeration system is found by using the enthalpy lines and the system graph drawn on a pressure-enthalpy diagram. Heat transfer includes heat rejected by the evaporating medium in the evaporator, heat absorbed by the refrigerant in the compressor, and heat absorbed by the condensing medium in the condenser. The amount of heat transfer in the different parts of a refrigeration system is found by applying the procedure:

1. As the refrigerant flows through an evaporator, it absorbs heat from the evaporating medium. The point that represents the properties of the refrigerant as it enters the evaporator is the intersection of the line that represents the properties of the refrigerant in the expansion device and the low-pressure line. See Figure 8-9. The enthalpy of the refrigerant that enters the evaporator is found on a line drawn vertically from this point to the enthalpy scale.

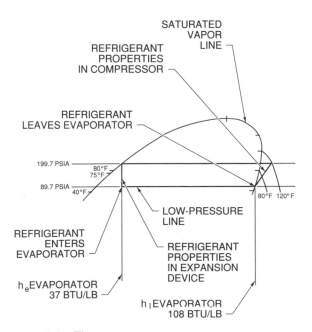

Figure 8-9. The amount of heat transfer at any point in an ideal refrigeration system is found by using the system graph and the enthalpy scale on a pressure-enthalpy diagram.

2. The properties of the refrigerant as it leaves the evaporator are found at the intersection of the low-pressure line and the saturated vapor line. The enthalpy of the refrigerant that leaves the evaporator is found on a line drawn vertically from this point to the enthalpy scale.

3. The difference between the enthalpy of the refrigerant that leaves the evaporator and the enthalpy of the refrigerant that enters the evaporator is the amount of heat absorbed by the refrigerant in the evaporator.

The heat that is added to the refrigerant in the evaporator is the refrigeration effect of a mechanical compression refrigeration system. Refrigeration effect is found by applying the formula:

$$RE = h_1 - h_e$$

where

RE = refrigeration effect (in Btu/lb)

h_1 = enthalpy of refrigerant leaving evaporator (in Btu/lb)

h_e = enthalpy of refrigerant entering evaporator (in Btu/lb)

Example: Finding Refrigeration Effect — Ideal Cycle

The enthalpy of the refrigerant that enters the evaporator in a refrigeration system is 37 Btu/lb. The enthalpy of the refrigerant as it leaves the evaporator is 108 Btu/lb. Find the refrigeration effect of the system.

$$RE = h_1 - h_e$$

$$RE = 108 - 37$$

$$RE = \textbf{71 Btu/lb}$$

4. As the refrigerant flows through the compressor, it is heated by the heat of compression. The point that represents the properties of the refrigerant as it enters the compressor is at the intersection of the line that represents the refrigerant properties in the compressor and the low-pressure line. See Figure 8-10. The point that represents the properties of the refrigerant as it enters the compressor is the same for the properties of the refrigerant that leaves the evaporator. The enthalpy of the refrigerant as it enters the compressor is found on a line drawn vertically from this point to the enthalpy scale.

The point that represents the properties of the refrigerant as it leaves the compressor is at the intersection of the line that represents the refrigerant properties in the compressor and the high-pressure line. The enthalpy of the refrigerant as it leaves the compressor is found on a line drawn vertically from this point to the enthalpy scale.

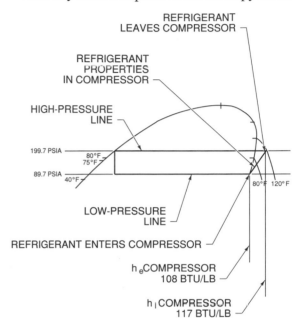

Figure 8-10. Heat of compression is the difference between the enthalpy of the refrigerant entering the compressor and the enthalpy of the refrigerant leaving the compressor.

Heat of compression is the amount of heat absorbed by the refrigerant in the compressor, which is the difference between the enthalpy of the refrigerant as it enters the compressor and the enthalpy of the refrigerant that leaves the compressor. Heat of compression is found by applying the formula:

$$HC = h_1 - h_e$$

where

HC = heat of compression (in Btu/lb)

h_1 = enthalpy of refrigerant leaving compressor (in Btu/lb)

h_e = enthalpy of refrigerant entering compressor (in Btu/lb)

Example: Finding Heat of Compression — Ideal Cycle

The enthalpy of a refrigerant as it leaves the compressor in a refrigeration system is 117 Btu/lb. The enthalpy of the refrigerant as it enters the compressor is 108 Btu/lb. Find the heat of compression.

$$HC = h_1 - h_e$$

$$HC = 117 - 108$$

HC = 9 Btu/lb

5. The condenser in a refrigeration system must reject all of the heat absorbed in the evaporator and compressor. The point that represents the properties of the refrigerant as it enters the condenser is located at the intersection of the line that represents the refrigerant properties in the compressor and the high-pressure line. The point that represents the properties of the refrigerant as it enters the condenser is the same for the properties of the refrigerant as it leaves the compressor. The enthalpy of the refrigerant as it enters the condenser is found on a line drawn vertically from this point to the enthalpy scale. See Figure 8-11.

The properties of the refrigerant as it leaves the condenser are found at a point located at the intersection of the high-pressure line and the line that represents refrigerant properties in the expansion device. The enthalpy of the refrigerant as it leaves the condenser is found on a line drawn vertically from this point to the enthalpy scale.

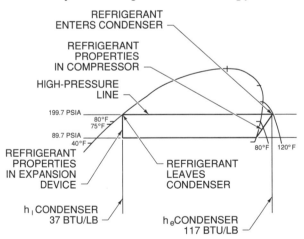

Figure 8-11. Heat of rejection is found by subtracting the enthalpy of the refrigerant entering the condenser from the enthalpy of the refrigerant leaving the condenser.

The *heat of rejection* is the amount of heat rejected by the condenser. Heat of rejection is found by subtracting the enthalpy of the refrigerant as it enters the condenser from the enthalpy of the re-

frigerant as it leaves the condenser. The amount of heat rejected by the condenser is found by applying the formula:

$$HR = h_e - h_1$$

where

HR = heat of rejection (in Btu/lb)

h_e = enthalpy of refrigerant entering condenser (in Btu/lb)

h_1 = enthalpy of refrigerant leaving condenser (in Btu/lb)

Example: Finding Heat of Rejection — Ideal Cycle

The enthalpy of a refrigerant as it enters the condenser is 117 Btu/lb. The enthalpy of a refrigerant as it leaves the condenser is 37 Btu/lb. Find the amount of heat rejected by the condenser.

$$HR = h_e - h_1$$

$$HR = 117 - 37$$

HR = 80 Btu/lb

The enthalpy of the refrigerant does not change as it flows through the expansion device. The line that represents the refrigerant properties in the expansion device is a vertical line, which indicates that there is no change in enthalpy. No change in enthalpy indicates that no heat is added to or rejected by the refrigerant.

ACTUAL CYCLE

In an actual cycle, the refrigerant that leaves the evaporator is superheated and the refrigerant that leaves the condenser is subcooled. The amount of superheat added to the refrigerant in the evaporator varies among different refrigeration systems. A refrigeration system should always add enough superheat to the refrigerant to ensure that only vapor arrives at the suction port of the compressor. Superheat ranges from 5°F to 30°F among different refrigeration systems. Superheat is found by measuring the operating temperature of a refrigeration system. Subcooling ranges from 5°F to 25°F among different refrigeration systems. Refrigeration systems are designed to provide enough subcooling to offset any temperature difference that occurs from pressure loss in the liquid line.

Plotting the low-pressure and high-pressure side lines of an actual operating cycle is the same as plotting the lines of an ideal cycle. If the system pressure on the low-pressure and high-pressure side of a refrigeration system and the type of refrigerant are known, the system graph can be plotted by applying the procedure:

1. To plot system pressure on a pressure-enthalpy diagram, gauge pressure must be converted to absolute pressure. A refrigeration system that contains R-22, for example, has a low-pressure side at 100 psig and a high-pressure side at 180 psig. The line that represents the low-pressure side of the refrigeration system is located on a horizontal line at 114.7 psia (100 psig + 14.7 = 114.7 psia). The low-pressure line is drawn from the left of the saturated liquid line to the right of the saturated vapor line. See Figure 8-12.

2. The line that represents the high-pressure side of the refrigeration system is located on a horizontal line at 194.7 psia (180 psig + 14.7 = 194.7 psia). The high-pressure line is drawn from the left of the saturated liquid line to the right of the saturated vapor line.

3. The refrigerant leaves the evaporator and enters the compressor. The temperature of the refrigerant is 80°F as it enters the compressor. Temperature is measured with a thermometer while the refrigeration system is operating. The point that represents the properties of the refrigerant as it enters the compressor is located at the intersection of the low-pressure line and the 80°F temperature line. A line drawn parallel with the nearest constant entropy line represents

the refrigerant properties in the compressor. The line ends at the point where it intersects the high-pressure line.

4. The restriction in the expansion device changes the pressure and temperature of the refrigerant. The temperature of the refrigerant is 98°F as it leaves the condenser. Temperature is measured with a thermometer while the refrigeration system is operating. The line that represents the refrigerant properties in the expansion device is drawn vertically from 98°F on the saturated liquid line to the high-pressure and low-pressure lines. The properties of the refrigerant as it leaves the condenser are found at the intersection on the high-pressure line. The properties of the refrigerant as it enters the evaporator are found at the intersection on the low-pressure line.

Example: Plotting a System Graph — Actual Cycle

A mechanical compression refrigeration system contains R-22. The low-pressure side is 80 psig and the high-pressure side is 180 psig. A thermometer reads 85°F on the outlet of the evaporator. The temperature of the refrigerant in the condenser as measured with a thermometer is 75°F. Plot the actual cycle on a pressure-enthalpy diagram.

1. The low-pressure line is located on the pressure-enthalpy diagram along the 94.7 psia line (80 psig + 14.7 = 94.7 psia). The low-pressure line is a horizontal line drawn from the left of the saturated liquid line to the right of the saturated vapor line. See Figure 8-13.

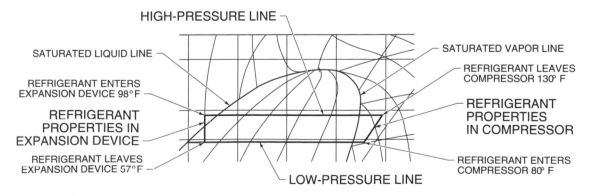

Figure 8-12. In an actual cycle, the system graph includes superheat from the evaporator and subcooling from the condenser.

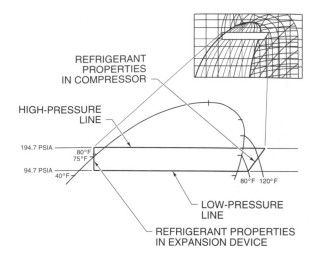

Figure 8-13. When plotting the system graph of an actual cycle, the actual temperature of the refrigerant as it enters the compressor and the actual temperature of the refrigerant as it enters the expansion device are used.

2. The high-pressure line is located on the pressure-enthalpy diagram along the 194.7 psia line (180 psig + 14.7 = 194.7 psia). The high-pressure line is drawn from the left of the saturated liquid line to the right of the saturated vapor line.

3. The line that represents the refrigerant properties in the compressor begins at the intersection of the low-pressure line and the 80°F temperature line. The line is drawn parallel with the closest constant entropy line until it intersects the high-pressure line.

4. The line that represents the refrigerant properties in the expansion device is a vertical line that extends from the 75°F temperature scale on the saturated liquid line to the high-pressure and low-pressure lines. The resulting figure is the system graph that represents the performance of an actual cycle of an operating refrigeration system.

Heat Transfer

The amount of heat transfer in the different parts of a refrigeration system is found by using the system graph and the enthalpy lines on a pressure-enthalpy diagram. Heat transfer includes heat rejected by the evaporating medium in the evaporator, heat absorbed in the compressor, and heat rejected in the condenser.

1. Refrigeration effect is the heat absorbed by the refrigerant as it flows through the evaporator. The point that represents the properties of the refrigerant as it enters the evaporator is located at the intersection of the line that represents the refrigerant properties in the expansion device and the low-pressure line. The enthalpy of the refrigerant as it enters the evaporator is found on a line drawn vertically from this point to the enthalpy scale. See Figure 8-14.

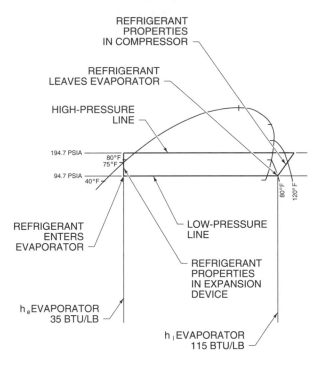

Figure 8-14. Refrigeration effect is the difference between the enthalpy of the refrigerant entering the evaporator and the enthalpy of the refrigerant leaving the evaporator.

2. The point that represents the properties of the refrigerant as it leaves the evaporator is located at the intersection of the low-pressure line and the line that represents the refrigerant properties in the compressor. The enthalpy of the refrigerant as it leaves the evaporator is found on a line drawn vertically from this point to the enthalpy scale. The difference between the enthalpy of the refrigerant as it enters the evaporator and the enthalpy of the refrigerant as it leaves the evaporator is the refrigeration effect, which is the amount of heat absorbed by the refrigerant in the evaporator.

Example: Finding Refrigeration Effect — Actual Cycle

The enthalpy of a refrigerant that leaves the evaporator is 115 Btu/lb. The enthalpy of the refrigerant that enters the evaporator is 35 Btu/lb. Find the refrigeration effect.

$RE = h_1 - h_e$

$RE = 115 - 35$

$RE = 80$ Btu/lb

3. Heat of compression is the thermal energy equivalent of the mechanical energy used to compress the refrigerant. The point that represents the properties of the refrigerant as it enters the compressor is located at the intersection of the line that represents the refrigerant properties in the compressor and the low-pressure line. The point that represents the properties of the refrigerant as it enters the compressor is the same for the properties of the refrigerant as it leaves the evaporator. The enthalpy of the refrigerant as it enters the compressor is found on a line drawn vertically from this point to the enthalpy scale. See Figure 8-15.

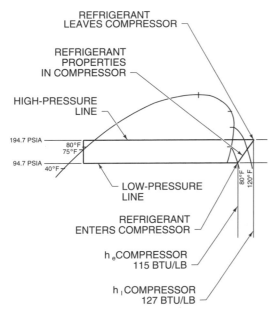

Figure 8-15. Heat of compression is the difference between the enthalpy of the refrigerant entering the compressor and the enthalpy of the refrigerant leaving the compressor.

4. The point that represents the properties of the refrigerant as it leaves the compressor is located at the intersection of the line that represents the refrigerant properties in the compressor and the high-pressure line. The enthalpy of the refrigerant as it leaves the compressor is found on a line drawn vertically from this point to the enthalpy scale. The difference between the enthalpy of the refrigerant as it enters the compressor and the enthalpy of the refrigerant as it leaves the compressor is the heat of compression, which is the amount of heat absorbed by the refrigerant in the compressor.

Example: Finding Heat of Compression — Actual Cycle

The enthalpy of a refrigerant that leaves the compressor is 127 Btu/lb. The enthalpy of the refrigerant that enters the compressor is 115 Btu/lb. Find heat of compression.

$HC = h_1 - h_e$

$HC = 127 - 115$

$HC = 12$ Btu/lb

5. Heat of rejection is the heat rejected by a refrigerant in the condenser. Heat of rejection must equal the heat absorbed in the evaporator plus the heat of compression. The point that represents the properties of the refrigerant as it enters the condenser is located at the intersection of the line that represents the refrigerant properties in the compressor and the high-pressure line. The enthalpy of the refrigerant as it enters the condenser is found on a line drawn vertically from this point to the enthalpy scale. See Figure 8-16.

The point that represents the properties of the refrigerant as it leaves the condenser is located at the intersection of the high-pressure line and the line that represents refrigerant properties in the expansion device. The enthalpy of the refrigerant as it leaves the condenser is found on a line drawn vertically from this point to the enthalpy scale. The heat of rejection, which is the amount of heat rejected by the condenser, is found by subtracting the enthalpy of the refrigerant as it enters the condenser from the enthalpy of the refrigerant as it leaves the condenser. It may also be found by adding the refrigeration effect to the heat of compression.

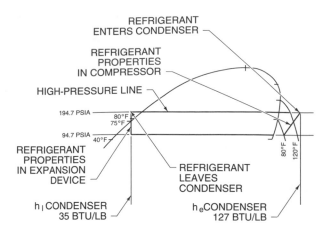

Figure 8-16. Heat of rejection is the difference between the enthalpy of the refrigerant leaving the condenser and the enthalpy of the refrigerant entering the condenser.

Example: Finding Heat of Rejection — Actual Cycle

The enthalpy of the refrigerant that enters the condenser is 127 Btu/lb. The enthalpy of the refrigerant that leaves the condenser is 35 Btu/lb. Find the heat of rejection.

$$HR = h_e - h_1$$

$$HR = 127 - 35$$

$$HR = \mathbf{92 \ Btu/lb}$$

6. The amount of superheat that is added in the evaporator is the difference between the enthalpy of the refrigerant at the point where the low-pressure line intersects the saturated vapor line and the enthalpy of the refrigerant at the point where the low-pressure line intersects the line that represents the refrigerant properties in the compressor. See Figure 8-17. The amount of subcooling that occurs in the condenser is the difference between the enthalpy of the refrigerant at the point where the high-pressure line intersects the saturated liquid line and the enthalpy of the refrigerant at the point where the high-pressure line intersects the line that represents the refrigerant properties in the expansion device.

REFRIGERATION SYSTEM PERFORMANCE

The system graph of an operating refrigeration system drawn on a pressure-enthalpy diagram is used for collecting data about a refrigeration system. The data obtained from a system graph is used to calculate performance characteristics, which indicate the efficiency of a refrigeration system. The performance characteristics of a refrigeration system are refrigeration effect, compression ratio, and cooling coefficient of performance.

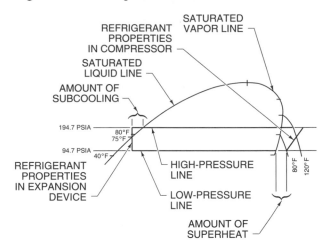

Figure 8-17. Superheat and subcooling occur during an actual refrigeration cycle.

Refrigeration Effect

Refrigeration effect is the amount of heat absorbed by a refrigerant in the evaporator of a refrigeration system. Refrigeration effect is the amount of cooling that a refrigeration system can produce. Refrigeration effect is calculated by subtracting the enthalpy of a refrigerant that enters the evaporator from the enthalpy of the refrigerant as it leaves the evaporator. The enthalpy at each point is found on a system graph drawn on a pressure-enthalpy diagram.

Compression Ratio

Compression ratio is the ratio of the pressure on the high-pressure side to the pressure on the low-pressure side of a refrigeration system. The pressures must be absolute pressure. Compression ratio is equal to the pressure on the high-pressure side divided by the pressure on the low-pressure side. Compression ratio is found by applying the formula:

$$C_r = \frac{P_H}{P_L}$$

where

C_r = compression ratio

P_H = pressure on high-pressure side (in psia)

P_L = pressure on low-pressure side (in psia)

Example: Finding Compression Ratio

A refrigeration system has a high-pressure side of 150 psig and a low-pressure side of 80 psig. Find the compression ratio.

1. Convert gauge pressure to absolute pressure.

psia = *psig* + 14.7

High-pressure Side

psia = 150 + 14.7

psia = **164.7 psia**

Low-pressure Side

psia = 80 + 14.7

psia = **94.7 psia**

2. Find the compression ratio.

$$C_r = \frac{P_H}{P_L}$$

$$C_r = \frac{164.7}{94.7}$$

$$C_r = \textbf{1.74}$$

A low compression ratio indicates efficient compressor operation. A high compression ratio requires that the compressor use more energy to raise the pressure of the refrigerant. Therefore, a refrigeration system with a high compression ratio uses more energy than a refrigeration system with a low compression ratio to produce the refrigeration effect.

On a system graph for a refrigerant with a high compression ratio, the high-pressure and low-pressure lines move apart as the heat of compression increases. Energy used to operate the refrigeration system increases as the difference between the pressures increases, but the refrigeration effect does not increase proportionally.

Coefficient of Performance

The *coefficient of performance* of a mechanical compression refrigeration system is the theoretical efficiency of the operation of the refrigeration system. COP is the ratio of the refrigeration effect achieved in a refrigeration system to the energy used to achieve the refrigeration effect. COP is found by dividing the refrigeration effect by the heat of compression. COP is found by applying the formula:

$$COP = \frac{RE}{HC}$$

where

COP = coefficient of performance

RE = refrigeration effect (in Btu/lb)

HC = heat of compression (in Btu/lb)

Example: Finding Coefficient of Performance

The refrigeration effect of a refrigeration system is 77 Btu/lb and the heat of compression is 15 Btu/lb. Find the COP of the refrigeration system.

$$COP = \frac{RE}{HC}$$

$$COP = \frac{77}{15}$$

$$COP = \textbf{5.13}$$

This is cooling COP and should not be confused with heating COP, which is calculated for a heat pump. A high cooling COP indicates efficient operation. Because the heat of compression is the thermal equivalent of the work that produces the refrigeration effect, a low COP indicates that more energy is used to produce the same effect.

Cooling COP is a performance rating that indicates actual efficiency. COP is used primarily by engineers when designing a refrigeration system. It is not commonly used when referring to refrigeration equipment. Seasonal energy efficiency rating (SEER) is used more often to refer to refrigeration systems. *Seasonal energy efficiency rating* is the cooling performance rating of a refrigeration system such as an air conditioner, which operates under normal conditions over a period of time. The efficiency of a unit is lower when the unit starts and stops often, such as during cool weather cycles. The efficiency increases when the cycles get longer, such as during warm weather. By using SEER ratings, the differences in efficiency are averaged out over a longer period of operating time.

Review Questions

## 8

1. What refrigerant properties can be found on a pressure-enthalpy diagram?

2. What is the relationship between absolute pressure and gauge pressure? How is gauge pressure converted to absolute pressure?

3. Define *saturation pressure*. Explain how it relates to saturation temperature.

4. Define *superheat* and *subcooling*. How are they related?

5. Define *enthalpy*. Where are the scales for enthalpy found on a pressure-enthalpy diagram?

6. Define *constant quality*. How is it related to the critical point?

7. Define *constant volume*. Where are the constant volume lines located on a pressure-enthalpy diagram?

8. Define *constant entropy*. Why is it an important part of a pressure-enthalpy diagram?

9. What is a system graph? What does it do?

10. What are the two main characteristics of the system graph of an ideal refrigeration cycle?

11. What does the low-pressure line on a system graph represent?

12. What does the high-pressure line on a system graph represent?

13. What does the vertical line that connects the left side of a system graph represent?

14. What does the diagonal line that runs along the right side of a system graph represent?

15. What causes the temperature decrease in the refrigerant as it passes through the expansion device?

16. What is the difference between an ideal refrigeration cycle and an actual refrigeration cycle?

17. At what two points in a refrigeration system is the refrigerant superheated?

18. At what point does subcooling occur in a refrigeration system?

19. Define *heat of rejection*.

20. What must the heat of rejection be equal to? What law of thermodynamics does this illustrate?

21. How does COP indicate the efficiency at which a refrigeration system operates?

22. Why is the SEER of a refrigeration system a better indication of efficiency than the COP of the refrigeration system?

Compression System—Low-pressure Side Chapter 9

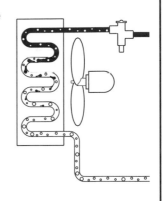

A mechanical compression refrigeration system consists of a low-pressure side and a high-pressure side. The low-pressure side of the refrigeration system is where the evaporating medium is cooled when the refrigerant absorbs heat from the evaporating medium in the evaporator. The cooled evaporating medium cools building spaces or objects. The low-pressure side of a mechanical compression refrigeration system consists of an expansion device, evaporator, refrigerant lines, and accessories. These components are used to control the refrigerant flow and heat transfer. Controlling heat transfer causes the refrigeration effect.

LOW-PRESSURE SIDE

The low-pressure side of a mechanical compression refrigeration system begins at the expansion device. The expansion device and the compressor suction pressure create a relatively low-pressure area between the expansion device and the compressor inlet. The expansion device reduces the pressure on the refrigerant by allowing the refrigerant to expand. Some of the refrigerant vaporizes in the expansion device as the pressure on the refrigerant decreases. This refrigerant vapor absorbs heat from the rest of the refrigerant, which becomes relatively cold. The relatively cold refrigerant flows through the evaporator and absorbs heat from the evaporating medium. The temperature of the evaporating medium decreases because of the heat loss. The evaporating medium is then used for cooling.

EXPANSION DEVICES

An expansion device is a valve or mechanical device that decreases the pressure in a mechanical compression refrigeration system. In the low-pressure side, the hot, high-pressure refrigerant vaporizes at a relatively low temperature. In the expansion device, part of the refrigerant vaporizes and cools the rest of the refrigerant. The refrigerant then absorbs

heat from the evaporating medium. This is the refrigeration effect.

The expansion device is located in the liquid line ahead of the evaporator. The refrigerant enters the expansion device at a high pressure and leaves the expansion device at a low pressure. The sudden pressure decrease that occurs as the refrigerant passes through the expansion device causes some of the refrigerant to vaporize. This vaporization removes heat from the remaining refrigerant and causes a temperature decrease. The pressure decrease in the expansion device causes a temperature decrease. See Figure 9-1.

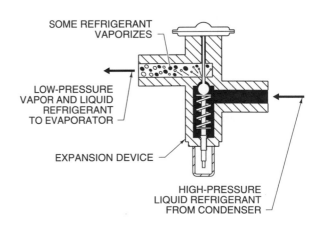

Figure 9-1. The expansion device causes a pressure decrease in a refrigeration system.

In the evaporator, the temperature of the refrigerant decreases to a temperature lower than the temperature of the evaporating medium. The evaporating medium is the air or water that is cooled by a refrigeration system. The evaporating medium is cooled as it flows over the evaporator coil because the refrigerant inside the evaporator coil is cooler than the evaporating medium.

A mechanical compression refrigeration system will not work properly if the expansion device does not work properly. Expansion devices control the flow of refrigerant by controlling the volume, pressure, or temperature of the refrigerant in the low-pressure side.

Volume Control

In a mechanical compression refrigeration system, restricting the volume of the refrigerant that flows through the liquid line causes a low-pressure area. Reducing the volume reduces the flow of refrigerant and increases the pressure of the refrigerant before it reaches the expansion device. After the expansion device, the refrigerant is at low pressure because of the compressor suction pressure.

Volume control expansion devices include capillary tubes, float valves, and hand valves. Capillary tubes are used on small package refrigeration systems. Float valves are used on refrigeration systems that have flooded evaporator coils. Hand valves are used on large refrigeration systems that are manually operated and maintained.

Capillary Tubes. Capillary tubes are the most common volume control expansion device. A *capillary tube* is a long, thin tube that resists fluid flow, which causes a pressure decrease. A capillary tube has a small diameter and may contain a wire that reduces the internal area of the tube. The small internal area and relatively long length of a capillary tube cause considerable resistance to the flow of refrigerant through the tube. This resistance inside the tube causes the pressure of the refrigerant to decrease when the refrigerant passes out of the tube. See Figure 9-2.

When a capillary tube is used as an expansion device, the volume of the refrigerant in the refrigeration system is critical. A capillary tube is a con-

stant volume control expansion device, which means that the device requires constant volume in a refrigeration system. Constant volume control indicates that the device cannot handle significant changes in the cooling load. Capillary tubes can be used as expansion devices in refrigeration systems that do not have significant load changes.

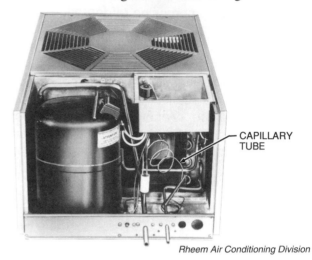

Figure 9-2. A capillary tube is a long, thin tube that reduces the pressure of a refrigerant as the refrigerant passes out of the tube.

An advantage to using a capillary tube as an expansion device is that refrigerant can flow through a capillary tube when the refrigeration system is OFF. This flow equalizes the pressures in the two sides of the refrigeration system. The equalized pressure is an advantage because the compressor does not have to start against a high pressure when the compressor is switched ON. A split-phase motor that has a low starting torque and has no starting capacitor can be used because of the low starting pressure. A split-phase motor without a starting capacitor is simpler and less expensive than a motor with a high starting torque and a starting capacitor.

Float Valves. A *float valve* is a valve controlled by a hollow ball that floats in a liquid. Depending on the level of the liquid, the float ball opens or closes a port. A float valve expansion device is located in the liquid line. Float valves are used to control the volume of liquid refrigerant. If the liquid level rises, the valve will close and stop the liquid flow. If the liquid level falls, the valve will open and allow more liquid flow. See Figure 9-3.

A float valve expansion device is used in large air conditioning units that have flooded evaporator coils. A *flooded evaporator coil* is an evaporator coil that is full of liquid refrigerant during normal operation. A float valve controls the liquid level in the coil. The float valve expansion device causes a pressure decrease on the low-pressure side of a refrigeration system.

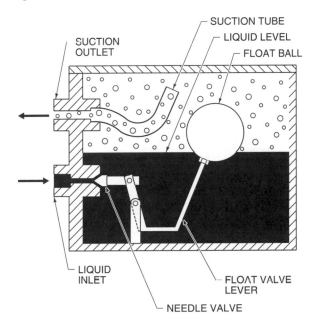

Figure 9-3. A float valve is controlled by a float ball that opens or closes a port in the liquid line.

Hand Valves. A *hand valve* is a needle valve that has fine threads on the valve stem and body. The needle adjusts against the valve seat and controls flow through the valve. Hand valves are used as expansion devices in refrigeration systems and are adjusted by hand to regulate the flow of refrigerant. Because of the fine threads on a hand valve, the valve can be adjusted precisely to allow the correct flow of refrigerant. See Figure 9-4.

A refrigeration system that has a hand valve expansion device requires constant supervision. The valve position is fixed when the valve is set. If the load on the refrigeration system changes, adjustments to the valve setting must be made manually. Hand valves, like float valves, are used on large commercial or industrial refrigeration systems. Hand valves are not used on small automatic refrigeration systems because they require supervision.

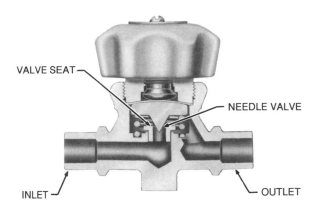

Superior Valve Company
Division of Amcast Industrial Corporation

Figure 9-4. A hand valve is an expansion device that is adjusted manually to regulate refrigerant flow.

Pressure Control

Pressure control expansion devices are used to control the pressure on the low-pressure side of a refrigeration system. Controlling the pressure on the low-pressure side controls the temperature of the refrigerant in the evaporator. Other pressure control devices are used to control compressor suction pressure or compressor crankcase pressure.

Automatic Expansion Valves. The most common pressure control device is the automatic expansion valve. An *automatic expansion valve* is a valve that is opened and closed by the pressure in the line ahead of the valve. An automatic expansion valve is a pressure-regulating valve that maintains a constant pressure in the evaporator. The valve controls the temperature of the refrigerant by controlling the pressure in the evaporator. An automatic expansion valve is adjustable for a range of pressures.

In an automatic expansion valve, a diaphragm is connected to a ball valve by a push rod. An adjustment spring sits above the diaphragm. An adjustment screw controls the pressure that the adjustment spring exerts on the diaphragm. The push rod connects the diaphragm to the ball valve. The ball valve opens or closes to control the flow of refrigerant. Pressure from the adjustment spring balances the pressure on the diaphragm from the refrigerant in the evaporator. See Figure 9-5.

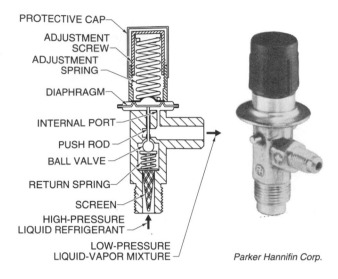

Parker Hannifin Corp.

Figure 9-5. An automatic expansion valve controls the temperature of the refrigerant by controlling the pressure in the evaporator.

Pressure from the refrigerant in the evaporator exerts pressure on the diaphragm through the internal port. A return spring also exerts pressure on the diaphragm. Pressure from the adjustment spring above the diaphragm balances this pressure. If the pressure of the refrigerant in the evaporator exceeds the pressure of the adjustment spring, the ball valve will close and reduce the refrigerant flow. If the refrigerant pressure in the evaporator is less than the adjustment spring pressure, the push rod will open the ball valve and allow more refrigerant into the low-pressure side of the refrigeration system.

If the cooling load on a refrigeration system increases, the pressure in the evaporator will increase. The pressure increase occurs because the refrigerant vaporizes more quickly in the evaporator, which increases the amount of vapor in the evaporator compared to the amount of liquid. Because the vapor has greater specific volume than the liquid, it exerts greater pressure. The pressure increase causes the valve to close, which reduces the flow of refrigerant. The reduced flow of refrigerant causes the pressure to fall back to the setpoint.

If the cooling load on the evaporator decreases, the refrigerant will vaporize slowly and the pressure on the low-pressure side of the refrigeration system will decrease. The pressure decrease causes the valve to open. The open valve allows more refrigerant flow and the pressure increases to the setpoint.

For example, the adjustment spring in an automatic expansion valve exerts 30 psi above the diaphragm. When the pressure from the evaporator exerts 30 psi below the diaphragm, the pressure of the adjustment spring and the ball valve are balanced. See Figure 9-6.

If the cooling load on the evaporator increases, the pressure in the evaporator and the pressure on the diaphragm will increase. This pressure is larger than the adjustment spring pressure, which closes the ball valve. The reduced refrigerant flow causes the reduced pressure in the refrigeration system.

If the cooling load on the evaporator decreases, the pressure in the evaporator and the pressure on

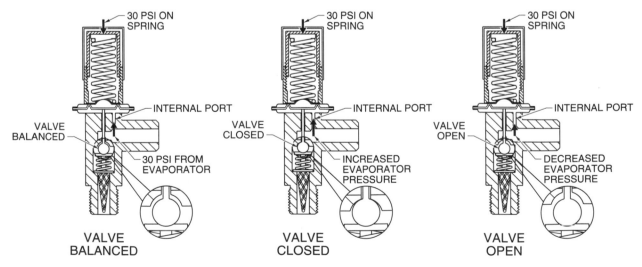

Figure 9-6. An automatic expansion valve is controlled by pressure of refrigerant, which opposes the compression of the adjustment spring.

the diaphragm will decrease. As the pressure decreases, the adjustment spring pressure is larger than the evaporator pressure and return spring pressure. The valve opens and the refrigerant flow rate increases. The increase in refrigerant flow causes increased pressure in the refrigeration system.

Automatic expansion valves are used on refrigeration systems that have constant cooling loads. An automatic expansion valve does not compensate for changes in the cooling load or for changes in the condensing medium temperature. Because of the way it works, an automatic expansion valve starves the evaporator for refrigerant if the refrigeration system is heavily loaded and floods the evaporator with refrigerant if the load is light. Automatic expansion valves are used most often on small air conditioning units or in systems that have slight load variations.

Temperature Control

Temperature control devices control the flow of refrigerant by controlling the temperature of the refrigerant in part of the refrigeration system. Two types of temperature control devices are the thermostatic expansion valve and the thermoelectric expansion valve.

Thermostatic Expansion Valves. Thermostatic expansion valves are the most commonly used temperature control expansion devices in refrigeration systems. A *thermostatic expansion valve* is an expansion valve that uses pressure to control the amount of superheat the refrigerant absorbs in the evaporator. The pressure is generated in a small, refrigerant-filled remote bulb, which is located on the suction line of the refrigeration system. Superheat is sensible heat absorbed by the refrigerant after it has completely vaporized. See Figure 9-7.

The remote bulb is connected to the expansion valve by a capillary tube. The pressure in the remote bulb is generated by the refrigerant in the bulb. The refrigerant evaporates or condenses depending on the temperature of the refrigerant in the suction line.

A thermostatic expansion valve is located in the liquid line of a refrigeration system ahead of the evaporator. A diaphragm in the top of the valve is connected to a ball valve inside the valve body by a push rod. The pressure above the diaphragm depends on the temperature of the refrigerant in the suction line. The pressure on the valve opens or closes the valve, which controls the flow of refrigerant into the evaporator.

Moving the adjustment screw on a thermostatic expansion valve changes the superheat setting of

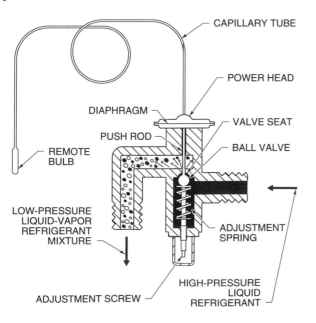
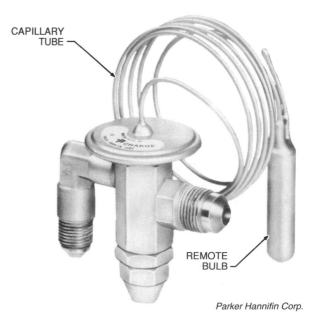

Parker Hannifin Corp.

Figure 9-7. A thermostatic expansion valve controls the amount of superheat that a refrigerant absorbs in the evaporator.

the valve. Adjusting the adjustment screw opens or closes the valve according to the temperature sensed by the remote bulb.

For example, a refrigeration system using R-22 operates with a high-pressure side of 260 psig and a low-pressure side of 80 psig. The thermostatic expansion valve is adjusted to provide 10°F superheat. At these conditions, the pressure exerted below the diaphragm is the low-pressure side pressure, which is 80 psig. At 80 psig, the temperature of the refrigerant in the evaporator is 45°F. With a superheat of 10°F, the remote bulb senses a temperature of 55°F. The pressure in the remote bulb, which is the pressure above the diaphragm, is 92.6 psig. The adjustment spring must be set at 12.6 psig (92.6 psig – 80 psig = 12.6 psig) to bring the valve into balance. See Figure 9-8.

If the load on the evaporator increases, the refrigerant flowing through the evaporator will vaporize rapidly and more superheat will be absorbed by the refrigerant. The remote bulb senses the added superheat. The pressure in the bulb and the pressure above the diaphragm increases. The pressure increase opens the valve, which allows more refrigerant flow. The added flow allows the refrigeration system to handle the increased load.

If the load on the evaporator decreases, the refrigerant flowing through the evaporator will vaporize slowly and cause a decrease in the amount of superheat. The remote bulb senses the reduced superheat in the suction line, which decreases the pressure above the diaphragm. The pressure decrease closes the valve, which allows less refrigerant to flow. The reduced refrigerant flow allows the refrigeration system to compensate for the reduced load.

A thermostatic expansion valve is an excellent expansion device for refrigeration systems that have varying cooling loads. The valve will automatically adjust to the load.

Thermoelectric Expansion Valves. A *thermoelectric expansion valve* is an expansion device that controls the flow of refrigerant in response to temperature sensed by a solid-state sensor. The valve is temperature-actuated and uses electric current as a control signal.

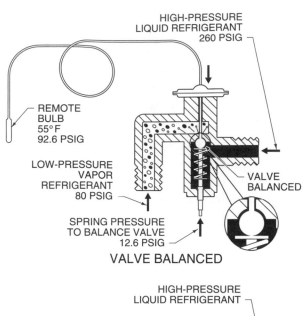

VALVE BALANCED

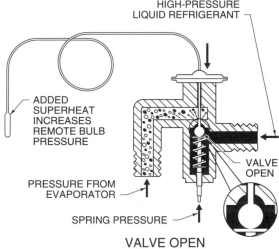

VALVE OPEN

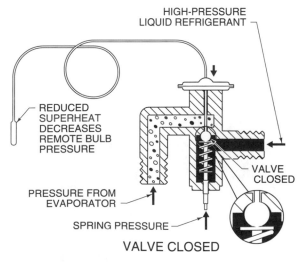

VALVE CLOSED

Figure 9-8. A thermostatic expansion valve is operated by pressure generated in a small, refrigerant-filled remote bulb located in the suction line.

A thermoelectric expansion valve looks like a thermostatic expansion valve, but contains a low-voltage electrical circuit that runs from an electric power source through an electronic temperature sensor instead of a capillary tube running to a remote bulb. The top of the valve has a small, electric heater and a bimetal operator. The bimetal operator is connected to a needle valve by a push rod. The electric heater is connected to electric terminals that are connected to the power source and temperature sensor. See Figure 9-9.

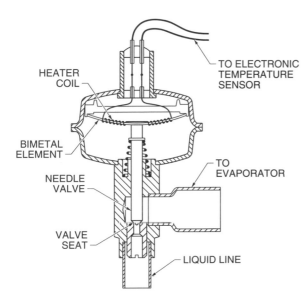

Figure 9-9. A thermoelectric expansion valve controls the flow of refrigerant in response to temperature sensed by a solid-state sensor.

The electric heater coil is connected to the electrical power source and sensor. When electric current flows through the electric circuit, heat generated by the heater bends the bimetal element. The valve opens when the bimetal element bends. When no electric current flows through the circuit, the bimetal element returns to its original position and the valve closes. A 24 V power source is normally used for a thermoelectric expansion valve. A thermistor is the sensor that is used with a thermoelectric expansion valve.

A thermistor is a temperature-sensing element that changes electrical resistance in response to a temperature change. A thermistor is attached to the suction line of a refrigeration system or any location where a temperature reading is desired. The thermistor is connected to the electrical circuit and ther-

moelectric valve. The valve controls the flow of refrigerant in response to temperature changes at the location of the thermistor.

Thermoelectric expansion valves have a longer response time than thermostatic expansion valves. Thermoelectric expansion valves can sense temperatures at a great distance from the valve. More than one thermistor can be used to switch from one control point to another. Some control functions require this arrangement.

EVAPORATORS

The evaporator is the component of a mechanical compression refrigeration system where the refrigerant vaporizes as it absorbs heat from the evaporating medium. The evaporator section in a refrigeration system consists of the evaporator, cabinet, blower, and accessories. See Figure 9-10. The two kinds of evaporators are air-cooled evaporators and coolers.

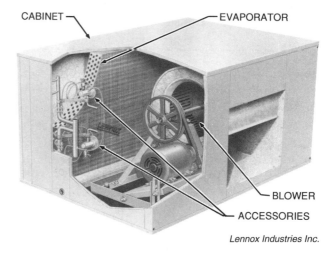

Lennox Industries Inc.

Figure 9-10. Refrigerant vaporizes and absorbs heat from the evaporating medium in the evaporator.

Air-cooled Evaporators

Air-cooled evaporators are evaporators that cool air as the air flows across the evaporator coils. Air is the evaporating medium in an air-cooled evaporator. The refrigerant flows through a coil and air is blown across the outside of the coil by the evaporator blower. The amount of heat transferred between the refrigerant in the coil and the air depends on the size of the coil. See Figure 9-11.

FINNED TUBES

Super Radiator Coils

Figure 9-11. An air-cooled evaporator coil cools air as the air flows across the evaporator coil.

Refrigerant enters the evaporator coil at the low-pressure side pressure and saturated temperature for that pressure. As the refrigerant flows through the coil, it absorbs heat from the air that flows across the outside of the coil. The added heat vaporizes the refrigerant. Coils used in refrigeration systems include bare-tube, finned-tube, and flat-plate coils.

Bare-tube Coils. A *bare-tube coil* is a coil of bare, copper tubes through which refrigerant flows. Air is blown across the outside of the coil. This type of coil is not used extensively because the rate of heat transfer between the refrigerant and air is not as great as that attained with finned-tube coils.

Finned-tube Coils. A *finned-tube coil* is a copper or aluminum tube with aluminum fins pressed on the tubing to increase the surface area of the coil. The increased surface area provides a larger heat transfer surface area. Heat is transferred from the air to the finned-tube coil and then to the refrigerant. Finned-tube coils are commonly used in air conditioning and refrigeration systems.

Flat-plate Coils. A *flat-plate coil* is a coil pressed or buried within flat plates of metal. This type of coil is used mostly in refrigeration applications that use water as the evaporating medium but not often for air conditioning systems that use air as the evaporating medium.

Blowers. A blower moves air across the outside of the evaporator coil and through building spaces. The air may flow either through ductwork or directly into a building. The type of blower used depends on the air distribution system. The two basic types of blowers are centrifugal blowers and propeller fans.

A centrifugal blower is normally used with air distribution systems that have ductwork. Return air ductwork from building spaces is connected to the inlet side of the blower. Supply air ductwork is connected to the outlet side of the evaporator coil. A centrifugal blower moves air efficiently against the resistance in a typical air distribution system.

In nonducted systems such as blower coils, cabinet air conditioners, or wall or window air conditioners, the evaporator blower is often a propeller fan. Propeller fans move air efficiently when resistance to air flow is low.

Cooling Capacity. *Cooling capacity* is the total heat transfer capacity of an evaporator coil expressed in Btu per hour. The cooling capacity of an evaporator coil is a function of the kind of refrigerant used, refrigeration effect, and refrigerant flow rate.

The refrigeration effect is the amount of heat in Btu per pound absorbed by the refrigerant as it flows through the evaporator coil. There is no pressure change in an evaporator, only heat is transferred. The refrigerant enters an evaporator as a liquid and leaves the evaporator as a vapor. The enthalpy (heat content) of the liquid refrigerant and the enthalpy of the refrigerant vapor are used to find refrigeration effect. The enthalpy values are found on a refrigerant property table for the refrigerant. The low-pressure side pressure (in psig) is used to find the line that contains the enthalpy values. On a refrigerant property table for R-22, for example, a low-pressure side pressure of 48.809 psig corresponds to an enthalpy of 17.317 Btu/lb as a liquid and an enthalpy of 106.65 Btu/lb as a vapor.

The *refrigerant flow rate* is the amount of refrigerant that flows through the refrigeration system per unit of time. Refrigerant flow rate is expressed by weight (mass) or volume. *Mass flow rate* is the rate of refrigerant flow expressed in pounds per minute. *Volumetric flow rate* is the rate of refrigerant flow expressed in cubic feet per minute. Volumetric flow rate is the mass flow rate expressed as volume. Volumetric flow rate is a function of the refrigeration effect of the system, cooling capacity, and density of the refrigerant. Density is the weight of a substance per unit of volume.

The flow rate of the refrigerant is the rate in Btu per hour at which the refrigerant flows through the coil. The evaporator coil cooling capacity equals the refrigeration effect of a refrigeration system multiplied by the flow rate of the refrigerant times 60 minutes. Evaporator coil cooling capacity is found by applying the formula:

$$cc = RE \times w \times 60$$

where

cc = cooling capacity of evaporator (in Btu/hr)

RE = refrigeration effect (in Btu/lb)

w = mass flow rate (in lb/min)

60 = constant

Example: Finding Cooling Capacity — Evaporator

A refrigeration system that contains R-22 has a low-pressure side pressure of 92.58 psig. The mass flow rate of R-22 is 21.47 lb/min. Find the cooling capacity of the evaporator.

1. Find the enthalpy of the refrigerant.

   Find the enthalpy of the refrigerant as a liquid and a vapor according to the low-pressure side pressure. On a refrigerant property table for R-22, the enthalpy of the refrigerant as a vapor is 109.12 Btu/lb at 92.58 psig and the enthalpy of the refrigerant as a liquid is 25.73 Btu/lb at 92.58 psig. See Appendix.

2. Find the refrigeration effect.

   $RE = h_1 - h_e$

   $RE = 109.12 - 25.73$

   $RE = 83.39$ Btu/lb

3. Find the cooling capacity of the evaporator.

   $cc = RE \times w \times 60$

   $cc = 83.39 \times 21.47 \times 60$

   $cc = 107,423$ Btu/hr

Actual Refrigeration Effect. Actual refrigeration effect of a refrigeration system is approximately 85% of the cooling capacity of the evaporator. Actual refrigeration effect indicates the amount of cooling produced by a refrigeration system. Actual refrigeration effect is found by applying the formula:

$$RE_a = cc \times .85$$

where

RE_a = actual refrigeration effect (in Btu/hr)

cc = cooling capacity of evaporator (in Btu/hr)

$.85$ = constant

Example: Finding Actual Refrigeration Effect — Evaporator

A refrigeration system contains R-22. The low-pressure side pressure is 111.26 psig. The mass flow rate of the refrigerant is 30.12 lb/min. Find the actual refrigeration effect of the refrigeration system.

1. Find the enthalpy of the refrigerant.

   Find the enthalpy of the refrigerant as a liquid and a vapor according to the low-pressure side pressure. On a refrigerant property table for R-22, the enthalpy of the refrigerant as a vapor is 109.84 Btu/lb at 111.26 psig and the enthalpy of the refrigerant as a liquid is 28.63 Btu/lb at 111.26 psig. See Appendix.

2. Find refrigeration effect of the evaporator.

   $RE = h_1 - h_e$

   $RE = 109.84 - 28.63$

   $RE = 81.21$ Btu/lb

3. Find the cooling capacity of the evaporator.

   $cc = RE \times w \times 60$

   $cc = 81.21 \times 30.12 \times 60$

   $cc = 146,762.71$ Btu/hr

4. Find the actual refrigeration effect.

   $RE_a = cc \times .85$

   $RE_a = 146,762.71 \times .85$

   $RE_a = 124,748.30$ Btu/hr

Coolers

Coolers, which are components of chillers, are evaporators that chill water. Water is the evaporating medium in a cooler. The chilled water is used to cool the air in building spaces. Water is contained in the shell (tank) of a cooler and refrigerant flows through a coil that runs through the shell. The water in a cooler is chilled as it comes in contact with the cold coil. Shell-and-tube coolers and shell-and-coil coolers are used commonly. See Figure 9-12.

Shell-and-Tube Cooler. A *shell-and-tube cooler* is a cooler that contains tubes that run from one end of the shell to the other. Refrigerant flows through the tubes. The evaporating medium, which is water or brine, is cooled as it circulates through the shell and around the cooled tubes. The tubes act as the evaporator of the refrigeration system and the shell holds the evaporating medium.

Shell-and-Coil Cooler. A *shell-and-coil cooler* is a cooler that contains a coil of copper tubing. Water circulates through the shell and refrigerant circulates through the coil. The coil acts as the evaporator of the refrigeration system and the shell holds the evaporating medium.

Both shell-and-tube and shell-and-coil coolers transfer heat efficiently between the water and refrigerant in a chiller. Only the wall of the copper tubing separates the refrigerant from the evaporating medium in a cooler.

Water Source. A makeup water line supplies the water used in a cooler. Makeup water can be supplied from the water system in a building, a well, or any other available source. The water in a chiller is circulated within a closed piping system that includes the cooler, circulating pump(s), piping, and terminal devices. The water is continually circulated and new water is added only as needed.

Cooling Capacity. The cooling capacity of a cooler is the total heat transfer capacity of a cooler expressed in Btu per hour. The cooling capacity of a cooler is a function of the kind of refrigerant used, refrigeration effect, and refrigerant flow rate. The procedure for finding the cooling capacity and actual refrigeration effect of an evaporator coil is used to find the cooling capacity and actual refrigeration effect of a cooler.

Example: Finding Actual Refrigeration Effect — Cooler

A refrigeration system that uses R-502 has a low-pressure side pressure of 72.83 psig. The flow rate of the refrigerant is 12.0 lb/min. Find the actual refrigeration effect of the cooler.

1. Find the enthalpy of the refrigerant.

Find the enthalpy of the refrigerant as a liquid and a vapor according to the low-pressure side pressure. On a refrigerant property table for R-502, the enthalpy of the refrigerant as a vapor is 81.41 Btu/lb at 72.83 psig and the enthalpy of the refrigerant as a liquid is 18.85 Btu/lb at 72.83 psig. See Appendix.

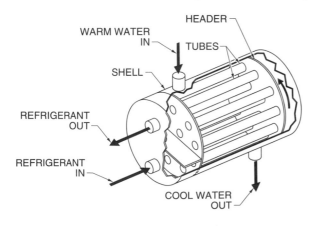

SHELL-AND-TUBE COOLER

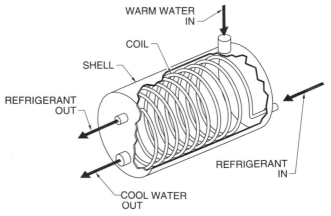

SHELL-AND-COIL COOLER

Figure 9-12. Refrigerant flows through tubes in a shell-and-tube cooler and through a coil in a shell-and-coil cooler.

2. Find refrigeration effect of the cooler.

$RE = h_1 - h_e$

$RE = 81.41 - 18.85$

$RE =$ **62.56 Btu/lb**

3. Find cooling capacity of the cooler.

$cc = RE \times w \times 60$

$cc = 62.56 \times 12.0 \times 60$

$cc =$ **45,043.20 Btu/hr**

4. Find actual cooling capacity of the cooler.

$RE_a = cc \times .85$

$RE_a = 45,043.20 \times .85$

$RE_a =$ **38,286.72 Btu/hr**

REFRIGERANT LINES

Refrigerant lines connect the components in a mechanical compression refrigeration system. Refrigerant lines include the suction, hot gas discharge, and liquid lines. The refrigerant line on the low-pressure side of the refrigeration system is the suction line. On the low-pressure side of the system, the suction line carries refrigerant vapor. Refrigerant lines are designed and installed so that the oil returns to the compressor after the refrigerant and oil circulate through the refrigeration system together.

Refrigeration systems that use halocarbon refrigerants have refrigerant lines that consist of copper or aluminum tubing. The size of the tubing ensures proper refrigerant flow to each section. Refrigeration systems that use ammonia or other corrosive material as a refrigerant have lines that consist of iron or steel piping. Fittings, valves, and other parts of a refrigeration system are made of the same material as the tubing or piping.

Suction Line

The suction line is located in the low-pressure side of a refrigeration system. The suction line runs from the evaporator coil to the compressor inlet port and carries the refrigerant from the evaporator coil to the compressor. The suction line has a relatively large diameter because it carries refrigerant vapor. A refrigerant has a greater specific volume as a vapor than it has as a liquid or as a liquid-vapor mixture.

When a thermostatic expansion valve is used on a refrigeration system, the remote bulb is attached to the suction line close to the evaporator coil. The remote bulb is attached to the outside of the line and does not affect the refrigerant flow in the line at the point of connection. See Figure 9-13.

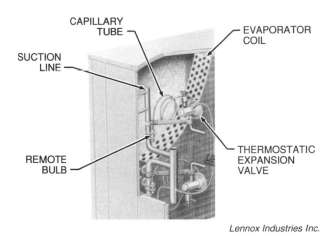

Lennox Industries Inc.

Figure 9-13. The remote bulb of a thermostatic expansion valve is located in the suction line close to the evaporator.

Distributor

Larger refrigeration systems have a distributor on the inlet of the evaporator coil. A *distributor* is a piping arrangement that splits the refrigerant flow into several separate return bends on the evaporator coil to evenly distribute the refrigerant into the coils. A distributor carries refrigerant from the expansion device to the evaporator coil. The tubes running from the distributor head to the coils are relatively small. Distributors are also known as distribution heads. See Figure 9-14.

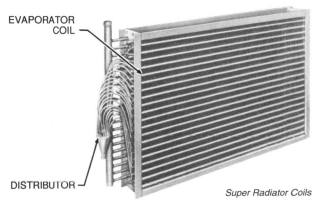

Super Radiator Coils

Figure 9-14. A distributor splits refrigerant flow into several separate return bends.

ACCESSORIES

Accessories are used for servicing, maintaining, and controlling the flow of refrigerant in a refrigeration system. Accessories include service valves, sight glasses, moisture indicators, filter-dryers, suction accumulators, and liquid receivers.

Low-pressure side accessories are located in the suction line and include filter-dryers and suction accumulators. Low-pressure side accessories are used to filter or dehydrate the refrigerant, or to control, clean or regulate the refrigerant as it flows through the line. Low-pressure side accessories are usually designed to work while the refrigeration system is operating. Service valves, sight glasses, and moisture indicators can be located in either the low-pressure or high-pressure side of a refrigeration system.

Cabinets

Cabinets are sheet metal boxes or panels that protect and support the parts of an evaporator. Cabinets are enclosures that direct the flow of air through the coil. Cabinets are usually weatherproofed, insulated, and painted a neutral color.

Service Valves

A *service valve* is a manually operated valve that contains two valve seats. One valve seat (front seat) is closed when the valve is turned completely in. The other valve seat (back seat) is closed when the valve is turned completely out. See Figure 9-15.

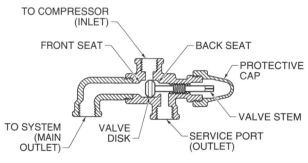

TO COMPRESSOR
(INLET)

FRONT SEAT — — BACK SEAT

— PROTECTIVE CAP

— VALVE STEM

TO SYSTEM   VALVE   — SERVICE PORT
(MAIN   DISK   (OUTLET)
OUTLET)

Figure 9-15. A service valve is a manually operated valve that has two valve seats.

A service valve has an inlet port and two outlet ports. One of the outlet ports is a service port that is used for connecting hoses to a gauge manifold for service procedures. Front seating the valve opens the service port to the inlet port of the valve. Back seating the valve completely closes the service port. When the valve is in an intermediate position, all three ports are open. This position is used for service procedures.

The stem of a service valve is turned in or out with a valve wrench. A *valve wrench* is a ratchet wrench that fits common valve stems. Valve stems are usually $\frac{1}{4}''$ to $\frac{3}{8}''$ in size.

Caps cover the valve stem and service port opening of service valves when the valves are not being used. These caps have sealing gaskets and should always be sealed when the refrigerant system is operating. The caps prevent refrigerant leaks that could occur around the valve stem or packing.

Most refrigeration systems, especially those with serviceable compressors, have service valves. Two service valves are usually used. One valve is located on the discharge port and the other valve is located on the suction port of the compressor.

Some refrigeration systems, especially refrigeration systems that have hermetic compressors, have a single service valve that is located on the liquid line. Refrigeration systems with liquid receivers may have one service valve called a king valve. A *king valve* is a regular service valve that is located directly on the liquid line at the discharge side of the liquid receiver. Service valves can be used as shutoff valves or for access to the refrigeration system for service procedures.

Sight Glasses

A *sight glass* is a fitting for refrigerant lines that contains a small, glass window. Refrigerant flow can be observed through the window. Normally, a sight glass located in the suction line that leaves the evaporator will show only refrigerant vapor with a small quantity of oil flowing through it. Bubbles or foam in the refrigerant lines indicate a shortage of refrigerant or some other problem in the refrigeration system. See Figure 9-16.

Moisture Indicators

A *moisture indicator* is a colored chemical patch located inside the glass window of a sight glass.

The color of the chemical patch indicates whether moisture is present in the refrigerant. The amount of moisture is important in a dry refrigeration system. A dry refrigeration system is a refrigeration system that contains a refrigerant that should have no moisture in it. If the system is dry, the chemical patch will typically be bright green. If moisture is present in the refrigerant, the chemical patch will typically turn pale yellow. Other colors may be used for indicating moisture.

Parker Hannifin Corp.

Figure 9-16. A sight glass is a fitting in a refrigerant line that is used to see the flow of refrigerant inside the lines.

Filter-Dryers

A *filter-dryer* is a combination filter and dryer located in the liquid line of a refrigeration system. A filter-dryer removes solid particles and moisture from a refrigerant. A filter-dryer contains a desiccant that absorbs the moisture that circulates with the refrigerant. A *desiccant* is a substance that acts as a drying agent. See Figure 9-17.

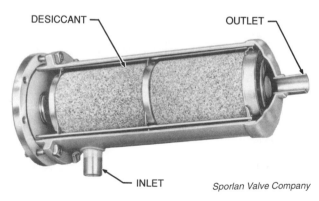

Sporlan Valve Company

Figure 9-17. Filter-dryers remove solid particles from refrigerant and absorb moisture in the refrigerant.

Particles such as slivers of metal from installation or service procedures contaminate the refrigerant, which reduces the efficiency of a refrigeration sys-

tem. A filter-dryer prevents contaminants from circulating through a refrigeration system. Carbon particles, for example, would damage a compressor if the particles flowed through the compressor.

Suction Accumulator

A *suction accumulator* is a metal container located in the suction line between the evaporator and the compressor that catches liquid refrigerant that passes through the evaporator coil before the refrigerant reaches the compressor. Inside a suction accumulator, a section of tubing runs down from the top, makes a U-bend and returns out of the top. The inlet end of the tubing section is connected to the refrigerant line that runs from the evaporator. An opening in the side of the tubing inside the tank allows liquid refrigerant to spill to the bottom of the tank. Another opening in the tubing near the top of the tank draws refrigerant vapor into the tube. This end of the tubing goes to the suction line that leads to the inlet of the compressor. See Figure 9-18.

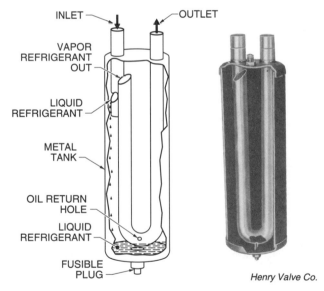

Henry Valve Co.

Figure 9-18. A suction accumulator stores liquid refrigerant so that liquid does not enter the compressor.

Liquid refrigerant must not enter the compressor. If refrigerant enters the compressor as a liquid, the pressure of the liquid will damage the valves and other parts of the compressor. The suction accumulator stores liquid refrigerant and allows the refrigerant to flow out only after it vaporizes.

Review Questions

## 9

1. What processes occur on the low-pressure side of a mechanical compression refrigeration system?

2. What causes the temperature of the refrigerant to decrease as it flows through the expansion device?

3. Name three expansion devices that control the refrigerant in a system by controlling volume.

4. Why is a capillary tube not used as an expansion device in refrigeration systems with varying loads?

5. A float valve is normally used in refrigeration systems with what type of evaporator?

6. An automatic expansion valve controls the pressure in the evaporator. How does this control the temperature?

7. An automatic expansion valve should not be used on a system with varying loads. Why?

8. List two types of expansion devices that control the refrigerant in a system by controlling the temperature. Explain how each device works.

9. How does a thermostatic expansion valve control refrigerant flow during load changes?

10. What does the sensor of a thermoelectric expansion device actually sense?

11. Why is an external electrical power circuit used with a thermoelectric expansion valve?

12. Why is a thermistor used with a thermoelectric expansion valve?

13. Explain how aluminum fins make an evaporator coil more efficient.

14. What is the main difference between a shell-and-coil evaporator and a shell-and-tube cooler?

15. A propeller fan is used with what kind of air distribution system? Why?

16. What refrigerant lines are used on the low-pressure side of a mechanical compression refrigeration system?

17. How does a service valve work? Explain.

18. How should the stem of a service valve be positioned to gain access to the service port?

19. Where is a sight glass located on the low-pressure side of a refrigeration system?

20. How does a suction accumulator collect liquid refrigerant? Why is this important?

Compression System — High-pressure Side

Chapter **10**

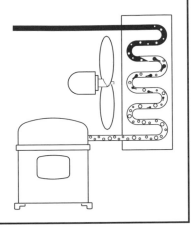

The high-pressure side of a mechanical compression refrigeration system is where the refrigerant vapor rejects heat to the condensing medium in the condenser. The refrigerant cools and condenses to a liquid. The heated condensing medium is exhausted or used for heating. The high-pressure side of a mechanical compression refrigeration system includes the compressor, condenser, refrigerant lines, and accessories. Accessories in the refrigerant lines filter the refrigerant and regulate the refrigerant flow.

HIGH-PRESSURE SIDE

The high-pressure side of a mechanical compression refrigeration system begins at the compressor. The low-pressure refrigerant vapor is compressed into high-pressure refrigerant vapor in the compressor. The compressor raises the pressure and temperature of the refrigerant so that heat can be removed from the refrigerant. The high-pressure refrigerant vapor flows through the hot gas line to the condenser.

The condenser is a heat exchanger where heat in the refrigerant is transferred to the condensing medium. A condenser is like an evaporator in that heat is transferred between a refrigerant in a coil and a medium that surrounds the coil. In a condenser, heat is transferred from the hot refrigerant vapor to the cool condensing medium. The heat in the condensing medium is then vented to the atmosphere or to a place where the heat can be used for heating. Refrigerant lines carry refrigerant through the high-pressure side of the refrigeration system. Accessories in the refrigerant lines filter the refrigerant and regulate the refrigerant flow.

COMPRESSORS

A compressor is a mechanical device that compresses refrigerant or other fluid to maintain a pres-

sure difference between the high-pressure side and low-pressure side of a mechanical compression refrigeration system. The compressor compresses the refrigerant, which becomes a high-pressure refrigerant vapor. The compressor circulates the refrigerant through the refrigeration system.

Compressors can only compress a substance when it is in the gas state. A refrigerant is a vapor as it flows through a compressor. When a substance is a vapor, the molecules in the substance are relatively far apart. The pressure of the vapor is low because it can spread apart and take up volume. A compressor reduces the space (volume) that the vapor can occupy. Mechanical action forces the molecules together, which increases the pressure and the temperature of the vapor. By rapidly decreasing the volume that a vapor can occupy, a compressor increases the pressure and temperature of the vapor.

Compressors are classified by mechanical action and construction. Mechanical action indicates how the compressor compresses a vapor. Construction indicates how the compressor can be serviced.

Mechanical Action

Mechanical action is the manner in which the compressor compresses a vapor or substance in the gas state. The categories of compressors are reciprocating, rotary, centrifugal, and screw compressors. Each of these compressors compresses refrigerant vapor in a different way.

Reciprocating Compressors. A *reciprocating compressor* is a compressor that uses mechanical energy to compress a vapor. A reciprocating compressor operates in the opposite manner of an internal combustion engine. An internal combustion engine produces mechanical energy when vapor expands. A reciprocating compressor uses mechanical energy to compress a vapor. See Figure 10-1.

Copeland Corporation

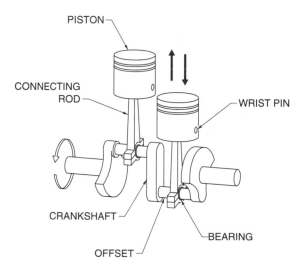

Figure 10-1. A piston is connected to a crankshaft by wrist pins, connecting rods, and bearings on the shaft in a reciprocating compressor.

In a reciprocating compressor, one or more pistons move back and forth inside closed cylinders. A *stroke* is the travel distance of a piston inside a cylinder. The movement of the piston in the cylinder compresses the vapor. The pistons move in the cylinders because of mechanical energy from a motor. The motor turns a crankshaft, which moves the pistons in a reciprocating motion. The pistons are connected to a crankshaft by wrist pins, connecting rods, and bearings on the crankshaft. Connecting rod bearings are connected to offsets on the crankshaft.

The *suction stroke* of a cylinder is the stroke that occurs when the piston moves down. As the piston moves down in the cylinder, the pressure in the cylinder decreases. Refrigerant vapor flows into the cylinder when the pressure in the cylinder is lower than the pressure of the refrigerant in the suction line. The pressure of the refrigerant is greater than the resistance of the suction valve spring. The suction valve opens and refrigerant vapor flows into the cylinder.

The *compression stroke* of a cylinder is the stroke that occurs after the piston completes its suction stroke and begins to move up in the cylinder toward the cylinder head. The *cylinder head* is the top part of the cylinder that seals the upper end of the cylinder. The pressure in the cylinder increases and the suction valve closes during the compression stroke. See Figure 10-2.

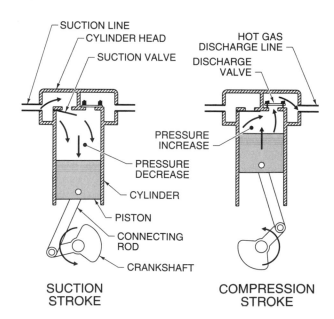

Figure 10-2. The suction stroke draws refrigerant vapor into a cylinder. The compression stroke compresses and pushes refrigerant vapor out of a cylinder.

When the pressure in the cylinder is high enough to overcome the pressure of the refrigerant vapor in the hot gas discharge line and the tension in the discharge valve spring, the discharge valve opens and the refrigerant vapor in the cylinder is forced into the hot gas discharge line. A *cycle* consists of one suction stroke and one compression stroke. Each piston completes a cycle for every revolution of the crankshaft.

The valves in the cylinder control the pressure of the refrigerant in a reciprocating compressor. Leaf valves and plate valves are used in compressors. A *leaf valve* is a valve that consists of a steel flapper. The flapper is fastened at one end and is held in place by the tension of the flapper itself or by springs acting on the flapper. A *plate valve* is a valve that consists of a floating plate held in place by springs. Tension in a spring holds either valve closed until the pressure of the refrigerant vapor opens the valve. See Figure 10-3.

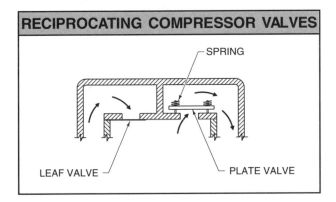

RECIPROCATING COMPRESSOR VALVES

SPRING

LEAF VALVE

PLATE VALVE

Figure 10-3. Leaf and plate valves control the pressure of the refrigerant in a cylinder.

A reciprocating compressor is a positive-displacement compressor. *Positive displacement* is the replacement of a fixed amount of a substance released from the cylinder with each cycle. Each cycle of a positive-displacement compressor moves a fixed amount of refrigerant. Because of positive-displacement, a reciprocating compressor produces higher pressures than other compressors and therefore operates efficiently in systems that require relatively high pressures. Reciprocating compressors are used most often in air conditioning systems. Reciprocating compressors can be adapted so that different refrigerants can be used. The volumetric capacity of a reciprocating compressor is a function of the

volume of the cylinder, speed of the crankshaft, and number of cylinders in the compressor.

Volumetric Capacity. *Volumetric capacity* of a compressor is the volume of refrigerant vapor that a reciprocating compressor moves. Volumetric capacity (volumetric displacement) of a compressor is based on the physical performance of the compressor and is a function of the volumetric capacity of each cylinder in the compressor, number of cylinders, and speed of the compressor. To find the volumetric capacity of the compressor, first find the volumetric capacity of the individual cylinders. The volumetric capacity for one cylinder is found by applying the formula:

$$V = \pi r^2 \times s$$

where

V = volumetric capacity of one cylinder (in cu in.)

π = 3.14 (constant)

r = radius of cylinder (in inches)

s = stroke length (in inches)

Example: Finding Volumetric Capacity — Cylinder

A compressor has two cylinders. Each cylinder has a radius of 2″ and a stroke length of 1.5″. Find the volumetric capacity of one of the cylinders.

$$V = \pi r^2 \times s$$
$$V = 3.14 \times (2 \times 2) \times 1.5$$
$$V = 3.14 \times 4 \times 1.5$$
$$V = \textbf{18.84 cu in.}$$

To find the volumetric capacity of the compressor, multiply the volume of one cylinder by the number of cylinders times the speed of the compressor and divide by 1728. Dividing by 1728, which is the number of cubic inches in a cubic foot, converts cubic inches into cubic feet. The volumetric capacity of a compressor is found by applying the formula:

$$V_C = \frac{V \times n \times N_c}{1728}$$

where

V_C = volumetric capacity of compressor (in cfm)
V = volume of one cylinder (in cu in.)

n = number of cylinders

N_c = speed of compressor (in rpm)

1728 = constant

Example: Finding Volumetric Capacity — Compressor

A compressor has two cylinders. Each cylinder has a radius of 1.5″ and a stroke length of 1.25″. The compressor operates at 1725 rpm. Find the volumetric capacity of the compressor.

1. Find volumetric capacity of one cylinder.

$$V = \pi r^2 \times s$$

$$V = 3.14 \times (1.5 \times 1.5) \times 1.25$$

$$V = 3.14 \times 2.25 \times 1.25$$

$$V = \textbf{8.83 cu in.}$$

2. Find the volumetric capacity of the compressor.

$$V_C = \frac{V \times n \times N_c}{1728}$$

$$V_C = \frac{8.83 \times 2 \times 1725}{1728}$$

$$V_C = \textbf{17.63 cfm}$$

Rotary Compressors. A *rotary compressor* is a positive-displacement compressor that uses a rolling piston (rotor) that rotates in a cylinder to compress refrigerant vapor. The piston or rotor is smaller than the cylinder and is mounted off-center on its shaft. The piston is located so it is in contact with one point on the cylinder wall at all times.

The pistons, rotors, and cylinders in rotary compressors must be machined accurately because the tolerances and clearances between the parts must be precise. The surfaces are sealed with a thin film of lubricating oil. Rotary compressors are used mostly in small air conditioning systems such as window air conditioners and small package units. Stationary vane and rotating vane compressors are two kinds of rotary compressors.

A *stationary vane compressor* is a positive-displacement compressor that has a spring-loaded vane in the side of the cylinder wall. The vane rides against the rolling piston. An electric motor turns the rolling piston. A crescent-shape opening is created between the piston and the cylinder wall as

the piston rotates inside the cylinder. As the piston rotates, the opening moves. A low-pressure area is created in the crescent-shape opening as the opening moves away from the stationary vane, and a high-pressure area is created in the opening as the opening moves toward the stationary vane. See Figure 10-4.

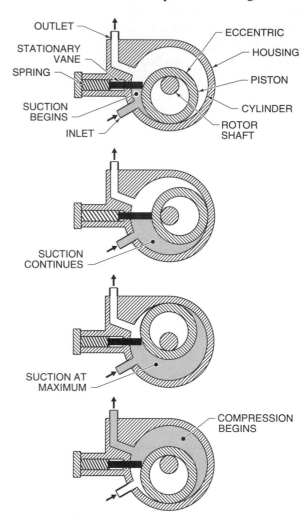

Figure 10-4. A stationary vane rotary compressor compresses refrigerant vapor as a piston rotates in a cylinder.

An inlet for refrigerant vapor is located in the sidewall of the cylinder on the side of the vane where the low-pressure area occurs. An outlet for the compressed refrigerant vapor is located on the sidewall of the cylinder on the side of the vane where the high-pressure area occurs.

A *rotating vane compressor* is a positive-displacement compressor that has multiple vanes that are located in the rotor. These vanes form a seal as they are forced against the cylinder wall by centrifugal force. As the rotor rotates, refrigerant is drawn into the crescent-shape opening between the rotor and the cylinder wall as the opening becomes larger and is forced out when the opening becomes smaller. See Figure 10-5.

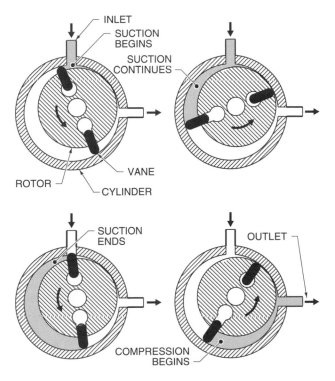

Figure 10-5. Vanes in the rotor of a rotating vane compressor are pushed outward by centrifugal force.

Centrifugal Compressors. A *centrifugal compressor* is a compressor that uses centrifugal force to move refrigerant vapor. A centrifugal compressor and a centrifugal blower work in the same way. An impeller wheel turns inside a housing. The inlet for refrigerant vapor is an opening located in the side of the housing near the center of the impeller wheel. The outlet of the compressor is an opening located on the outer perimeter of the housing. The impeller wheel turns at a high speed and draws refrigerant into the center of the impeller wheel through the intake opening. The refrigerant is compressed as it flows through the vanes on the impeller wheel and is thrown off the tips of the impeller wheel by centrifugal force. The speed of the refrigerant increases as it is thrown to the perimeter of the impeller wheel. As the refrigerant leaves the outer rim of the impeller wheel, the speed is converted

to pressure because the refrigerant is forced into a smaller opening. The high-pressure refrigerant leaves the housing through the discharge opening. See Figure 10-6.

A centrifugal compressor is not a positive-displacement compressor. A centrifugal compressor delivers large volumes of refrigerant but does not create a large pressure difference between the refrigerant that enters and the refrigerant that leaves the compressor. Because centrifugal compressors move large volumes of refrigerant, they can be used on large refrigeration systems. The low pressure difference can be compensated for by using a refrigerant that boils at fairly low pressures.

High pressures are produced by multistage centrifugal compressors. A *multistage centrifugal compressor* is a centrifugal compressor that has more than one impeller wheel. The impeller wheels are designed so that the refrigerant flows from the outer rim of one impeller wheel and into the center of another impeller wheel. The pressure of the refrigerant increases in stages as the refrigerant flows through each impeller wheel. A multistage centrifugal compressor moves large volumes of refrigerant at relatively high pressures.

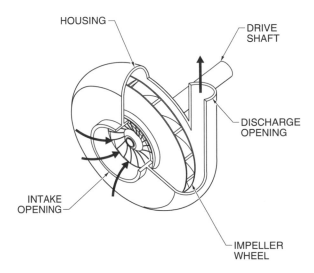

Figure 10-6. Refrigerant is thrown off the tips of an impeller wheel by centrifugal force.

Screw Compressors. A *screw compressor* is a compressor that contains a pair of screw-like helical gears that interlock as they turn. A screw compressor is also known as a helical gear compressor. The gears are located in a tight-fitting housing and are turned by an electric motor. As the gears turn, refrigerant is drawn into the suction inlet and is forced through the housing by the interlocking lobes of the gears. The refrigerant is compressed as the opening between the gears becomes smaller. The compressed refrigerant is then discharged through the outlet. See Figure 10-7.

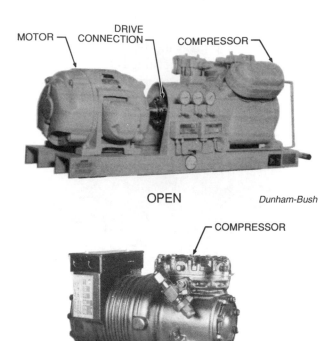

OPEN   *Dunham-Bush*

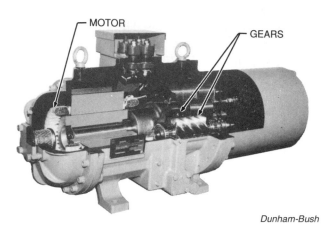

Dunham-Bush

Figure 10-7. A screw compressor forces refrigerant into spaces between two rotating, interlocking gears.

A screw compressor is an efficient positive-displacement compressor. The parts of a screw compressor must be machined to close tolerances, and proper lubrication is necessary for maintaining a seal between the moving parts. The lubrication system of the compressor keeps all moving parts covered in oil during operation. Screw compressors are usually used on large refrigeration and air conditioning systems.

Compressor Construction

The compressor is the component of a mechanical refrigeration system that circulates the refrigerant. If the flow of refrigerant stops, the refrigeration system may be damaged. Maintaining a compressor is important for proper operation. The efficiency of a mechanical compression refrigeration system depends on the compressor. Reciprocating, rotary, centrifugal, and screw compressors are available in open, semi-hermetic, or hermetic configurations. See Figure 10-8.

SEMI-HERMETIC   *Copeland Corporation*

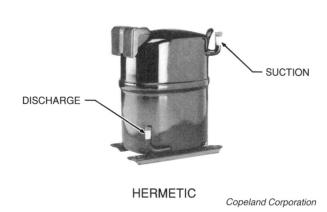

HERMETIC   *Copeland Corporation*

Figure 10-8. Compressors are classified as open, semi-hermetic, and hermetic.

Open Compressors. An *open compressor* is a compressor that has all of the components except for the motor inside one housing. A *housing* is the protective enclosure for moving parts. The motor that operates the compressor is mounted externally. The compressor is driven by either a direct drive or belt drive connection.

Open compressors can be easily serviced on a job site. The compressor crankshaft extends outside the compressor housing. The shaft passes through an opening that is sealed by a shaft seal to prevent refrigerant and lubricating oil from leaking. Various types of seals are used, but all seals are subject to wear and require maintenance and service. Lubricating oil for open compressors is contained in the crankcase. See Figure 10-9.

Open compressors are available in all sizes but are often used in large refrigeration systems that produce many tons of cooling. Small open compressors are often used in refrigeration systems. Centrifugal compressors and screw compressors are usually open compressors.

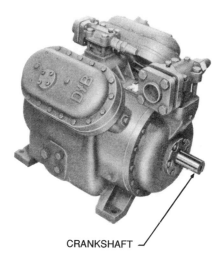

CRANKSHAFT

Dunham-Bush

Figure 10-9. The housing of an open compressor does not contain the compressor motor.

Semi-hermetic Compressors. A *semi-hermetic compressor* is a compressor that has all of the components and a motor located inside a housing. Semi-hermetic compressors are also known as serviceable compressors because the housing can be opened and the components can be serviced on a job site.

An extension of the motor drive shaft is the crankshaft. Because the motor and compressor are located in the same housing, the refrigerant that circulates through the system can be used to cool the compressor motor windings. Lubricating oil is contained in the crankcase of a semi-hermetic compressor. See Figure 10-10.

Semi-hermetic compressors are available in all sizes but are most commonly used on air conditioning systems of 60,000 Btu/hr (5 tons of cooling) cooling capacity or larger. Semi-hermetic compressors are not costly to repair because they can be serviced at a job site, but the initial cost is greater than other types of compressors.

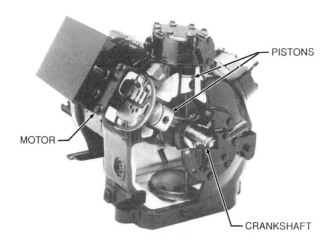

PISTONS

MOTOR

CRANKSHAFT

Tecumseh Products Company

Figure 10-10. The housing of a semi-hermetic compressor contains the compressor motor.

Hermetic Compressors. A *hermetic compressor* is a compressor that has all of the components and the motor sealed in a metal housing. A hermetic compressor cannot be serviced on a job site because it cannot be opened without causing damage to the compressor. If the motor or any part of the compressor fails, the entire compressor will have to be replaced. See Figure 10-11.

Hermetic compressors are usually used in air conditioners that produce 5 tons of cooling or less. The motor in a hermetic compressor is cooled by refrigerant that circulates through the air conditioning system. Lubricating oil for both the motor and compressor is contained in the compressor crankcase. Hermetic compressors are normally used in small air conditioning systems such as window air conditioners or through-the-wall air conditioners. The air conditioning systems may be package systems or split systems. A *package system* is an air conditioning system that has all components enclosed in one cabinet. A *split system* is an air conditioning system that has the evaporator and condenser enclosed in separate cabinets.

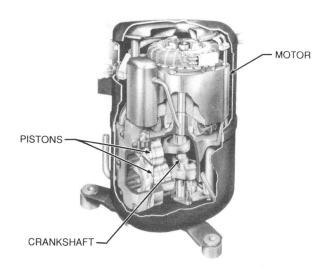

PISTONS

CRANKSHAFT

MOTOR

Tecumseh Products Company

Figure 10-11. The sealed housing of a hermetic compressor contains the motor and other compressor components.

Compressor Performance

Compressor performance is the cooling capacity produced by the amount of refrigerant the compressor moves through the refrigeration system. A certain refrigeration effect can be produced when a compressor circulates the required amount of refrigerant at the required pressures.

Refrigerant Flow Rate. *Refrigerant flow rate* is the amount of refrigerant that flows through the refrigeration system per unit of time. The compressor controls the flow of refrigerant. Refrigerant flow rate is expressed by weight (mass) or volume.

Mass flow rate is the rate of refrigerant flow expressed in pounds per minute. The mass flow rate is the weight of the refrigerant that must be circulated per minute by the compressor to provide the required cooling capacity. The mass flow rate of the refrigerant is found by dividing the required cooling capacity by the refrigeration effect of the system multiplied by 60. Multiplying by 60 converts the hours in Btu per hour to minutes in Btu per minute. The mass flow rate of a refrigerant is found by applying the formula:

$$w = \frac{cc}{RE \times 60}$$

where

w = mass flow rate (in lb/min)

cc = cooling capacity (in Btu/hr)

RE = refrigeration effect (in Btu/lb)

60 = constant

Example: Finding Mass Flow Rate

A refrigeration system that contains R-22 has a low-pressure side pressure of 121.45 psig and a high-pressure side pressure of 260.03 psig. Find the mass flow rate for the system that will provide a 90,000 Btu/hr cooling capacity.

1. Find the enthalpy of the refrigerant.

   On a refrigerant property table for R-22, the enthalpy of the refrigerant as a vapor is 110.18 Btu/lb at 121.45 psig. The enthalpy of the refrigerant as a liquid is 45.69 Btu/lb at 260.03 psig. See Appendix.

2. Find the refrigeration effect.

   $RE = h_1 - h_e$

   $RE = 110.18 - 45.69$

   $RE = 64.49$ Btu/lb

3. Find the mass flow rate.

   $$w = \frac{cc}{RE \times 60}$$

   $$w = \frac{90,000}{64.94 \times 60}$$

   $w = 23.1$ lb/min

Volumetric flow rate is the rate of refrigerant flow expressed in cubic feet per minute. Volumetric flow rate is the mass flow rate expressed as volume. Volumetric flow rate is a function of the refrigeration effect of the system, cooling capacity, and specific volume of the refrigerant. Specific volume is the volume of a substance per unit of the substance. The volumetric flow rate of the refrigerant is found by multiplying the refrigerant mass flow rate by the specific volume of the refrigerant. The refrigerant volumetric flow rate is found by applying the formula:

$$Q = w \times v$$

where

Q = volumetric flow rate (in cfm)

w = mass flow rate (in lb/min)

v = specific volume (in cu ft/lb)

Example: Finding Volumetric Flow Rate

A refrigeration system that contains R-22 operates with a low-pressure side of 68.5 psig and a high-pressure side of 337.4 psig. Find the volumetric flow rate that must be maintained by the compressor to provide a cooling capacity of 60,000 Btu/hr.

1. Find the enthalpy of the refrigerant.

   On a refrigerant property table for R-22, the enthalpy of the refrigerant as a vapor is 107.94 Btu/lb at 68.5 psig. The enthalpy of the refrigerant as a liquid is 52.55 Btu/lb at 337.4 psig. See Appendix.

2. Find the specific volume.

   The specific volume of R-22 as a vapor at 68.5 psig and 107.94 Btu/lb is .66 cu ft/lb. See Appendix.

3. Find the refrigeration effect.

   $RE = h_e - h_1$

   $RE = 107.94 - 52.55$

   $RE = \textbf{55.39 Btu/lb}$

4. Find the mass flow rate.

   $$w = \frac{cc}{RE \times 60}$$

   $$w = \frac{60,000}{55.39 \times 60}$$

   $w = \textbf{18.05 lb/min}$

5. Find the volumetric flow rate.

   $Q = w \times v$

   $Q = 18.05 \times .66$

   $Q = \textbf{11.91 cfm}$

CONDENSERS

A condenser is a heat exchanger that removes heat from high-pressure refrigerant vapor. The heat absorbed by the refrigerant on the low-pressure side of the system is transferred to the condensing medium. The heat from the condensing medium is then vented to the atmosphere or used for heating.

Hot, high-pressure refrigerant vapor enters the condenser coil from the hot gas discharge line. The temperature of the refrigerant is greater than the temperature of the condensing medium flowing across the outside of the coil. Because of the temperature difference, heat passes from the refrigerant to the condensing medium.

The three basic types of condensers used on refrigeration systems are air-cooled, water-cooled, and evaporative condensers. Air-cooled condensers use air for the condensing medium. Water-cooled condensers use water for the condensing medium. Evaporative condensers use both air and water for condensing mediums.

Air-cooled Condensers

An *air-cooled condenser* is a condenser that uses air as the condensing medium. Heat is transferred from the refrigerant to the air. An air-cooled condenser consists of a frame or cabinet that holds the various components, a blower, a condenser coil, a compressor, refrigerant lines, and accessories. See Figure 10-12.

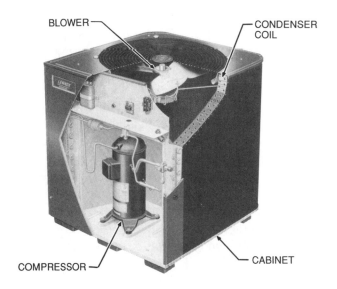

Lennox Industries Inc.

Figure 10-12. An air-cooled condenser includes a frame or cabinet, blower, condenser coil, and compressor.

Cabinets. The components of an air-cooled condenser are enclosed in a cabinet. A cabinet is necessary because air-cooled condensers are usually located outdoors. In many systems the condenser cabinet also contains the compressor. The cabinet is usually steel and is painted with a weather-resistant paint. The condenser coil and blower are arranged in the cabinet so that outdoor air is drawn into the cabinet by the blower, blown across the coil, and exhausted back to the outdoors. The coil and blower(s) may be arranged for either horizontal or vertical air flow.

Blowers. The blower in an air-cooled condenser usually consists of one or more propeller fans. An air-cooled condenser is designed to be located outdoors so that outdoor air can be used as the condensing medium. Propeller fans move the air in a condenser because there is little resistance to the flow of air. The condenser blower normally handles a greater volume of air than the evaporator of a system. The condenser blower has a larger volume because the temperature of the air that enters the condenser is higher, and the amount of heat to be rejected by the condenser is greater than the amount of heat absorbed in the evaporator.

Condenser Coils. A condenser coil is the heat exchanger in an air-cooled condenser. A condenser coil absorbs heat from the refrigerant and transfers it to the air. Bare-tube, finned-tube, and flat-plate coils are used in both evaporators and condensers. Finned-tube coils are used most commonly. Finned-tube coils are copper or aluminum tubes with aluminum fins pressed on the tubing to increase the surface area of the coil. The increased surface area provides a larger heat transfer surface area. Heat is transferred from the refrigerant to the finned-tube coil to the air. See Figure 10-13. The amount of heat transferred between the refrigerant in the condenser coil and the air in the condenser depends on the size of the coil.

Superheat and Subcooling. When the refrigerant vapor leaves the compressor, it is at superheated temperature for the refrigerant. *Superheated temperature* is a temperature higher than the saturated vapor temperature for the pressure on the refrigerant. Superheat is the sensible heat absorbed by the refrigerant after it has vaporized. Most of the superheat is lost as the refrigerant flows through the hot gas discharge line before it reaches the condenser. The refrigerant enters the condenser coil as a

saturated vapor. In the condenser coil, heat flows from the refrigerant to the air. As heat is removed, the refrigerant condenses to a liquid.

Dunham-Bush

Figure 10-13. A finned-tube coil is made of copper or aluminum tubing with aluminum fins pressed on the tubing to extend the surface area.

The condenser coil must be large enough to condense all of the refrigerant to a liquid before the refrigerant leaves the coil. In the last few rows of the condenser coil, additional heat is absorbed from the refrigerant. Removing this heat subcools the refrigerant. Subcooling is lowering the temperature of the refrigerant below the saturated liquid temperature for the pressure on the refrigerant. The refrigerant leaves the condenser as a high-pressure, subcooled liquid. See Figure 10-14.

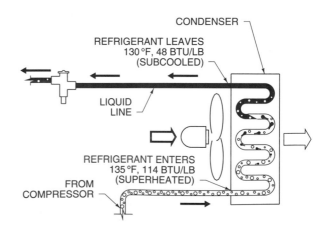

Figure 10-14. The refrigerant leaves the condenser as a high-pressure, subcooled liquid.

Heat Transfer. Heat is transferred from the refrigerant to the condensing medium in the condenser. Heat of rejection is the amount of heat rejected by the condenser. Heat of rejection is the difference between the enthalpy (heat content) of the vapor refrigerant as it enters the condenser and the enthalpy of the liquid refrigerant as it leaves the condenser. The difference between the heat of rejection and the refrigeration effect is the heat of compression and heat from the compressor motor. In the condenser heat transfer occurs, but there is no pressure change. The refrigerant enters a condenser as a vapor and leaves as a liquid.

Heat rejection rate is the rate at which heat is transferred from the refrigerant in the condenser to the condensing medium. The heat rejection rate is a function of the mass flow rate of the refrigerant and the heat of rejection. The heat rejection rate is found by multiplying the heat of rejection by the mass flow rate of the refrigerant times 60 minutes. Multiplying by 60 converts pounds per minute to Btu per hour. The heat rejection rate is found by applying the formula:

$$HR_r = HR \times w \times 60$$

where

HR_r = heat rejection rate (in Btu/hr)

HR = heat of rejection (in Btu/lb)

w = mass flow rate (in lb/min)

60 = constant

Example: Finding Heat Rejection Rate — Air-cooled Condenser

A refrigeration system that contains R-22 has a high-pressure side pressure of 337.5 psig. The mass flow rate is 18.05 lb/min. Find the heat rejection rate of the refrigerant in the condenser.

1. Find the enthalpy of the refrigerant.

   On a refrigerant property table for R-22, the enthalpy of the refrigerant as a liquid is 52.55 Btu/lb at 337.5 psig. The enthalpy of the refrigerant as a vapor is 112.47 Btu/lb at 337.5 psig. See Appendix.

2. Find the heat of rejection.

   $$HR = h_e - h_l$$
   $$HR = 112.47 - 52.55$$
   $$HR = \textbf{59.92 Btu/lb}$$

3. Find heat rejection rate.

   $$HR_r = HR \times w \times 60$$

$$HR_r = 59.92 \times 18.05 \times 60$$
$$HR_r = \textbf{64,893.36 Btu/hr}$$

Temperature Increase. As air passes across the outside of the condenser coil, heat flows from the hot refrigerant vapor to the ambient air. This heat transfer increases the temperature of the air. The temperature increase is a function of the volumetric flow rate of the air, heat rejection rate, and a constant. The constant is 1.08, which represents the heat absorbed by the refrigerant in the compressor. The volumetric flow rate is expressed in cubic feet per minute of air. The temperature increase of the air across the coil is found by applying the formula:

$$\Delta T = \frac{HR_r}{1.08 \times Q}$$

where

ΔT = temperature increase (in °F)

HR_r = heat rejection rate (in Btu/hr)

1.08 = constant

Q = volumetric flow rate of air (in cfm)

Example: Finding Temperature Increase — Air

A refrigeration system that contains R-22 has a high-pressure side pressure of 195.97 psig. The condenser blower moves 3000 cfm of air across the outside of the condenser coil. The mass flow rate of the refrigerant is 21.3 lb/min. Find the temperature increase of the air that leaves the condenser.

1. Find the enthalpy of the refrigerant.

   On a refrigerant property table for R-22, the enthalpy of the refrigerant as a liquid is 39.23 Btu/lb at 195.97 psig. The enthalpy of the refrigerant as a vapor is 111.81 Btu/lb at 195.97 psig. See Appendix.

2. Find the heat of rejection.

   $$HR = h_e - h_l$$
   $$HR = 111.81 - 39.23$$
   $$HR = \textbf{72.58 Btu/lb}$$

3. Find heat rejection rate.

   $$HR_r = HR \times w \times 60$$
   $$HR_r = 72.58 \times 21.3 \times 60$$
   $$HR_r = \textbf{92,757.24 Btu/hr}$$

4. Find the temperature increase.

$$\Delta T = \frac{HR}{1.08 \times Q}$$

$$\Delta T = \frac{92,757.24}{1.08 \times 3000}$$

$$\Delta T = \textbf{28.63} \ \degree\textbf{F}$$

Water-cooled Condensers

A *water-cooled condenser* is a condenser that uses water as the condensing medium. Heat is transferred from the refrigerant to the water. A water-cooled condenser consists of a frame or cabinet, compressor, condenser coil, refrigerant lines, and accessories. See Figure 10-15.

Most water-cooled condensers are part of a package refrigeration system. The condenser, evaporator, and blower are contained in a sheet metal cabinet and are installed as one unit. Package refrigeration systems may be used inside a building for zone cooling without any part of the refrigeration system being located outside the building. A water supply and a waste water connection are required. The three basic kinds of condensers used in water-cooled refrigeration systems are double pipe, shell-and-tube, and shell-and-coil condensers.

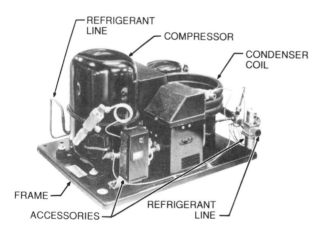

Tecumseh Products Company

Figure 10-15. A water-cooled condenser contains a frame, compressor, condenser coil, refrigerant lines, and accessories.

Double Pipe Condenser. A *double pipe condenser* is a condenser that contains a small tube that runs through the center of a larger tube. Double

pipe condensers are also known as tube-in-tube condensers. Refrigerant circulates through the inner tube and water circulates through the outer tube. Heat is transferred from the refrigerant in the inner tube to the water in the outer tube. The heat transfer condenses the refrigerant vapor in the inner tube. The heat is carried away by the water in the outer tube as the water leaves the condenser. See Figure 10-16.

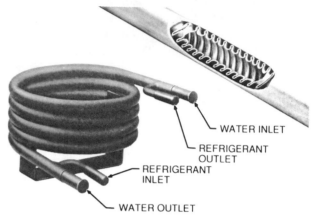

Turbotec Products, Inc.

Figure 10-16. A double pipe condenser is a condenser that contains a small tube that runs through the center of a larger tube.

Shell-and-Tube Condenser. A *shell-and-tube condenser* is a condenser that consists of a shell with headers, which are connected by tubes. The headers isolate a small section of the shell. Refrigerant from the system circulates through the shell, and water circulates through the tubes. Heat is transferred from the refrigerant in the shell to the water in the tubes.

Shell-and-tube condensers are manufactured in both vertical and horizontal models. Vertical shell-and-tube condensers are often open to the atmosphere. Water flows down through the tubes by gravity into a pan. Horizontal shell-and-tube condensers are usually used on smaller refrigeration systems. The water circulates through a closed piping system from a cooling tower.

Shell-and-Coil Condenser. Shell-and-coil condensers are similar to shell-and-tube condensers. A *shell-and-coil condenser* is a condenser that consists of a shell with headers, which are connected by a coil. The headers isolate a small section of the shell. Water flows through the coil. Heat is transferred from the refrigerant in the shell to the water.

Water Source. The water source for a water-cooled condenser is usually a cooling tower. The water piping for a water-cooled condenser includes make-up water piping, piping from the tower to the condenser, and piping from the condenser back to the tower. See Figure 10-17. Makeup water for the cooling tower is supplied from an outside system such as city or well water.

A *cooling tower* is an evaporative water cooler that uses natural evaporation to cool water. Air circulates through the tower by natural convection or is blown through the tower by fans located in the tower. The heat absorbed by the water in the condenser is rejected to the air in the cooling tower. Cool water from the bottom of the tower is then circulated back to the condenser for reuse.

Heat Transfer. Heat rejection rate is the rate at which heat is transferred from the refrigerant in the condenser to the condensing medium. The heat rejection rate of the refrigerant flowing through a water-cooled condenser is a function of the mass flow rate of the refrigerant and the heat of rejection. The heat rejection rate is a function of the mass flow rate of the refrigerant and the heat of rejection. Heat of rejection is the difference in enthalpy of the refrigerant as it enters and leaves the condenser. The heat rejection rate is a function of the heat of rejection and the mass flow rate.

Example: Finding Heat Rejection Rate — Water-cooled Condenser

A refrigeration system that has a water-cooled condenser contains R-12. The high-pressure side pressure is 99.73 psig and the mass flow rate is 19.05 lb/min. Find the heat rejection rate of the refrigerant flowing through the condenser.

1. Find the enthalpy of the refrigerant.

   On a refrigerant property table for R-12, the enthalpy of the refrigerant as a liquid is 29.03 Btu/lb at 99.73 psig. The enthalpy of the refrigerant as a vapor is 86.88 Btu/lb at 99.73 psig. See Appendix.

2. Find the heat of rejection.

   $HR = h_e - h_l$

   $HR = 86.88 - 29.03$

   $HR =$ 57.85 Btu/lb

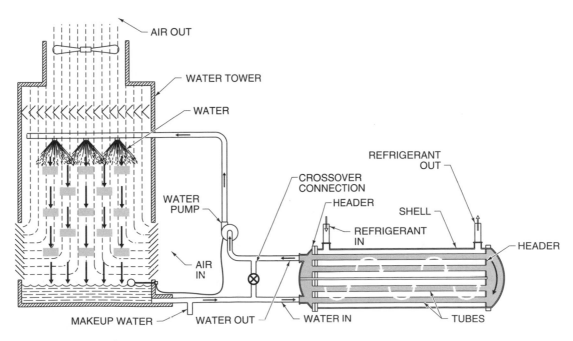

Figure 10-17. The water piping for a water-cooled refrigeration system includes makeup water piping and piping that connects the cooling tower to the condenser.

3. Find heat rejection rate.

$HR_r = HR \times w \times 60$

$HR_r = 57.85 \times 19.05 \times 60$

$HR_r = \textbf{66,122.55 Btu/hr}$

Temperature Increase. The heat rejected by the refrigerant in the condenser causes a temperature increase in the water flowing through a water-cooled condenser. The temperature increase of the water is a function of the volumetric flow rate of the water, heat rejection rate, and a constant. The constant is 500, which represents the heat absorbed by the refrigerant in the compressor. The volumetric flow rate is expressed in gallons per minute. The temperature increase of the water is found by applying the formula:

$$\Delta T = \frac{HR_r}{500 \times Q}$$

where

ΔT = temperature increase (in °F)

HR_r = heat rejection rate (in Btu/hr)

500 = constant

Q = volumetric flow rate (in gpm)

Example: Finding Temperature Increase — Water

A refrigeration system that has a water-cooled condenser contains R-12. The high-pressure side pressure is 46.74 psig and the mass flow rate of the refrigerant is 13.05 lb/min. The volumetric flow rate of the water through the condenser is 15 gpm. Find the temperature increase of the water flowing through the condenser.

1. Find the enthalpy of the refrigerant.

    On a refrigerant property table for R-12, the enthalpy of the refrigerant as a liquid is 19.63 Btu/lb at 46.74 psig. The enthalpy of the refrigerant as a vapor is 83.00 Btu/lb at 46.74 psig. See Appendix.

2. Find the heat of rejection.

$HR = h_e - h_l$

$HR = 83.00 - 19.63$

$HR = \textbf{63.37 Btu/lb}$

3. Find heat rejection rate.

$HR_r = HR \times w \times 60$

$HR_r = 63.37 \times 13.05 \times 60$

$HR_r = \textbf{49,618.71 Btu/hr}$

4. Find the temperature increase.

$$\Delta T = \frac{HR_r}{500 \times Q}$$

$$\Delta T = \frac{49,618.71}{500 \times 15}$$

$$\Delta T = \frac{49,618.71}{7500}$$

$$\Delta T = \textbf{6.62°F}$$

Evaporative Condensers

An *evaporative condenser* is a condenser that uses water in the condenser coil, air blown past the coil, and evaporation of water from the outside surface of the condenser coil to remove heat from the refrigerant. An evaporative condenser is more efficient than an air-cooled or water-cooled condenser because a large amount of latent heat is removed by evaporation. The wet surface of the outside of the coil transfers heat more efficiently than a dry surface. See Figure 10-18.

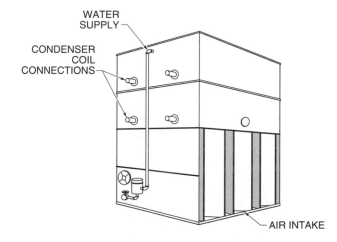

Figure 10-18. An evaporative condenser is more efficient than an air-cooled or water-cooled condenser because of the large amount of latent heat removed by evaporation.

An evaporative condenser contains a condenser coil, blower, cabinet, compressor, and water distribution system. The condenser coil, blower, and cabinet of an evaporative condenser are similar to the components used in air- and water-cooled condensers. The water distribution system keeps the outside of the coil wet. The water distribution system consists of a water pan in the bottom of the cabinet that collects runoff water, water piping with distributors across the top of the coil, and a pump that circulates water from the water pan to the distributor heads at the top of the coil. Makeup water is supplied to an evaporative condenser from an outside source.

Operation. The blower draws ambient air into the cabinet and blows the air past the coil. As refrigerant circulates through the condenser coil, water drips on the outside of the coil and air is blown past the coil. Most of the water on the outside of the coil evaporates as the water absorbs heat from the refrigerant. Some cooling also occurs due to the water and air itself.

Heat Transfer. Heat rejection rate is the rate at which heat is transferred from the refrigerant in the condenser to the condensing medium. In an evaporative condenser, water and air both act as condensing mediums. The heat rejection rate of the refrigerant flowing through an evaporative condenser is a function of the mass flow rate of the refrigerant and the heat of rejection.

Example: Finding Heat Rejection Rate — Evaporative Condenser

A refrigeration system that contains R-12 has a high-pressure side pressure of 117.03 psig. The mass flow rate is 14.7 lb/min. Find the heat rejection rate of the refrigerant flowing through the condenser.

1. Find the enthalpy of the refrigerant.

   On a refrigerant property table for R-12, the enthalpy of the refrigerant as a liquid is 31.47 Btu/lb at 117.03 psig. The enthalpy of the refrigerant as a vapor is 87.77 Btu/lb at 117.03 psig. See Appendix.

2. Find the heat of rejection.

   $HR = h_e - h_l$

   $HR = 87.77 - 31.47$

   $HR = \textbf{56.30 Btu/lb}$

3. Find heat rejection rate.

   $HR_r = HR \times w \times 60$

   $HR_r = 56.30 \times 14.7 \times 60$

   $HR_r = \textbf{49,656.60 Btu/hr}$

REFRIGERANT LINES

The high-pressure side of a refrigeration system has two refrigerant lines. The two refrigerant lines on the high-pressure side of a refrigeration system are the hot gas discharge line and the liquid line. See Figure 10-19.

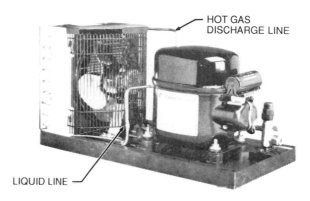

Copeland Corporation

Figure 10-19. The two refrigerant lines on the high-pressure side of a refrigeration system are the hot gas discharge line and the liquid line.

Hot Gas Discharge Line

The hot gas discharge line is the refrigerant line that runs from the compressor to the condenser. The hot gas discharge line carries hot gas from the compressor to the condenser. *Hot gas* is the hot, high-pressure refrigerant vapor that has been compressed and heated by the compressor. Because the refrigerant vapor is at a relatively high pressure in the hot gas discharge line, the line is relatively small.

Liquid Line

The liquid line is the refrigerant line that runs from a condenser to an expansion device. The liquid line carries high-pressure liquid refrigerant from the condenser to the liquid receiver (if the system has one) and on to the expansion device. The liquid line is usually the smallest line in the system. Because the refrigerant is in the liquid state in the line, the refrigerant requires less volume.

Sizing

Properly sized refrigerant lines are large enough to return oil to the compressor and to admit only the amount of refrigerant that will meet the cooling requirement. Refrigerant lines are sized for the lowest possible installation and operating costs.

The refrigerant lines must be installed and sized so that oil is carried along with the refrigerant and is returned to the compressor. The crankcase of the compressor housing in most compressors is flooded with refrigerant. The refrigerant cools the parts of the compressor. The refrigerant cools the motor windings of a hermetic compressor. The crankcase also contains lubricating oil for the compressor. As the refrigerant vapor and oil mix in the crankcase, some oil is carried with the refrigerant vapor through the discharge side of the system.

Liquid refrigerant can carry oil more easily than vapor refrigerant can carry oil. The design of the piping that carries liquid refrigerant is not as important as the design of the piping that carries vapor refrigerant. To help oil return to the compressor, all horizontal piping is installed with a slight downward slope in the direction of flow. The slope is $\frac{1}{2}''$ for every 10′ of piping. The piping is also installed without sags because oil collects in low places. Vertical suction risers have traps at the bottom of each set of risers and at about 10′ intervals on high risers.

Pressure Drop

Pressure drop is the decrease in pressure that occurs as refrigerant flows through the various connections and sections of the refrigerant lines. Pressure drop is caused by friction between the refrigerant and the piping walls and by turbulence in the refrigerant as it flows through the piping. Pressure drop should be as low as possible. Excessive pressure drop decreases system capacity and increases requirements for power from the compressor.

Pressure drop can be expressed in either pounds per square inch gauge or in degrees Fahrenheit. Pressure drop expressed in degrees Fahrenheit is the

temperature drop equivalent for the same pressure drop in pounds per square inch gauge. Temperature drop varies for different pressures and refrigerants. Each refrigerant has a different temperature drop for a given pressure drop. Tables show the temperature drops for given pressure drops for commonly used refrigerants. See Figure 10-20.

RECOMMENDED PRESSURE AND TEMPERATURE DROPS IN REFRIGERANT LINES				
Liquid Line				
Refrigerant	°F drop		psig drop	
R-12	2		3.50	
R-22	2		5.50	
R-502	2		5.70	
Hot Gas Discharge Line				
Refrigerant	°F drop		psig drop	
R-12	8.75		5.00	
R-22	13.75		5.00	
R-502	14.25		5.00	
Suction Line				
Evaporating Temperature				
	45°F		20°F	
Refrigerant	°F drop	psig drop	°F drop	psig drop
R-12	2	2.00	2	1.35
R-22	2	3.00	2	2.20
R-502	2	3.30	2	2.40

Figure 10-20. Temperature drop varies for different pressures and refrigerants.

Excessive pressure drop in the liquid line must be avoided. If excessive pressure drop occurs because a liquid line is sized too small, the pressure in the liquid line can drop below the saturated pressure of the refrigerant before the liquid refrigerant reaches the expansion device. If the pressure in the liquid line drops below the saturated pressure of the refrigerant before the liquid refrigerant reaches the expansion device, the liquid refrigerant will start to flash to a vapor.

Some refrigerant should flash to a vapor when exiting an expansion device. Expansion devices are designed to control the flow of liquid refrigerant. Vapor refrigerant in the liquid line ahead of the expansion device causes erratic operation of the expansion device and poor control of the refrigerant entering the evaporator.

Pressure Drop Chart

In a refrigeration system, the proper amount of refrigerant must circulate through the lines to meet the cooling requirement of the system. A *pressure drop chart* is a chart used to size refrigerant lines to produce the required pressure drop for each section of refrigerant lines. A pressure drop chart allows a design technician to select the proper size piping for the different refrigerant lines in a refrigeration system. The sizing is based on the cooling requirements and a given pressure drop in the line.

A pressure drop chart has a scale at the top right that represents the cooling capacity of the system. The cooling capacity scale is calibrated in thousands of Btu per hour up to 1 ton and in tons of cooling from 1 ton to 100 tons. The cooling capacity scale is used to enter the cooling load required for the system. See Appendix.

Diagonal lines at the lower right corner of the chart represent the evaporator coil temperature, condenser coil temperature, and size of the liquid line. See Figure 10-21. The evaporator coil temperature lines are used to size suction lines. The condenser coil temperature lines are used to size discharge lines. The line that represents the size of the liquid line is used to size the liquid line in a mechanical compression refrigeration system.

Diagonal lines on the left of the chart represent pipe sizes for refrigerant lines. Pipe sizes are marked on the pipe size lines on the chart from $\frac{3}{8}''$ to $6\frac{1}{8}''$. The pipe size lines cross vertical lines that run up from the pressure drop graph at the bottom of the chart. The pressure drop scale is located on the bottom of the pressure drop graph. The pressure drop graph has three horizontal lines that represent condenser coil temperatures. Diagonal lines run from the bottom condenser coil temperature line to the top condenser coil temperature line. The scale for the pressure drop that occurs in the lines at the given condenser coil temperatures is located at the bottom or the top of the condenser coil temperature lines.

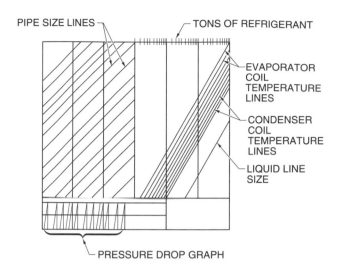

Figure 10-21. A pressure drop chart is a chart used to size refrigerant lines to produce the required pressure drop for each section of piping.

To use a pressure drop chart for sizing refrigerant lines, apply the procedure:

1. Find the cooling capacity of the system on the scale on the upper right of the chart. Draw a vertical line from the cooling capacity through the evaporator coil temperature lines, the condenser coil temperature lines, and the line that represents the liquid line. The evaporator coil temperature lines are for sizing the suction line, and the condenser coil temperature lines are for sizing the discharge line.

A refrigeration system has a cooling capacity of 8 tons, an evaporator coil temperature of 20°F, and a condenser coil temperature of 80°F. On the chart, a vertical line is drawn from 8 tons cooling capacity until the line intersects the 20°F evaporator coil temperature line, the 80°F condenser coil temperature line, and the liquid line. See Figure 10-22.

2. Draw a horizontal line from each of these three points to the pipe size lines.

3. Find the point on the pressure drop graph that represents the required condenser coil temperature and draw a horizontal line to the selected pressure drop for the system.

If the condenser coil temperature of a refrigeration system is 80°F and a 3 psi pressure drop is selected, a horizontal line will be drawn from the 80°F condenser coil temperature line on the pressure drop graph until it intersects the 3 psi pressure drop line.

4. Draw a vertical line from the intersection of the condenser coil temperature line and the pressure drop line on the pressure drop graph to the horizontal lines drawn from the evaporator coil temperature line, condenser coil temperature line, and the liquid line.

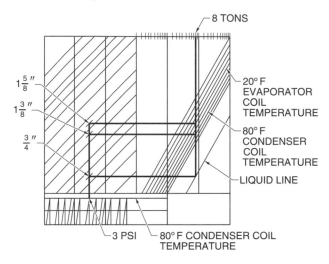

Figure 10-22. A vertical line is drawn from the cooling capacity of the system to the evaporator coil temperature lines, condenser coil temperature lines, and the liquid line.

The intersection of the vertical line from the pressure drop graph and the horizontal line from the evaporator coil temperature represents the proper size of piping to use for the suction line. The intersection of the vertical line from the pressure drop graph and the horizontal line from the condenser coil temperature represents the proper size of piping to use for the discharge line. The intersection of the vertical line from the pressure drop graph and the horizontal line from the liquid line represents the proper size of piping to use for the liquid line.

In a refrigeration system with 20°F evaporator coil temperature, 80°F condenser coil temperature, and a 3 psi pressure drop, the suction line should be 1⅝″, the discharge line should be 1⅜″, and the liquid line should be ¾″.

Example: Using a Pressure Drop Chart

A refrigeration system has a cooling capacity of 60,000 Btu/hr (5 tons). The evaporator coil temperature is 40°F and the condenser coil temperature

is 100°F. The pressure drop for the system is 2 psi. Find the proper size pipe for the suction line, discharge line, and liquid line.

1. Draw a vertical line from 5 tons of cooling on the cooling capacity scale through the 40°F evaporator coil temperature line and the 100°F condenser coil temperature line to the liquid line. See Figure 10-23.

2. Draw a horizontal line from the 40°F evaporator coil temperature line to the pipe size line. Draw a second horizontal line from the 100°F condenser coil temperature line to the pipe size line. Draw a third horizontal line from the liquid line to the pipe size lines on the left side of the chart.

3. Draw a horizontal line from the 100°F condenser coil temperature line on the pressure drop graph to the 2 psi pressure drop.

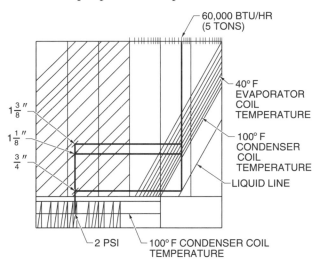

Figure 10-23. The size of the suction line, discharge line, and liquid line are found on a pressure drop chart.

4. Draw a vertical line from the intersection of the condenser coil temperature line and the pressure drop line to the horizontal line drawn from the evaporator coil temperature line, condenser coil temperature line, and the liquid line.

The intersection of the vertical line from the pressure drop graph and the horizontal line from the evaporator coil temperature line represents the proper size of piping to use for the suction line. The suction line should be 1⅜″. The intersection of the vertical line from the pressure drop graph and the horizontal line from the condenser coil temperature line represents the proper size of piping to use for the discharge line. The discharge line should be $1\frac{1}{8}''$. The intersection of the vertical line from the pressure drop graph and the horizontal line from the liquid line represents the proper size of piping to use for the liquid line. The liquid line should be $\frac{3}{4}''$.

Velocity

Velocity is the rate of motion in a given direction. Velocity of a refrigerant is the speed at which the refrigerant travels through a pipe. Velocity is expressed in feet per minute. The velocity of the refrigerant in the refrigerant lines must be high enough to push lubricating oil in the line along with the refrigerant so the oil returns to the compressor. The pressure drop must be kept low.

If the velocity of the refrigerant in any section of pipe exceeds certain limits, noise will be generated. Tables of maximum line sizes are used to check the velocity of the refrigerant in the system. The size is checked to ensure that there is enough pressure to push the oil along. If the velocity is too high or too low for a pipe sized for a given pressure drop, velocity and pressure drop should be considered when finding the actual size.

Tables of maximum suction line sizes show the maximum size pipe that can be used to ensure oil return to the compressor. See Appendix. The horizontal lines represent the capacity of the compressor. See Figure 10-24.

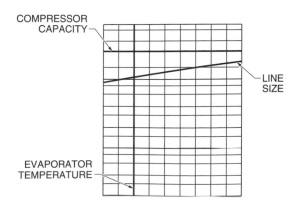

Figure 10-24. Tables of maximum suction line sizes show the maximum size pipe that can be used to ensure oil return to the compressor.

The scale that shows the capacity of the compressor runs along the left side of the table. The vertical lines on the table represent evaporator temperature. The scale that shows the evaporator temperature is located along the bottom of the table. Diagonal lines represent pipe sizes. The scale that shows the pipe sizes is located along the right side of the table. To use a table of maximum suction line sizes to check for proper oil return to the compressor, apply the procedure:

1. Find the capacity of the compressor on the left side of the table and draw a horizontal line across the table from that point.

If the compressor in a refrigeration system that contains R-12 has a capacity of 20,000 Btu/hr, a horizontal line is drawn from the 20,000 Btu/hr scale on the left of the chart. See Figure 10-25.

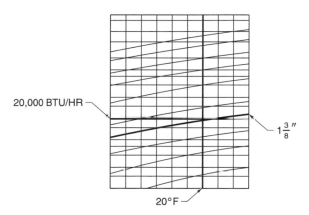

Figure 10-25. A horizontal line is drawn from the compressor capacity on the left of the chart.

2. Find the evaporator temperature at the bottom of the table and draw a vertical line from this point. The point where the lines intersect represents the maximum size pipe that can be used to ensure oil return.

If the evaporator temperature is 20°F, a vertical line is drawn from the 20°F scale on the bottom of the chart until the line intersects the 20,000 Btu/hr compressor capacity line.

The proper pipe size should be below the intersection of the lines and is indicated by the sloping line closest to the intersection of the two lines. If the pipe size is above the point of intersection, a smaller pipe is required to ensure proper oil return.

A refrigeration system that contains R-12 that has a compressor with a capacity of 20,000 Btu/hr and an evaporator temperature of 20°F should have a $1\frac{3}{8}''$ suction line to ensure proper oil return.

Each line in a system requires a different table. The kind of refrigerant in the system must also be considered. Each refrigerant has a different table.

Example: Using a Table of Maximum Suction Line Sizes

A refrigeration system that contains R-12 has a compressor that has a capacity of 60,000 Btu/hr. The evaporator temperature of the refrigeration system is 40°F. Find the maximum size pipe that can be used for the suction line to ensure proper oil return.

1. Draw a horizontal line from 60,000 Btu/hr. See Figure 10-26.

2. Draw a vertical line from 40°F. The maximum size pipe that can be used to ensure proper oil return is $1\frac{5}{8}''$.

ACCESSORIES

Accessories are used in the refrigerant lines to service or maintain a refrigeration system or to provide a specific function in controlling or regulating the flow of the refrigerant in the system. Accessories in the high-pressure side of a refrigeration system are located in the hot gas discharge line or the liquid line. The most commonly used accessory on the high-pressure side is the liquid receiver.

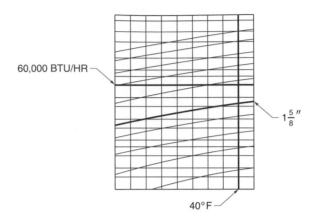

Figure 10-26. To find the maximum pipe size that can be used for the suction line, draw a horizontal line from the 60,000 Btu/hr compressor capacity point.

Liquid Receiver

A *liquid receiver* is a storage tank for refrigerant that is located in the liquid line of a refrigeration system. See Figure 10-27. The liquid receiver stores excess liquid refrigerant while the refrigeration system is operating. A liquid receiver allows for

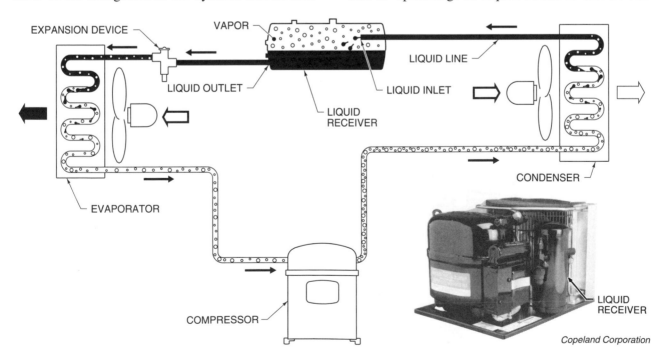

Copeland Corporation

Figure 10-27. The liquid receiver stores excess liquid refrigerant while a refrigeration system is operating.

changes in the volume of the refrigerant that occur when the cooling load changes or when the temperature of the condensing medium changes. If the temperature of the condensing medium increases or decreases, the volume of the refrigerant in the condenser coil will change. These changes affect the volume of the refrigerant. The storage capacity of the liquid receiver compensates for variations in the volume of the refrigerant. Liquid receivers are located only in the high-pressure side refrigerant lines.

Some of the refrigerant that flows from the condenser may still be a vapor. In a liquid receiver, liquid refrigerant stays in the bottom and the refrigerant vapor rises to the top. The liquid line from the compressor is connected at the top of the device. The liquid line from the receiver tank to the expansion valve is connected at the bottom. Only liquid refrigerant flows out of the tank to the expansion device. A liquid receiver in a refrigeration system ensures that only liquid refrigerant flows through the liquid line to the expansion device under all operating conditions.

If a liquid receiver is not part of the refrigeration system, a sight glass can be used to observe vapor in the liquid line. A sight glass in the liquid line between the condenser and the expansion device shows whether all of the refrigerant has condensed in the condenser coil. Other accessories such as service valves and moisture indicators can be located in either the low-pressure or high-pressure side of a refrigeration system.

Review Questions

1. What is the main function of the compressor in a mechanical compression system?

2. Name four types of compressors as characterized by mechanical action.

3. Define *positive displacement* as it refers to a compressor.

4. What connects the pistons to the crankshaft in a reciprocating compressor?

5. Name the two valves that are used most commonly in reciprocating compressors.

6. How many pistons are in a stationary vane rotary compressor? Why?

7. What is the purpose of oil in a rotary compressor?

8. List the two types of rotary compressors. What is the main difference between them?

9. Is a centrifugal compressor a positive-displacement compressor? Explain.

10. How are higher pressures produced with centrifugal compressors?

11. Explain how a screw compressor compresses fluid.

12. Define *volumetric capacity*.

13. What is the main identifying feature of an open compressor?

14. What cools the compressor motor on a semi-hermetic compressor?

15. What is the main identifying feature of a hermetic compressor?

16. Define *compressor performance*. Explain why compressor performance is important to system performance.

17. Define *volumetric flow rate*. In what unit of measure is it expressed?

18. Define *refrigeration effect*. What part of the refrigeration system does refrigeration effect describe?

19. What is the main function of a condenser?

20. List and describe the three basic types of condensers used in refrigeration systems.

21. Describe a finned-tube coil.

22. Define *heat rejection rate*.

23. Define *heat of compression*.

24. List and describe three types of water-cooled condensers.

25. Why could an evaporative condenser be called a combination air- and water-cooled condenser?

Air Conditioning Systems

Chapter 11

Air conditioning systems produce a refrigeration effect to maintain comfort within building spaces. Refrigeration systems produce a refrigeration effect for applications other than human comfort. Forced air conditioning and hydronic air conditioning are the two kinds of air conditioning systems. Forced air conditioning systems include an air conditioner, blower, supply and return ductwork, registers, grills, and controls. Hydronic air conditioning systems include a chiller, circulating pump, supply and return piping, terminal devices, and controls.

AIR CONDITIONING

Air conditioning is the process of cooling the air in a building to provide a comfortable temperature. An *air conditioner* is the component in a forced air conditioning system that cools the air. A *chiller* is the component in a hydronic air conditioning system that cools water, which cools the air. An air conditioning system is the equipment that produces a refrigeration effect and distributes cool air or water to building spaces. Air conditioning systems are classified by evaporating medium, condensing medium, physical arrangement, and cooling capacity.

Evaporating Mediums

An evaporating medium is the fluid that is cooled when heat is transferred in the evaporator from the evaporating medium to cold refrigerant. Air and water are evaporating mediums. The evaporator of an air conditioner cools air. The cooler of a chiller cools water.

Air. Air is used as the evaporating medium when an air conditioning system is located close to building spaces. Cool air is distributed to building spaces through ductwork. Air is readily available and inexpensive to use. Blower operation is the main

expense when using air as an evaporating medium. Exhausting air after it has been used is not difficult. Moving air over a long distance or moving a large amount of air requires a large amount of energy.

A relatively large quantity of air is required to carry heat. At standard conditions, air has a density of .0753 lb/cu ft and a specific heat of .24 Btu/lb. *Specific heat* is the amount of heat that is required to raise the temperature of 1 lb of a substance 1°F. Raising the temperature of 1 lb of air 1°F requires .24 Btu. Large ductwork is required to move large quantities of air. Large quantities of air are required to carry large amounts of heat.

Water. Water is used as an evaporating medium when an air conditioning system is located at a distance from building spaces. Cool water circulates from a cooler to terminal devices. At the terminal devices, the water cools the air and lowers the temperature of the air in the building spaces.

At standard conditions, water has a density of 62.32 lb/cu ft and a specific heat of 1.0 Btu/lb. Raising 1 lb of water 1°F requires 1 Btu. Because water is more dense than air, water holds heat better. The water in a small pipe holds the same amount of heat as the air in a large duct.

When used in a hydronic air conditioning system, water is treated with chemicals to prevent scale

171

buildup in the system. Scale buildup causes decreased efficiency. Some hydronic air conditioning systems may contain a solution of water and antifreeze compound. Chemically treated water can carry less heat than untreated water, but chemically treated water performs relatively well as an evaporating medium.

Condensing Mediums

A condensing medium is the fluid in the condenser of a refrigeration system that carries heat away from the refrigerant. Air and water are condensing mediums. An air-cooled condenser in an air conditioning system cools air. A water-cooled condenser in a chiller cools water. An evaporative condenser uses both air and water as the condensing medium.

Air. Air is used as the condensing medium for an air conditioning system when the condenser can be located outdoors. Air flow to the condenser cannot be restricted. The normal ambient temperature of outdoor air should be cooler than the condensing temperature of the refrigerant in the air conditioning system. When the temperature of the air is lower than the temperature of the refrigerant, the air can carry heat away from the condenser. An air-cooled condenser has a large capacity, which allows a large amount of air to circulate through the condenser.

Water. Water is used as a condensing medium when the condenser cannot be located outdoors. The initial cost of the equipment for a water-cooled condenser is high. Eliminating the waste water from a water-cooled condenser is also expensive. Water may be supplied to the condenser directly from a public or private water supply system or may be recirculated from a cooling tower. Because supply water temperature does not vary greatly, a water-cooled condenser works well regardless of changing outdoor temperatures. Waste water from a water-cooled condenser is piped to a cooling tower for cooling and reuse or dumped into a storm sewer. Three kinds of cooling towers are forced draft, induced draft, and hyperbolic. See Figure 11-1.

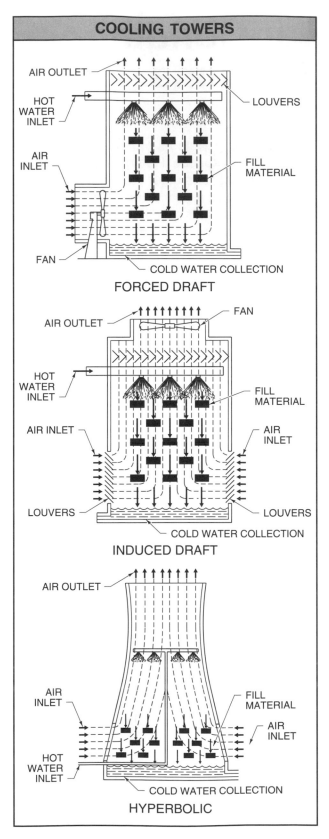

Figure 11-1. Cooling towers cool water for reuse in a water-cooled air conditioning system.

A cooling tower is an evaporative heat exchanger that removes heat from water. Water from a water-cooled condenser is cooled in a cooling tower for reuse in a hydronic air conditioning system. A cooling tower is a large structure that contains louvered panels for air flow. Natural or forced convection cause air flow in the cooling tower. A reservoir or tank in the tower holds hot water from the condenser. The water is sprayed over a fill material inside the tower. The fill material breaks the streams of water into small, cascading droplets. The water droplets are cooled by evaporation as air passes through the louvers in the tower. The water that flows out of the bottom of the tower is much cooler than the water sprayed in at the top.

A *forced draft cooling tower* is a cooling tower that has a fan located at the bottom of the tower that forces a draft through the tower. An *induced draft cooling tower* is a cooling tower that has a fan located at the top of the tower that induces a draft by pulling the air through the tower. A *hyperbolic cooling tower* is a cooling tower that has no fan. Natural draft moves air through a hyperbolic cooling tower.

Air and Water. An evaporative condenser uses both air and water as condensing mediums. In an evaporative condenser, air passes over the condenser coil and water is sprayed on the coil. The refrigerant flowing through the coil rejects heat because of the air passing over the coil, the water flowing over the coil, and the water evaporating from the surface of the coil. Evaporative condensers remove heat more efficiently than air-cooled or water-cooled condensers. Evaporation makes evaporative condensers more efficient. See Figure 11-2.

By combining the two evaporating mediums and the two condensing mediums, four different refrigeration systems are possible. Air conditioners and chillers can have either an air-cooled condenser or a water-cooled condenser. An air conditioner uses air as the evaporating medium. A chiller uses water as the evaporating medium. These four systems can be arranged to fit almost any application.

Physical Arrangement

Physical arrangement is the location of the various components of an air conditioning system in relation to the other components. Air conditioning systems have different arrangements depending on the size and layout of a building. Split systems, package units, and combination units are examples of different physical arrangements.

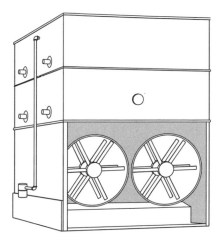

Figure 11-2. An evaporative condenser uses evaporation to remove heat from refrigerant.

Split Systems. A split system is an air conditioning system that has separate cabinets for the evaporator and the condenser. The cabinets are connected by refrigerant lines, plumbing, and electrical conductors. The evaporator of a split system is located inside a building, and the condenser is located outdoors. If the evaporator is located inside the furnace cabinet, the same blower can be used to move cool air and warm air. Split air conditioning systems have the evaporator in one cabinet and the condenser and compressor in another cabinet. Some split air conditioning systems have three parts, which are the compressor, the evaporator, and the condenser. See Figure 11-3.

The major components of a split system must be compatible. Each component of a split system is designed to have the same capacity as the other components in the system. Equipment specification sheets describe the equipment and the combinations of equipment that are compatible. Refrigerant lines connect the components of smaller split systems. Electrical conductors are field-installed and must meet NEC® requirements.

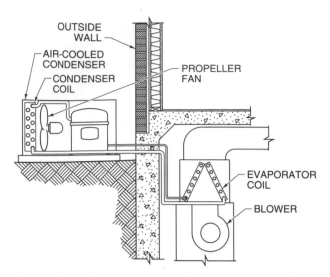

Figure 11-3. The evaporator and condenser of a split air conditioning system are enclosed in separate cabinets.

Package Units. A *package unit* is a self-contained air conditioner that has all of the components contained in one sheet metal cabinet. A package unit requires electrical power and control connections and must be connected to an air distribution system.

Controls for a package unit are installed by the manufacturer. Controls operate the components of a package unit. The thermostat and other remote controls must be installed and connected on-site when the unit is installed.

Air-cooled package units use air as the condensing medium and are normally installed outdoors. Air-cooled package units take in ambient air from outdoors and exhaust the air outdoors. An air distribution system that consists of ductwork runs from an air-cooled package unit to building spaces.

A *rooftop unit* is an air-cooled package unit that is located on the roof of a building. Rooftop units save building space. See Figure 11-4. If the ductwork in a building is installed under the floor or if access to the ductwork from the roof is difficult, a unit may be located on the ground near the building.

Water-cooled package units use water as the condensing medium. Water-cooled package units can be installed wherever supply water and a waste water outlet are available. Water-cooled package units are often located inside a building where using an air-cooled unit would be difficult. Water-cooled package units often cool a zone of a building. A zone is a specific section of a building that requires

separate temperature control. Zones are also known as modules. A water-cooled package unit is located in a zone, which requires less piping than an entire building. Most water-cooled package units are single-zone units, which cool a specific zone in a building. A water-cooled package unit with ductwork supplies cooling to specific building spaces. A large building may be divided into several zones and may have several package water-cooled units. This method of cooling is known as modular cooling.

The Trane Company

Figure 11-4. All of the components of a package unit are contained in one sheet metal cabinet.

Combination Units. Buildings that require cooling in the summer and heating in the winter require combination units. A *combination unit* is an air conditioner that contains the components for cooling and heating in one sheet metal cabinet. Power lines, fuel line connections, and control connections are provided. Most combination units are air-cooled and are located outdoors.

The cabinet of a combination unit is divided into two sections. The heating unit, evaporator, and blower are in one section, and the condenser and compressor are in the other section. See Figure 11-5. Inlet and outlet louvers for air, which is the condensing medium, are located in the condenser-compressor section. An intake louver for air and an outlet for flue gas are also located on the combination unit cabinet.

Combination units can be located on the roof of a building or on the ground near the building. The location of the unit depends on the location of the ductwork in the building. Most combination units are installed on a roof. One combination unit may be used on a building or several combination units may be used for modular applications. Because a combination unit controls heating and cooling, air distribution systems for heating and cooling are used. Single duct or double duct systems are used with combination units.

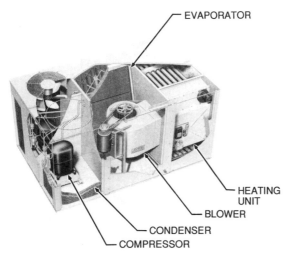

Lennox Industries Inc.

Figure 11-5. The cabinet of a combination unit is divided into two sections.

A *single duct system* is an air distribution system that consists of a supply duct that carries both cool air and warm air and a return duct that returns the air. When the thermostat calls for cooling, the combination unit distributes cool air to the building spaces through the supply duct. When the thermostat calls for heating, the combination unit distributes warm air to the building spaces through the supply duct. The return duct returns the air to the heating and cooling unit.

A *double duct system* is an air distribution system that consists of a supply duct that carries cool air and a supply duct that carries heated air. Mixing boxes (damper packages) mix the air at take-offs for each zone or building space. A *mixing box* is a sheet metal box that is attached to the cool air duct and the warm air duct. Openings in the side of the box connect with the ducts. Dampers in the box are controlled to introduce both warm and cool air and mix it in the proper proportion to provide a certain temperature.

Cooling Capacity

Air conditioners produce a given cooling capacity, which is expressed in Btu per hour or in tons of cooling. A ton of cooling equals 12,000 Btu/hr. A ton of cooling is the amount of heat in Btu required to melt a ton of ice over a 24-hour period. Air conditioners are rated in Btu per hour of cooling at certain standard conditions and are given a nominal size in tons of cooling. *Nominal size* is the cooling capacity of an air conditioner in Btu per hour rounded to the nearest ton or half-ton of cooling. See Figure 11-6.

Small air conditioners have cooling capacities of less than 1 ton of cooling to about 2 tons of cooling. Window and through-the-wall air conditioners are small package units. Small package units cool one room at a time and can be connected to the electrical system in most buildings.

Medium-size air conditioners have cooling capacities of 2 tons to 7.5 tons of cooling. Medium-size air conditioners are connected to ductwork and cool entire buildings such as homes or small commercial buildings. Medium-size air conditioners are either package or split systems.

Large air conditioners are available with cooling capacities of 20 tons to 60 tons of cooling. Large air conditioners are available as package units and as components in built-up systems.

A chiller is a water-cooled refrigeration system. Chillers are connected to a piping system that distributes the cool water to terminal devices. The water in the terminal devices cools the air in building spaces. Chillers have cooling capacities of 500 tons to 750 tons of cooling.

Refrigeration Process

Air conditioners use either the mechanical compression refrigeration process or absorption refrigeration process. Each refrigeration process removes heat from air or water and exhausts the heat where it is not objectionable. In the mechanical compression process, a compressor controls the pressure of the refrigerant in the system. In the absorption process, an absorbent chemical controls the pressure of the refrigerant in the system.

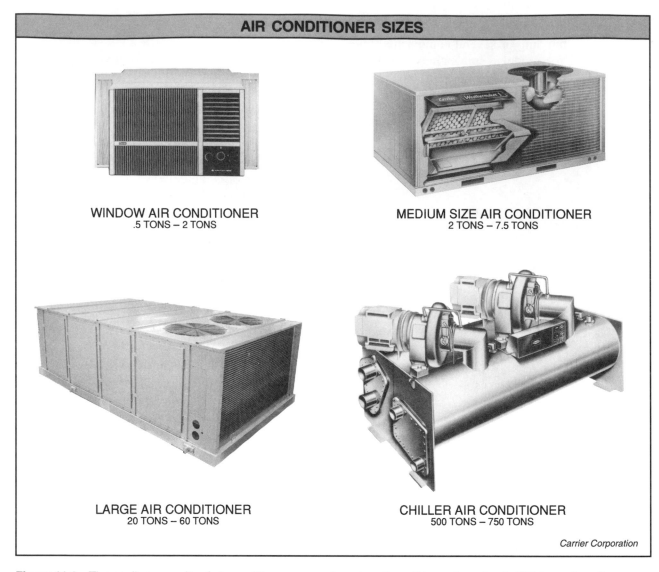

AIR CONDITIONER SIZES

WINDOW AIR CONDITIONER
.5 TONS – 2 TONS

MEDIUM SIZE AIR CONDITIONER
2 TONS – 7.5 TONS

LARGE AIR CONDITIONER
20 TONS – 60 TONS

CHILLER AIR CONDITIONER
500 TONS – 750 TONS

Carrier Corporation

Figure 11-6. The cooling capacity of air conditioners range from less than .5 tons of cooling to 750 tons of cooling.

Mechanical Compression Refrigeration. In the mechanical compression refrigeration process, a compressor increases the pressure and temperature of a refrigerant so the refrigerant can reject heat to the condensing medium as it condenses. A mechanical compression refrigeration system includes an expansion device, evaporator, compressor, condenser, blower or circulating pump, refrigerant lines, and controls.

The expansion device causes a pressure decrease in the refrigerant. The evaporator transfers heat from the evaporating medium to the refrigerant in the system. The compressor increases the pressure and temperature of the refrigerant. The condenser trans-

fers heat from the refrigerant to the condensing medium. The blower or circulating pump moves the condensing medium through the system. Refrigerant lines connect the components, and controls regulate the operation of the components.

Absorption Refrigeration. In the absorption refrigeration process, an absorbent controls the pressure and temperature of a refrigerant. An absorption refrigeration system contains an absorber, generator, condenser, orifice, evaporator, and controls. An absorption system may also contain a heat exchanger and a circulating pump.

In the absorber, absorbent combines with refrigerant to form a strong refrigerant-absorbent solution. In the generator, the solution absorbs heat and separates into refrigerant and absorbent. In the condenser, the refrigerant rejects heat and condenses to a high-pressure liquid. In the orifice, the pressure decreases. In the evaporator, the low-pressure refrigerant evaporates. The refrigerant absorbs heat from the evaporating medium that flows through the evaporator. This cools the evaporating medium, which is used for cooling.

Controls

Air conditioning systems are designed to produce a refrigeration effect at a fixed rate. Air conditioning systems are selected to provide enough cooling output to compensate for the warmest outdoor temperature. Since the outdoor temperature is normally cooler than the expected maximum temperature, a typical air conditioning system produces more cooling than required most of the time. Controls cycle the components of an air conditioning system ON and OFF to produce the required refrigeration effect consistently, safely, and automatically.

Power Controls. Power controls control the flow of electricity to an air conditioning system. Power controls are located in the electrical circuit between the power source and the air conditioning system. Power controls should be installed by a licensed electrician per article 424 of the National Electrical Code®. Power controls include disconnects, fuses, and circuit breakers.

At least two disconnects are located in the electrical power circuit to an air conditioning system. One disconnect is located in the electrical service panel where the circuit originates, and the other disconnect is located near the air conditioner. Manual disconnects are used on most air conditioners. A manual disconnect is a disconnect that is opened or closed manually by an operator. See Figure 11-7.

Fuses or circuit breakers are electric overcurrent protection devices that protect the electrical conductors in a circuit from an overcurrent condition, which is excessive current flow. If an overcurrent condition occurs, a fuse will burn out. The burned-out fuse breaks the circuit and shuts OFF the current

flow before the circuit is damaged. Small air conditioners may have circuit breakers for combination disconnects and fusing. Large air conditioners have cartridge fuses for overcurrent protection.

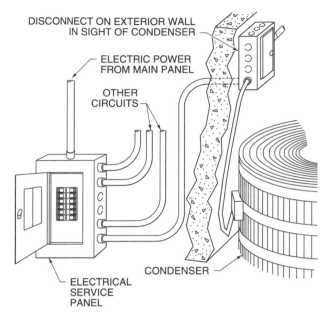

Figure 11-7. Power controls control the flow of electricity to an air conditioner.

Operating Controls. Operating controls cycle equipment ON or OFF. Each component of an air conditioner must function separately when combined with other components for proper performance. Operating controls include transformers, thermostats, blower controls, relays, contactors, magnetic starters, and solenoids. Operating controls on air conditioners and furnaces are similar. Operating controls regulate the operation of air conditioner components. See Figure 11-8.

A transformer is an electric device that decreases the voltage in an electrical circuit to the voltage required in the control circuit of an air conditioner. The transformer on an air conditioner is wired on the primary side to the electrical power source and is wired on the secondary side to the control circuits. Most control circuit transformers are step-down transformers with a secondary voltage of 24 V.

A thermostat is a temperature-actuated electric switch that turns an air conditioner ON or OFF in response to temperature changes in building spaces. When the temperature at the cooling thermostat rises above a setpoint, the thermostat closes a switch to complete the electrical circuit. The system turns ON when the circuit is complete. When the temperature at the thermostat falls below a setpoint, the thermostat opens a switch to open the electrical circuit. The system turns OFF when the circuit is open. The setpoint is the temperature at which the switch in the thermostat opens and closes. The setpoint on most thermostats can be changed manually.

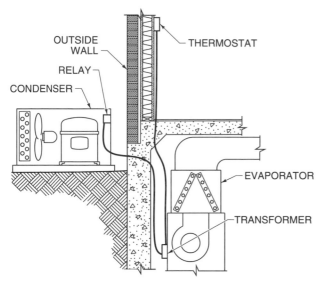

Figure 11-8. Operating controls regulate the operation of the components in an air conditioner.

The sensor in many thermostats is a bimetal element. A bimetal element is a sensor that consists of two different kinds of metal that are bonded together into a strip or coil. The metals expand at different rates when heated. When a bimetal element bends, it trips an electric switch. The switch controls the flow of electric current in the control circuit.

A cooling anticipator improves temperature control in a building space. A *cooling anticipator* is a small heating element that is wired into the control circuit inside the thermostat case. The heater is wired in parallel with the thermostat switch. In this position, the element is heated when the switch is open. This source of false heat causes the thermostat to call for cooling slightly before the room temperature would call for cooling.

A relay is an electric device that controls the flow of electric current in one circuit with another circuit. Relays are used in control circuits where a low-voltage circuit is used to control a line voltage cir-

cuit. Electromechanical relays, contactors, and magnetic starters are used in air conditioners.

An electromechanical relay is an electric device that uses a magnetic coil to open or close one or more sets of contacts. One set of contacts is fixed, and another set is mounted on a movable metal arm that is controlled by the coil. The moving contact is held in the initial position by a spring. When the coil is energized, the contacts open or close depending on the action desired. The magnetic coil is connected in the control circuit and the contacts are connected in the load circuit. When the coil is energized by the control circuit being energized, the contacts close and the load circuit is completed.

Relays are identified by poles and by the position of the contacts when the control circuit is de-energized. SP indicates single-pole. DP indicates double-pole. The position of the contacts are identified as normally open (NO) or normally closed (NC). Relays are available with many combinations of poles, but the contact positions are limited to NO, NC, or a combination of the two. Throw indicates the number of closed contact positions per pole. Circuits are identified as single throw (ST) or double throw (DT).

A contactor is a heavy-duty relay. A contactor is a device that opens and closes contacts in an electrical circuit. A contactor consists of a coil and contacts that are designed to operate with high electric current. Contactors may be used for controlling compressors or a large motor.

A magnetic starter is a contactor with an overload relay added to it. Overload relays are electric switches that are controlled by electric current flow or the ambient air temperature. Overload relays protect the motor against overheating or overloading.

When a motor is overloaded, it overheats. Overload occurs when a motor is connected to a load that is larger than the capacity of the motor. When a motor is overloaded, it draws more electric current than it is designed to carry. Overheating causes the insulation on the wiring in the motor to break down. Magnetic starters are used in electric motor circuits on motors that do not have internal overloads.

A bimetal overload relay consists of a set of contacts that are actuated by a bimetal element. If the temperature around the bimetal element rises higher than a setpoint, the element will bend and open the contacts. When the temperature drops below the set-point, the contacts close. In certain applications a reset button must be pressed to restart the motor. In other applications the motor restarts when the bimetal element cools.

The condenser blower motor on small air conditioners is wired in parallel with the compressor motor so that the condenser blower and compressor motor run at the same time. The relay, contactor, or magnetic starter that controls the compressor also controls the condenser blower. In large air conditioners, the condenser blower may have a separate contactor or magnetic starter, but the control coil is wired in parallel with the control coil on the compressor starter. The compressor and condenser blower still run at the same time.

The evaporator blower is controlled by a separate relay, contactor, or magnetic starter. The relay is actuated by a signal from the thermostat. Most cooling thermostats are wired to the blower and are actuated whenever the thermostat calls for cooling or when the blower switch on the thermostat is set for constant operation. If the evaporator blower and compressor relay are wired in parallel, the evaporator blower relay may also be actuated whenever the compressor relay is actuated.

A solenoid is similar to a relay in that it controls one electrical circuit with another. In a solenoid, an electromagnet positions a movable core that opens or closes a set of contacts. The contacts are wired into a circuit so the circuit opens and closes as the switch is energized and de-energized. A solenoid coil is energized by the control circuit of an air conditioner. The contacts on a solenoid are wired into the load circuit.

Safety Controls. Safety controls on an air conditioner shut the unit OFF to prevent damage to the components in case one of the components malfunctions. Because the components of an air conditioner are interrelated, the failure of one component can cause damage to the others. The safety controls in an air conditioner include pressure switches.

Pressure switches are electric switches that contain contacts and a spring-loaded lever arrangement. The lever opens and closes the contacts. The lever moves according to the amount of pressure acting on a diaphragm or bellows element. A section of

narrow tubing is connected to the refrigerant line at the point in the system where the pressure is controlled. The other end of the tube is connected to the diaphragm or bellows. Pressure is transmitted to the diaphragm or bellows through the tubing. A change in the pressure of the refrigerant in the system moves the diaphragm or bellows element. This movement of the diaphragm or bellows is transmitted to a lever arm by mechanical linkage. The switch is wired in series in the electrical conductors that operate the compressor.

Pressure switches are automatic reset switches. The contacts in the pressure switch automatically close when the pressure in the system returns to normal. On most pressure switches, the setpoint temperature and differential can be adjusted manually.

Refrigerant pressure switches are actuated by refrigerant pressure in the system and may sense high or low pressure. Oil pressure switches are actuated by the pressure of the oil in the compressor.

A high-pressure refrigerant switch is connected to the high-pressure side of an air conditioner. The high-pressure refrigerant switch is connected to a refrigerant line near the hot gas discharge outlet of the compressor. The high-pressure refrigerant switch shuts the system OFF if the refrigerant pressure becomes too high. High pressure in an air conditioner can damage the compressor and burn out the compressor motor. See Figure 11-9.

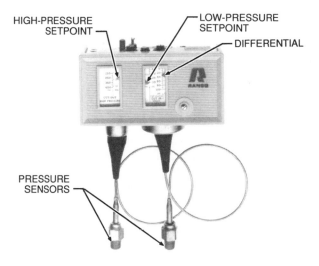

Figure 11-9. High- and low-pressure switches shut a compressor OFF when the refrigerant pressure becomes too high or too low.

A low-pressure refrigerant switch is connected into the low-pressure side of an air conditioner. The low-pressure refrigerant switch is connected to a refrigerant line near the suction inlet of the compressor or to the crankcase of the compressor. The low-pressure switch shuts the system OFF if the refrigerant pressure becomes too low. Low pressure can damage hermetic and semi-hermetic compressor motors. Low pressure in the system causes a low temperature. A low temperature in the evaporator coil can freeze and damage the coil.

An oil pressure switch is similar to a refrigerant pressure switch. An oil pressure switch ensures that there is lubricating oil in the system when the system is operating. An oil pressure switch is connected into the lubricating line of an air conditioner. Large air conditioners have positive-pressure lubrication systems. Positive-pressure lubrication systems contain an oil pump that distributes lubricating oil to compressor bearings. Failure of the lubrication system causes failure of the compressor.

The oil pressure switch sensors are connected to the oil line close to the oil pump discharge. The electric switch on an oil pressure switch shuts the compressor OFF and may actuate a bell or light to indicate oil pressure failure. See Figure 11-10.

Ranco Inc.

Figure 11-10. An oil pressure switch ensures that lubricating oil is in the compresor when the compressor motor is operating.

FORCED AIR CONDITIONING SYSTEMS

Forced air conditioning systems use air as the evaporating medium. Air circulates through ductwork to building spaces. Forced air conditioning systems are used primarily where the air conditioning equipment can be located close to building spaces and where only certain building spaces require cooling. Because the required ductwork may be fairly large, it is not practical to run ductwork over long distances.

A forced air conditioning system consists of an air conditioner, blower, supply and return ductwork, registers, grills, and controls. See Figure 11-11.

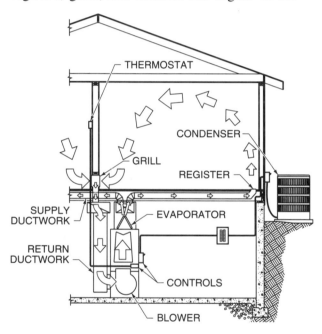

Figure 11-11. A forced air conditioning system consists of an air conditioner, blower, supply and return ductwork, registers, grills, and controls.

Air Conditioners

An air conditioner is the component in a forced air conditioning system that cools the air. An air conditioner uses either the mechanical compression or absorption refrigeration process to provide a refrigeration effect. The air conditioner contains a blower, filter, and controls. The supply and return air ductwork, registers, and grills have specifications for each application to provide the best air distribution for each building space.

Blowers

A blower moves air across the evaporator coil, circulates it through supply ductwork into the building spaces, and returns the air through return ductwork from the building spaces. Centrifugal blowers are used in air conditioning systems.

A centrifugal blower consists of a blower wheel mounted on a shaft inside a sheet metal scroll. Return air enters the wheel through openings on the sides of the sheet metal scroll. A discharge outlet is located on the perimeter of the scroll at a right angle to the axis of the blower wheel.

As an electric motor rotates the blower wheel, air is thrown out of the blower by centrifugal force from the tips of the vanes on the blower wheel. Air is drawn into the center of the wheel through the openings in the sides of the scroll because a low-pressure area is created inside the wheel.

Ductwork

Forced air conditioning systems include supply air ductwork and registers and return air ductwork and grills. The supply air ductwork runs from the blower in the air conditioner to building spaces. Registers are located on the supply branches and are sized and located to provide the proper amount of air required in each building space. See Figure 11-12.

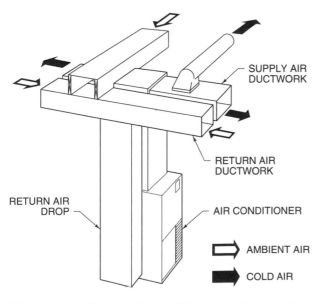

Figure 11-12. Forced air conditioning systems contain supply ductwork and registers and return ductwork and grills.

Return air ductwork returns air from the building spaces to the air conditioner for reuse. Grills are sized and located to provide proper circulation of the air through the building.

HYDRONIC AIR CONDITIONING SYSTEMS

A hydronic air conditioning system uses water or a solution of water and antifreeze as the evaporating medium. Terminal devices use the evaporating medium to cool air in building spaces. The relatively high specific weight and high specific heat of water make it an excellent evaporating medium. The pipes that carry water in a hydronic air conditioning system are much smaller than ducts that carry air in a forced air conditioning system.

Hydronic air conditioning systems are used where the air conditioning equipment is located centrally and the building spaces are located remotely. Hydronic air conditioning systems are often located in one central plant that cools several other buildings, such as campus school buildings. Hydronic air conditioning systems are also used in multistory buildings where the cooling equipment is located on one floor and the cool water is distributed to terminal devices located on other floors.

Because water carries heat better than air, chillers are used for larger air conditioning systems. Chillers are available with almost any cooling capacity and can be used for almost any size air conditioning application. Large air conditioning systems that produce cooling capacities of many hundreds of tons of cooling are built-up systems. Built-up systems are designed by engineers for specific buildings. A typical hydronic air conditioning system contains a chiller, a piping system, circulating pumps, terminal devices, and controls. See Figure 11-13.

Chillers

A chiller is the component in a hydronic air conditioning system that cools the water. A chiller uses either the mechanical compression or absorption refrigeration process to cool the water. Mechanical compression systems are used often for air conditioning applications. Absorption systems are used on large commercial or industrial applications.

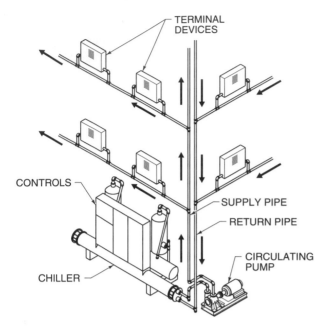

Figure 11-13. Hydronic air conditioning systems contain a chiller, piping system, circulating pump, terminal devices, and controls.

Piping Systems

Piping systems distribute water from the chiller to the terminal devices. Piping systems are classified by the arrangement of the piping loop that carries the water. A typical piping system has a supply line that carries water from the chiller to the terminal devices and a return line that carries water back from the terminal devices to the chiller. Piping systems include one-, two-, three-, and four-pipe systems. Piping systems for hydronic air conditioning systems are similar to the piping systems on hydronic heating systems.

One-pipe. Series and primary-secondary systems are one-pipe systems. In a one-pipe series system, the water in the system circulates through one pipe and flows through each terminal device in turn. Terminal devices are connected in series on the loop.

In a one-pipe primary-secondary system, the terminal devices are connected in parallel with the main supply pipe. Water flows from the supply pipe through the terminal devices and back into the supply pipe. The flow rate in each terminal device can be controlled within the loop. One-pipe systems are used in small- to medium-size applications.

Two-pipe. A two-pipe system has a supply pipe and a return pipe that run to each terminal device. The terminal devices are arranged in parallel. Water flows from the supply pipe through each terminal device and back to the chiller through the return pipe. Direct-return and reverse-return systems are two-pipe systems.

In a two-pipe direct-return system, supply water flows into each terminal device from the supply pipe. The water then enters the return pipe and flows directly back to the chiller. The return water flows in the opposite direction of the water in the supply pipe.

In a two-pipe reverse-return system, supply water flows into each terminal device from the supply pipe. The water then enters the return pipe and flows in the same direction as the water in the supply pipe. The return water continues around the piping loop from one terminal device to the next.

Three-pipe. A three-pipe system has two supply pipes and one return pipe. A three-pipe system is used for heating some building spaces and cooling other building spaces simultaneously. One supply pipe is connected to a boiler and the other is connected to a chiller. See Figure 11-14.

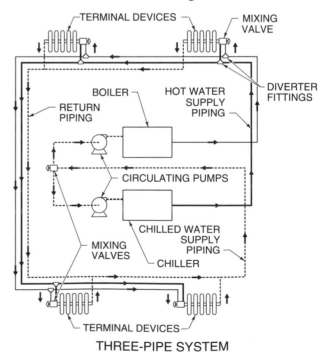

THREE-PIPE SYSTEM

Figure 11-14. A three-pipe system has two supply pipes and one return pipe.

Both supply pipes are connected to each terminal device. Return water flows in a common pipe that is connected to the heating and cooling supply pipes. Mixing valves that are controlled by system switches and thermostats regulate the flow of water from the supply pipes to the terminal devices. On a call for heat, water from the boiler flows through one supply pipe and into the terminal device. On a call for cooling, water from the chiller flows through the other supply pipe and into the terminal device.

Three-pipe systems provide good control of air temperature but are more expensive to install than two-pipe systems. Three-pipe systems are also expensive to operate because the hot and cold return water is mixed.

Four-pipe. A four-pipe system uses separate piping for heating and cooling. The terminal devices are connected to both heating and cooling pipes. Water flow to the terminal devices is controlled by mixing and diverting valves, which allow either hot or cold water to flow to any terminal device. Four-pipe systems are expensive to install but provide excellent control of air temperature. Four-pipe systems are more economical to operate than three-pipe systems. See Figure 11-15.

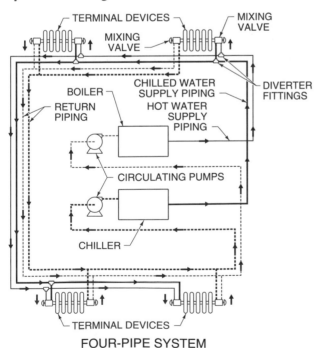

FOUR-PIPE SYSTEM

Figure 11-15. A four-pipe system has completely separate piping for heating and cooling.

Circulating Pumps

A circulating pump moves water from the chiller through the piping system to the terminal devices of a hydronic air conditioning system. A circulating pump may be connected to either the supply or return piping, but a circulating pump is usually connected to the return piping close to the chiller. The number and size of circulating pumps depends on the volume of water and resistance to flow in the system. If a hydronic air conditioning system has many piping loops feeding different zones in a building, a pump will be located on each loop.

Centrifugal pumps are often used on hydronic air conditioning systems. A centrifugal pump is a circulating pump that has a rotating impeller wheel inside a cast iron or steel housing. The housing has an opening parallel with the impeller wheel shaft and another opening on the outer perimeter of the housing at right angles to the impeller wheel shaft. The opening parallel with the impeller wheel shaft is the water inlet. The return water piping is connected at the water inlet. The opening on the outer perimeter of the housing is the water outlet. The supply water piping is connected at the water outlet. See Figure 11-16.

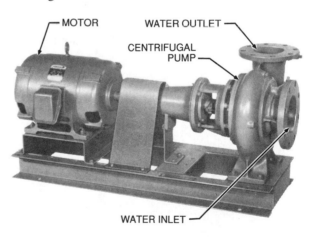

Figure 11-16. An impeller wheel rotates inside the cast iron or steel housing of a centrifugal pump.

The impeller wheel of a circulating pump rotates rapidly. As the impeller wheel spins, water is thrown out of the outlet opening by centrifugal force. This action creates a low-pressure area that draws water into the center of the impeller wheel.

Terminal Devices

In a hydronic air conditioning system, a terminal device holds cold water from the chiller. Terminal devices receive a supply of cold water from the main supply pipe and return the water to the main return pipe. Unlike terminal devices for hydronic heating systems, terminal devices for hydronic air conditioning systems do not use thermal radiation for heat transfer. Blowers move air through the terminal device to cool the air. Terminal devices for hydronic air conditioning systems are cabinet air conditioners, unit ventilators, duct coils, and unit air conditioners.

Most terminal devices are combinations of water coils and other components. The water coils serve as heat exchangers and the other components control the flow of water and air through the device. Water coils are finned-tube coils, which are used for hydronic heating. The amount of cooling produced depends on the size of the coil, the flow rate of water through the coil, and the flow rate of air across the outside of the coil.

Cabinet Air Conditioners. Cabinet air conditioners are similar to cabinet heaters. A cabinet holds a water coil, blower, and filter. Cabinet air conditioners have a louvers along the bottom of the cabinet for air intake and louvers along the top of the cabinet for exhaust. Cabinet air conditioners are installed on the floor along the outside walls of a room. See Figure 11-17.

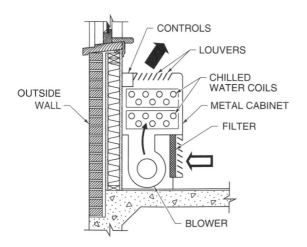

Figure 11-17. Cabinet air conditioners have a water coil, blower, and filter inside one cabinet.

Unit Ventilators. Unit ventilators are for air conditioning applications and heating applications. Unit ventilators differ from cabinet air conditioners in that they have an opening to bring in air from outdoors. Unit ventilators can be used for cooling and ventilating a room simultaneously. Unit ventilators are often used in buildings such as schools and apartments that hold a large number of people and require constant ventilation. See Figure 11-18.

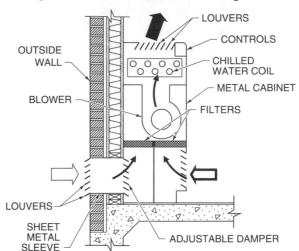

Figure 11-18. Unit ventilators have an opening to the outdoors that admits fresh air.

Duct Coils. A *duct coil* is a terminal device that is located in a duct. Air is supplied to the duct from a blower that may be remotely located. Duct coils are often used when a central chiller provides cooling to separate zones in a building. Each zone in the building has duct coils and a control system. See Figure 11-19.

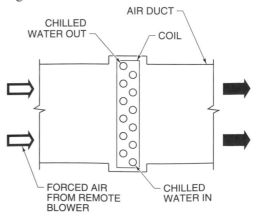

Figure 11-19. Duct coils are terminal devices installed in an air duct.

Unit Air Conditioner. A *unit air conditioner* is a self-contained air conditioner that contains a coil and a blower in one cabinet. Some unit air conditioners do not require ductwork. Unit air conditioners are similar to unit heaters. A propeller fan moves the air out of the outlet through louvers. See Figure 11-20.

Another type of unit air conditioner contains a centrifugal blower and is connected to ductwork. This type of unit air conditioner is designed to be used in ductwork.

Unit air conditioners can be used as unit coolers without any ductwork or can be used where each unit air conditioner and its ductwork are a separate system within a building. Unit air conditioners are often used for space cooling when appearance and space constraints are not important. Unit air conditioners are often used in individual rooms or zones when controlling an individual space is important.

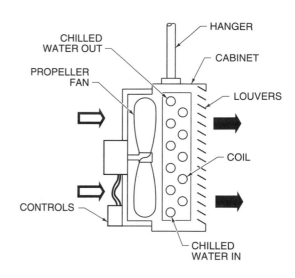

Figure 11-20. Unit air conditioners are self-contained air conditioners that contain a coil and a blower in one cabinet.

Review Questions

11

1. Define *air conditioning*. How does it differ from refrigeration?

2. List four ways in which air conditioning systems are classified.

3. Why are pipes in hydronic air conditioning systems smaller than air ducts in forced air conditioning systems?

4. Define *condensing medium*.

5. List and describe three kinds of condensers used in air conditioning systems.

6. What is the main characteristic of a mechanical compression refrigeration system?

7. List and describe the function of the seven main parts of an absorption refrigeration system.

8. How does an air conditioning system provide the correct amount of cooling for a building when the cooling load varies?

9. List and describe three classifications of controls found in a typical air conditioning system.

10. What is the main function of the operating controls on an air conditioning system?

11. What is the difference between a relay and a contactor?

12. What is the difference between a contactor and a magnetic starter?

13. What is the function of an overload relay in an electrical motor circuit?

14. What is the primary purpose of the safety controls in an air conditioning system?

15. What safety function does a pressure switch perform on an air conditioner?

16. Describe the difference between a package unit and a split system.

17. What are the three main functions of the blower in an air conditioning system?

18. List the four parts of a forced air conditioning system that are not part of the air conditioner itself.

19. List and describe the four different kinds of piping systems for hydronic distribution systems.

20. How do the terminal devices in a hydronic air conditioning system cool the air in a building space?

Heat Pumps

Chapter **12**

A heat pump is a mechanical compression refrigeration system that moves heat from one area to another. When a heat pump is in the cooling mode, it moves heat from inside a building to outdoors. When a heat pump is in the heating mode, it moves heat from outdoors to the inside of a building. Reversing the flow of refrigerant in the heat pump causes the change from cooling mode to heating mode.

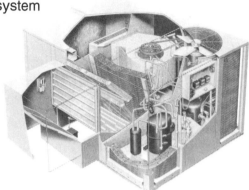

HEAT PUMPS

A *heat pump* is a mechanical compression refrigeration system that contains devices and controls that reverse the flow of refrigerant. Reversing the flow of refrigerant switches the relative position of the evaporator and condenser. A heat pump can absorb heat from inside a building and reject the heat outdoors and can absorb heat from the outdoors and reject the heat inside a building. A reversing valve reverses the refrigerant flow.

In the cooling mode, a heat pump operates like an air conditioning system. Air conditioning systems move heat from inside a building to the outdoors. When a heat pump is in the cooling mode, the indoor unit is the evaporator and the outdoor unit is the condenser. In the heating mode, the heat pump moves heat from the outdoors to the inside of a building. When a heat pump is in the heating mode, the indoor unit is the condenser and the outdoor unit is the evaporator.

As a heat pump operates, refrigerant flows from the compressor to a reversing valve. The refrigerant flows from the reversing valve to either the coil in the indoor unit or the coil in the outdoor unit. The direction of refrigerant flow depends on whether the system is in the cooling or heating mode.

Heat Sink

Unlike a furnace or boiler, a heat pump does not produce heat. A heat pump transfers heat from one area to another. A heat sink must be available to a heat pump. A *heat sink* is heat that is contained in a substance. This heat can be absorbed by a heat pump and used for heating.

The temperature of a substance is the intensity of heat in the substance. Any material that has a temperature above absolute zero (–460°F) contains heat. The higher the temperature of a material, the more heat it contains. A substance that has a temperature above absolute zero can be used as a heat sink for a heat pump. Because of availability, water and air are used as heat sinks for heat pumps.

Air. Air is used most often as a heat sink for a heat pump. Because the temperature of outdoor air varies during different times of the year and during different times of the day, the amount of heat available from outdoor air varies. Air above 15°F can be used as a heat sink.

Water. Water used as a heat sink comes from lakes, streams, wells, and industrial cooling processes. When used as a heat sink, the temperature of the water should be as high as possible. Groundwater is water that is found naturally in the ground. Groundwater is used as a heat sink for heat pumps. The average temperature of groundwater is from 45°F to 55°F, which makes water a good heat sink for a heat pump.

Classifications

Heat pumps are classified according to the cooling medium and heat sink. The four classifications of heat pumps are air-to-air, air-to-water, water-to-air, and water-to-water. The heat pump used for an application depends on the location of the heat pump and the availability of the cooling medium and heat sink. The coil of a heat pump that uses air as a heat sink is located outdoors. Heat pumps that use air as a heat sink are split systems. The coil of a heat pump that uses water as a heat sink may be located either outdoors or indoors.

Air-to-Air. An air-to-air heat pump system consists of a compressor, reversing valve, outdoor unit, expansion devices with bypass circuits, indoor unit, and refrigerant lines. See Figure 12-1. An *outdoor unit* is a package component that contains a coil heat exchanger and a blower. Depending on the direction of refrigerant flow, the outdoor unit acts as the evaporator or condenser of the heat pump.

The outdoor unit on an air-to-air heat pump contains a coil heat exchanger and a blower.

A *bypass circuit* is refrigerant line that contains a check valve on each side of an expansion device. The check valves in a bypass circuit are arranged to allow refrigerant to pass around the expansion device when the refrigerant flow is opposite to the flow direction of the expansion device. Only capillary tube expansion devices produce a pressure decrease when refrigerant flows in either direction. An *indoor unit* is a package component that contains a coil heat exchanger and a blower. Depending on the direction of refrigerant flow, the indoor unit acts as the evaporator or condenser of the heat pump. The indoor unit on an air-to-air heat pump contains a coil heat exchanger and a blower.

An air-to-air heat pump uses outdoor air as a heat sink when the system operates in the heating mode. Return air from the building is blown across the indoor coil and used to heat building spaces. Air-to-air heat pumps are available with cooling capacities from 1 ton to 5 tons of cooling. Air-to-air heat pumps are used in residences and small commercial buildings. Air-to-air heat pumps are available as both split systems and package heat pumps.

Air-to-Water. An air-to-water heat pump system consists of a compressor, reversing valve, outdoor unit, expansion devices with bypass circuits, indoor unit, and refrigerant lines. The outdoor unit contains

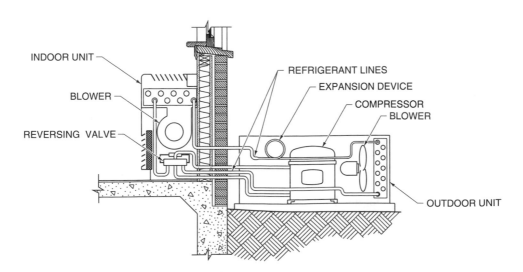

Figure 12-1. An air-to-air heat pump cools or heats air and uses air as a heat sink.

a coil heat exchanger. The indoor unit contains a coil heat exchanger and blower. An air-to-water heat pump uses outdoor air to cool or heat water that cools or heats the air in building spaces. Outdoor air is used as a heat sink when the heat pump operates in the heating mode.

Air-to-water heat pumps are used to heat and cool large buildings that use a hydronic heating and cooling system. The air-to-water heat pump system is a built-up system. Built-up systems contain factory-built components, which are assembled on a job site.

Water-to-Air. A water-to-air heat pump system consists of a compressor, reversing valve, outdoor unit, expansion devices with bypass circuits, indoor unit, and refrigerant lines. The outdoor unit contains a coil heat exchanger.

The indoor unit contains a finned-tube heat exchanger and blower. A water-to-air heat pump uses water as a heat sink and uses air to cool or heat building spaces. Water used as a heat sink in a water-to-air heat pump is taken from lakes, streams, wells, and industrial cooling processes. See Figure 12-2.

Water-to-air heat pumps are often used as package heat pumps in applications where an outdoor unit cannot be located outdoors. The outdoor unit is located inside the building. The supply water is piped to the outdoor unit, and the waste water is piped away. Small water-to-air heat pumps are often used to control the temperature of small spaces in high-rise buildings, such as hotel rooms.

Water-to-Water. A water-to-water heat pump system consists of a compressor, reversing valve, outdoor unit, expansion devices with bypass circuits, indoor unit, and refrigerant lines. The outdoor unit and the indoor unit each contain a coil heat exchanger. Neither unit contains a blower.

Water from outside the system is used as the heat sink for a water-to-water heat pump. Water from lakes, streams, wells, and industrial cooling processes may be used as a heat sink. The heat pump cools or heats water, which is used to cool or heat air in building spaces.

Water-to-water heat pump systems are used for cooling and heating the air in large buildings. Most of the systems are built-up systems, which must be assembled on-site. Centrifugal or screw compressors are often used in water-to-water heat pump systems.

COMPONENTS

Heat pumps contain a compressor, outdoor unit, indoor unit, expansion device, and refrigerant lines that are found on refrigeration systems. A reversing valve and expansion devices with bypass circuits allow the refrigerant to expand regardless of the direction of flow. A compressor moves refrigerant in one direction. A reversing valve reverses the flow of refrigerant in the refrigerant line.

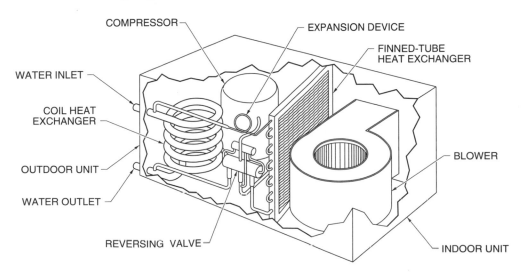

Figure 12-2. A water-to-air heat pump cools or heats water, which is used to cool or heat air in building spaces.

Reversing Valves

A *reversing valve* is a four-way valve that reverses the flow of refrigerant in a heat pump. A reversing valve consists of a piston and a cylinder that has four refrigerant line connections. The piston moves from one end of the cylinder to the other, which changes the direction of refrigerant flow by opening and closing outlets. See Figure 12-3.

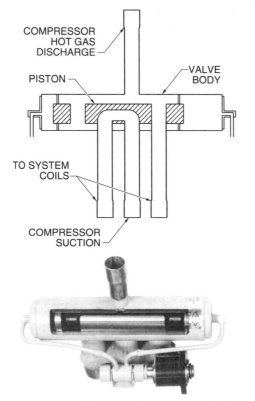

ALCO Controls Division Emerson Electric Company

Figure 12-3. A reversing valve consists of a cylinder with a piston that moves from one end of the cylinder to the other.

When the piston is at one end of the cylinder, refrigerant flows into the valve through the hot gas discharge line and is directed out of one of the outlets to the indoor or outdoor unit. When the piston moves to the other end of the cylinder, refrigerant flows into the valve through the same inlet port and is discharged to another outlet. The refrigerant is then directed to the opposite unit in the heat pump. In either position, the suction line to the compressor returns the refrigerant from the reversing valve to the compressor. See Figure 12-4.

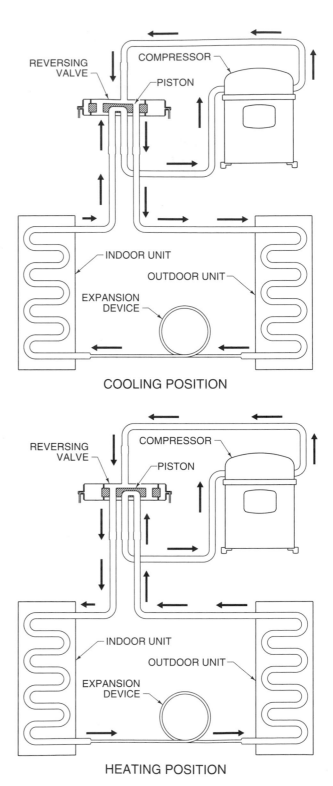

Figure 12-4. The position of the piston in the cylinder of a reversing valve determines where the refrigerant will flow.

A reversing valve is controlled by a solenoid- operated pilot valve. A *solenoid-operated pilot valve* is a small valve in the refrigerant line that contains a solenoid. The solenoid is operated by a low-voltage electric signal from a thermostat. The solenoid-operated pilot valve controls the reversing valve. The reversing valve is operated by refrigerant pressure. A solenoid-operated pilot valve is located near the reversing valve.

When an electric signal actuates the solenoid on the pilot valve, the solenoid moves a small piston back and forth in the pilot valve cylinder. A capillary tube runs from the pilot valve cylinder to the suction line on the reversing valve. Two other capillary tubes run from the pilot valve cylinder to each end of the reversing valve cylinder. See Figure 12-5.

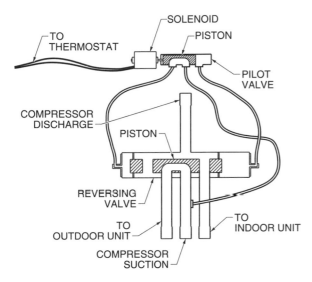

Figure 12-5. Capillary tubes connect the pilot valve to the reversing valve.

When the solenoid moves the pilot valve piston to the left side of the pilot valve cylinder, the capillary tube that runs to the left side of the reversing valve opens for refrigerant flow between the left side of the reversing valve and the pilot valve. The refrigerant flows from the pilot valve to the compressor suction line. The flow of refrigerant reduces the pressure in the left side of the reversing valve. The refrigerant pressure on the right side of the reversing valve pushes the reversing valve piston to the left side of the cylinder. See Figure 12-6.

When the solenoid moves the pilot valve piston to the right side of the pilot valve cylinder, the capillary tube that runs to the right side of the reversing valve opens for refrigerant flow between the right side of the reversing valve and the pilot valve. The refrigerant flows from the pilot valve to the com-

pressor suction line. The flow of refrigerant reduces the pressure in the right side of the reversing valve. The refrigerant pressure on the left side of the reversing valve pushes the reversing valve piston to the right side of the cylinder.

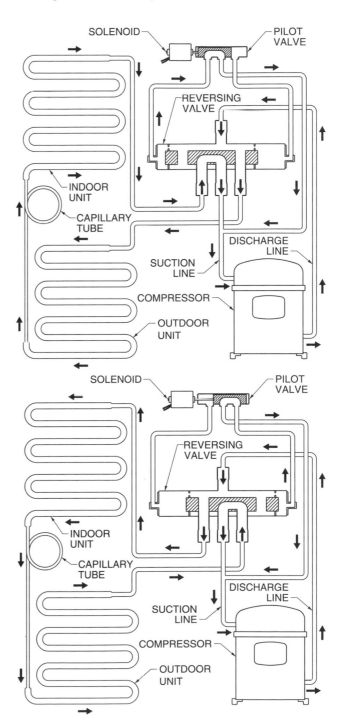

Figure 12-6. When a pilot valve releases refrigerant to a heat pump system, the pressure of the refrigerant moves the piston of the reversing valve.

Expansion Devices

The coil in the indoor unit and the coil in the outdoor unit each require an expansion device because the refrigerant in a heat pump flows in either direction in the heat pump. In small package heat pumps where the coils are close together, a capillary tube may be used as an expansion device. Some package heat pumps have refrigerant control valves, which are a combination check valve and orifice, as expansion devices. Large heat pumps, especially split systems, use thermostatic expansion valves as expansion devices.

Capillary Tubes. A capillary tube produces a pressure decrease regardless of the direction of refrigerant flow. Filters are located at each end of the capillary tube because the refrigerant must be filtered. The filters remove small contaminant particles but allow liquid refrigerant to pass. When a capillary tube expansion device is used in a heat pump system, a bypass circuit may be installed around the capillary tube because the refrigerant must be filtered. The capillary tube has a tube inside it to compensate for the higher pressures that occur in the heating mode. A capillary tube is used only on small package or split heat pump systems that have cooling capacities from 1 ton to 5 tons of cooling.

Refrigerant Control Valves. A *refrigerant control valve* is a combination expansion device and check valve. A refrigerant control valve has a ball inside the check valve that moves back and forth inside the valve housing. The ball moves in the same direction as the refrigerant flow. The ball partially closes the orifice in the end of the housing by moving to that end of the valve. One orifice is larger than the other to provide a pressure decrease for the refrigerant as refrigerant flows through the valve. Refrigerant control valves are simple and relatively trouble-free. See Figure 12-7.

Thermostatic Expansion Valves. Large heat pumps and split systems have thermostatic expansion valves, which are used as expansion devices. A thermostatic expansion valve is a valve that is controlled by pressure.

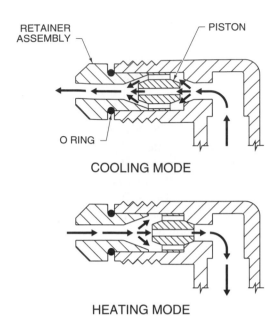

Figure 12-7. A refrigerant control valve is an expansion device that controls the flow of refrigerant in either direction.

In a thermostatic expansion valve, pressure from a remote bulb exerted against a diaphragm determines the size of the outlet. A pushrod connects the diaphragm in the top of the valve to a ball inside the valve body. If pressure on the diaphragm decreases, the ball will close the valve. If pressure on the diaphragm increases, the ball will open the valve. The pressure is exerted by a refrigerant-filled remote bulb. The bulb is attached by a capillary tube to the refrigerant line that leaves the outdoor coil. The temperature of the refrigerant leaving the outdoor coil is sensed by the refrigerant in the bulb. See Figure 12-8.

If the temperature of the refrigerant leaving the outdoor coil increases, the refrigerant in the bulb will vaporize and increase the pressure in the capillary tube. This pressure increase opens the thermostatic expansion valve, which increases the flow of refrigerant to the outdoor coil. If the temperature of the refrigerant leaving the outdoor coil decreases, the refrigerant in the bulb will condense and decrease the pressure in the capillary tube. This decrease in pressure closes the thermostatic expansion valve, which decreases the flow of refrigerant to the coil.

In large heat pumps, a thermostatic expansion valve is located on each coil. The valves are necessary because the indoor coil is the evaporator in the cooling mode and the outdoor coil is the evaporator in the heating mode. A thermostatic expansion valve is located near each coil at the end of the liquid line.

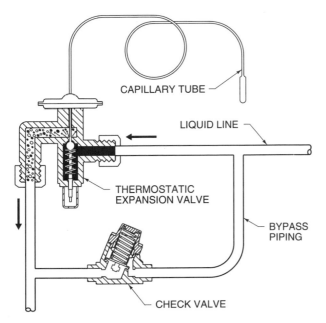

Figure 12-8. Large heat pumps and split systems have two thermostatic expansion valves. Bypass piping with a check valve allows refrigerant flow around the valve when flow is in the opposite direction.

Because refrigerant cannot be reversed in a thermostatic expansion valve, bypass piping is installed around each expansion device. The bypass piping has a check valve, which prevents the liquid refrigerant from flowing back through the expansion valve by allowing the refrigerant to flow through the bypass. The thermostatic expansion valve at the outdoor coil is bypassed when the system operates in the cooling mode, and the expansion valve at the indoor coil is bypassed when the system operates in the heating mode.

AUXILIARY HEAT

An auxiliary heat source is used during cold weather because a heat pump that is sized for cooling a building in summer may not have a heat output that is adequate for heating the building in winter. The heat output of a heat pump is directly related to the amount of heat available in a heat sink.

Auxiliary Heat Source

In cold weather the temperature of air falls and the amount of heat required to heat a building increases. If a heat pump is sized to manage the heat requirement of a building at the coldest temperatures, it will be oversized for cooling. To prevent this problem, the heat pump is sized to provide the heat required to heat a building to an outdoor air temperature of about 35°F. An auxiliary heat source increases the heat output of the heat pump at temperatures below 35°F. Sources of auxiliary heat for heat pumps are resistance heating elements and gas fuel- or fuel oil-fired heaters.

Resistance Heating Elements. Resistance heating elements are used most often as an auxiliary heat source for a heat pump. Electricity is a convenient energy source for auxiliary heat because most heat pumps are powered by electricity. See Figure 12-9.

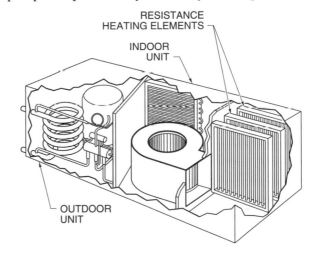

Figure 12-9. Resistance heating elements are often used as an auxiliary heat source for heat pump systems.

The resistance heating elements used for auxiliary heat in heat pumps are similar to those used in electric furnaces. The elements consist of a grid of wire coils in porcelain insulators suspended in a frame. The coils are made of nichrome wire, which has high resistance to the flow of electricity. When the coils are connected to an electrical circuit and the power is turned ON, the coils get very hot. Air is heated as it is blown across the coils. The resistance heating elements in a heat pump are installed downstream from the indoor coil.

Gas Fuel- or Fuel Oil-fired Heaters. Gas fuel- or fuel oil-fired heaters can provide auxiliary heat for heat pumps. Gas fuel- or fuel oil-fired heaters are used where a heat pump is added to an existing heating system, and the indoor coil can be placed on the supply side of the furnace. See Figure 12-10.

When a heat pump coil is used on the outlet side of an existing heating system, the controls must shut the heat pump down when the auxiliary heat is ON. The indoor coil should not be ON when the furnace is producing heat. Added heat causes excessive pressure in the heat pump system.

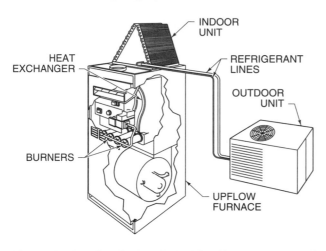

Figure 12-10. Gas fuel- or fuel oil-fired heaters are used where a heat pump is added to an existing heating system.

CONTROLS

Heat pumps have power, operating, and safety controls that are similar to the controls on a refrigeration system. Heat pumps include a heat pump thermostat, defrost controls, and an auxiliary heat source that has separate controls.

Thermostats

A *heat pump thermostat* is a component that incorporates a system switch, heating thermostat, and cooling thermostat. The heat pump thermostat switches the heat pump from the cooling mode to the heating mode and controls system operation in either mode. In the heating mode, the thermostat controls the reversing valve, compressor, and auxiliary heat source.

Different kinds of thermostats are used for controlling heat pumps. Some thermostats are switched manually from the cooling to heating mode while others switch automatically. Most heat pumps require a multibulb, multistage thermostat. A *multibulb thermostat* contains more than one mercury bulb switch. A *multistage thermostat* contains several mercury bulb switches that make and break contacts in stages. When a multibulb, multistage thermostat is connected to two different devices, the devices are energized at different times depending on changing temperatures.

For example, a thermostat used to control a heat pump may have two stages. In the cooling mode, the first-stage mercury bulb controls the reversing valve, and the second-stage mercury bulb controls the compressor. When the temperature increases, the first-stage mercury bulb switches the system to the cooling mode, and the second-stage mercury bulb energizes the compressor to provide cooling.

For the same heat pump system in the heating mode, the first-stage mercury bulb controls the compressor, and the second-stage mercury bulb controls auxiliary heat. As the temperature falls, the first-stage mercury bulb switches the compressor ON to provide heat. If the temperature continues to fall, the second-stage mercury bulb will switch the auxiliary heat ON. See Figure 12-11.

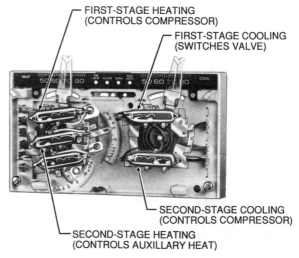

Honeywell Inc.

Figure 12-11. Most heat pump systems require a multibulb, multistage thermostat.

An auxiliary heat source is normally controlled by a second-stage mercury bulb in the heat pump thermostat. The first-stage bulb operates the heat pump. If the heat from the heat pump does not satisfy the thermostat, the second-stage bulb will close and turn ON the auxiliary heat source.

Defrost Controls

Defrost controls regulate the defrost cycles. A *defrost cycle* is a mechanical procedure that consists of reversing refrigerant flow in a heat pump to melt frost or ice that builds up on the outdoor coil. When an air-to-air heat pump operates in the heating mode, the outdoor coil functions as an evaporator. The temperature of the outdoor coil is usually below 32°F. At this temperature, moisture in the outdoor air freezes and forms frost or ice on the surface of the outdoor coil. The frozen moisture blocks air flow across the outdoor coil.

When the flow of refrigerant in the heat pump system is reversed by temporarily switching the system to the cooling mode, hot refrigerant gas flows through the outdoor coil. The cooling mode is activated when the position of the reversing valve is switched. The outdoor fans are shut OFF, and the auxiliary heat is turned ON. This action raises the temperature of the hot gas, which flows through the outdoor coil and melts the frost or ice. The air blown into the building is cooled slightly during a defrost cycle.

Most defrost control systems are based on time cycles. A timer in the defrost control system calls for defrost at regular time intervals. If defrost is required, an initiation sensor signals a need for defrost. The initiation sensor may sense either the temperature of the air leaving the outdoor coil or the temperature of the refrigerant leaving the outdoor coil by sensing pressure in the outdoor coil. If defrost is not needed when the timer signals, the initiation sensor will prevent the system from going into a defrost cycle.

During a defrost cycle, a termination sensor senses the temperature of the outdoor coil either directly or through refrigerant pressure. If the temperature or pressure indicates that the frost or ice has melted, the cycle will be terminated.

Time. Time is often used to initiate a defrost cycle. A timer in the defrost control system runs con-

tinuously. At intervals of 45, 60, or 90 minutes, the timer calls for a defrost. The initiation sensor, which is located on the outdoor coil, determines if a defrost cycle is needed and activates the defrost control system. See Figure 12-12.

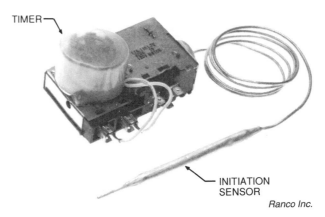

TIMER

INITIATION SENSOR

Ranco Inc.

Figure 12-12. Time is the most commonly used method of initiating a defrost cycle.

Temperature. The temperature of the air that enters and leaves the outdoor coil is used to determine when a defrost cycle should begin and end. If frost or ice begins to build up on the outdoor coil, the temperature difference across the coil will decrease. A temperature-actuated defrost control is mounted on the discharge side of the coil. A *temperature-actuated defrost control* consists of a remote bulb thermostat and an electric control switch. The switch is wired into the defrost control. The temperature-actuated defrost control sends a signal to the defrost control, which monitors temperature. If the temperature indicates a need for defrost, a defrost cycle will be initiated. As the frost or ice on the coil is melted by the hot gas, the temperature of the refrigerant leaving the coil increases. This temperature increase is sensed by the thermostat, and the temperature-actuated defrost control sends a signal to the defrost control to terminate the defrost cycle. See Figure 12-13.

Pressure. The pressure of the refrigerant in a refrigeration system can be used to determine when a defrost cycle should begin and end. At saturation conditions, the refrigerant pressure in an outdoor coil represents the temperature of the refrigerant because of the pressure-temperature relationship.

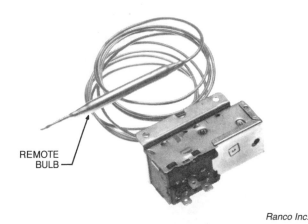

Ranco Inc.

Figure 12-13. A temperature-actuated defrost control mounted on the coil in the outdoor unit senses the temperature of the coil.

The temperature of the outdoor coil is monitored by monitoring refrigerant pressure. A tapping valve is placed on one of the refrigerant lines in the outdoor coil, usually on a return bend. A *tapping valve* is a valve that pierces a refrigerant line. A tube from the tapping valve leads to a pressure switch. If the coil begins to frost, the pressure of the refrigerant in the coil will decrease. If the pressure decreases enough to indicate a freezing temperature, the pressure switch will activate the defrost control. See Figure 12-14.

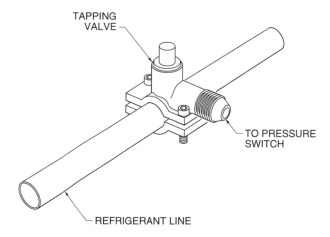

Figure 12-14. The pressure of the refrigerant in a system may determine the proper time to begin and end a defrost cycle.

Auxiliary Heat Control

Auxiliary heat for a heat pump is controlled by the thermostat. On a call for heat, the reversing valve

piston is in the heating mode position. The first-stage heating bulb of the thermostat operates the compressor, which provides heat from the heat pump. If heat pump output does not raise the temperature in the building to the thermostat setpoint, the second-stage heating bulb will call for auxiliary heat. The auxiliary heat source is actuated by the heating relay when the second-stage bulb closes.

To ensure that the heat pump heats as much as possible and that auxiliary heat is used only when necessary, an outdoor thermostat is often used with the regular thermostat to control the amount of auxiliary heat produced. The outdoor thermostat has a set of contacts that close when the temperature falls below the outdoor thermostat setpoint. If the outdoor temperature is above the setpoint temperature, the auxiliary heat will not turn ON.

The auxiliary heat is also controlled by the defrost control system. When the heat pump switches to a defrost cycle, the auxiliary heat is turned ON to heat the air that is blown into the building.

OPERATION

When a heat pump operates in the cooling mode, the refrigerant vapor is compressed in the compressor. The compressed refrigerant flows through the hot gas discharge line to the reversing valve. The refrigerant flows through the reversing valve to the outdoor coil. In the outdoor coil, the refrigerant condenses to a liquid as it rejects heat to the condensing medium. The refrigerant flows through the liquid line where the pressure of the refrigerant decreases as the liquid refrigerant flows through an expansion device. The refrigerant then flows to the indoor coil.

In the indoor coil, the refrigerant vaporizes as it rejects heat to the evaporating medium. The refrigerant leaves the indoor coil through the suction line and returns to the reversing valve. The refrigerant flows through the reversing valve to the compressor. See Figure 12-15.

A heat pump contains R-22. When operating in the cooling mode, the heat pump has a high-pressure side pressure of 265 psig and a low-pressure side pressure of 80 psig. In the cooling mode, the outdoor coil is the high-pressure side and the indoor coil is the low-pressure side.

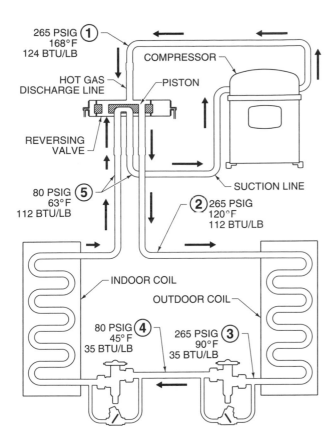

Figure 12-15. When a heat pump is in the cooling mode, the refrigerant vapor flows from the compressor, through the reversing valve, and into the outdoor coil.

1. The refrigerant leaves the compressor with a pressure of 265 psig, a temperature of 168°F, and an enthalpy of 124 Btu/lb.

2. As the refrigerant enters the outdoor coil, it is still at a pressure of 265 psig. The temperature has decreased to 120°F, and the enthalpy is 112 Btu/lb. The temperature and enthalpy decreased because of sensible heat loss in the hot gas line between the compressor and the outdoor coil.

3. The refrigerant leaves the outdoor coil with a pressure of 265 psig, a temperature of 90°F, and an enthalpy of 35 Btu/lb. The temperature decreases because the refrigerant is subcooled before it leaves the coil in the outdoor unit. The enthalpy decreases because the refrigerant rejects heat to the air as the refrigerant flows through the outdoor coil.

The refrigerant leaves the outdoor coil and enters the expansion device with the same properties. The refrigerant has a pressure of 265 psig, a temperature of 90°F, and enthalpy of 35 Btu/lb. After the refrigerant flows through the expansion device, the pressure has decreased to 80 psig.

4. The temperature is 45°F, and the enthalpy is still 35 Btu/lb. The pressure decrease is caused by the restriction in the expansion device. The temperature decreases because some of the refrig erant vaporizes.

5. The refrigerant leaves the indoor coil with a pressure of 80 psig, temperature of 63°F, and enthalpy of 112 Btu/lb. The temperature and enthalpy increase because the refrigerant in the coil in the indoor unit absorbs heat from the air that surrounds the coil.

The refrigerant leaves the indoor coil and enters the compressor with the same properties. The pressure, temperature, and enthalpy increase as the refrigerant flows through the compressor. The temperature and enthalpy increase because of the heat of compression and heat from the compressor and motor.

When a heat pump operates in the heating mode, the refrigerant flows from the reversing valve to the indoor coil. The refrigerant flows through the indoor unit, liquid line, and expansion device. From the expansion device, the refrigerant flows through the outdoor unit and returns to the reversing valve.

For example, when a heat pump that contains R-22 operates in the heating mode, the high-pressure side pressure is 260 psig and the low-pressure side pressure is 33 psig. In the heating mode, the indoor unit is the high-pressure side and the outdoor unit is the low-pressure side. See Figure 12-16.

1. As the refrigerant leaves the compressor, it has a pressure of 260 psig, a temperature of 200°F, and an enthalpy of 130 Btu/lb of refrigerant.

2. The refrigerant enters the indoor unit with a pressure of 260 psig, a temperature of 120°F, and an enthalpy of 113 Btu/lb. The temperature and enthalpy decrease because of heat losses in the hot gas discharge line and the reversing valve.

3. The refrigerant leaves the indoor coil with a pressure of 260 psig, a temperature of 60°F, and an enthalpy of 27 Btu/lb. The temperature and enthalpy decrease because heat is rejected to the air in the indoor coil.

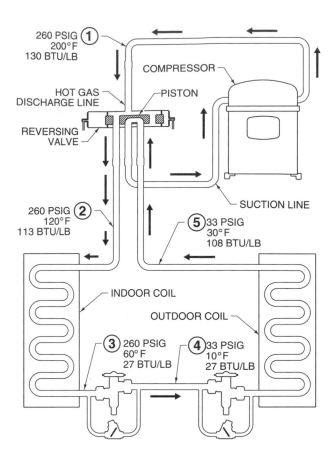

Figure 12-16. When a heat pump is in the heating mode, the refrigerant vapor flows from the compressor, through the reversing valve, and into the indoor coil.

4. The refrigerant enters the expansion device at the outdoor coil with approximately the same conditions as the refrigerant that leaves the indoor coil. The refrigerant leaves the expansion device at the outdoor coil and enters the evaporator. The refrigerant has a pressure of 33 psig, a temperature of 10°F, and an enthalpy of 27 Btu/lb. The pressure and temperature decrease because of the expansion device. The pressure decreases because of the restriction in the expansion device. The temperature decreases because some of the liquid refrigerant vaporizes.

5. The refrigerant leaves the outdoor coil at a pressure of 33 psig, a temperature of 30°F, and an enthalpy of 108 Btu/lb. The increase in temperature and enthalpy are due to heat absorbed from the heat sink in the outdoor coil.

The refrigerant enters the compressor at about the same conditions as it leaves the outdoor coil.

As the refrigerant flows through the compressor, the pressure, temperature, and enthalpy increase. The temperature and enthalpy increase because of the heat of compression and heat from the compressor motor.

HEAT PUMP SELECTION

A heat pump is selected for a given application based on the available source of heat, the type of heat pump for the application, and the cooling and heating capacity of the heat pump.

Sizing

A heat pump selected for cooling and heating a building located in a warm climate is sized to accommodate the cooling load for the building. In a warm climate, the heating capacity of a heat pump is great enough to handle the heating load.

A heat pump selected for cooling and heating a building located in a cold climate may require auxiliary heat to meet the required heating capacity. In this case, the heating capacity of the unit at the design temperature for the area should be checked against the heat loss of the building at the same design temperature to determine the amount of auxiliary heat required.

Balance Point Temperature. *Balance point temperature* is the temperature at which the output of a heat pump balances the heat loss of a building. The balance point temperature for a heat pump used to heat and cool a particular building is found on a graph. The graph has the heat loss of the building at different outdoor temperatures plotted against the heat output of the heat pump plotted against the same outdoor temperatures. See Figure 12-17.

The heat loss of the building is plotted from the point that represents the total heat loss of the building at the outdoor design temperature. A line is drawn from this point to the base line of the graph at the outdoor air temperature at which the heat loss would be 0 Btu/hr. The heat loss is usually about 5°F less than the indoor design temperature for the building.

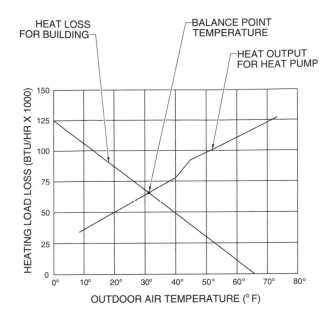

Figure 12-17. Balance point temperature is the temperature at which the output of a heat pump balances the heat loss of a building.

After the building heat loss line is drawn on the graph, the heat output for a heat pump is drawn by using heat output data from the heat pump data sheet. The data sheet shows that the heat output for a heat pump varies with outdoor temperature.

The heat output for a heat pump is found for several different outdoor temperatures, and the points are plotted on the graph. A line is drawn to connect the points. The line represents the heat output for the heat pump over a range of temperatures. Lines can be drawn for several sizes of heat pumps on the same graph. Some heat pump manufacturers provide output graphs drawn for their heat pumps.

The point on the graph where the heat loss line for the building crosses the heat output line for the

heat pump is the balance point temperature for the heat pump. At this temperature, the heat output from the heat pump balances the heat loss from the building. See Figure 12-18.

At temperatures above the balance point temperature, a heat pump heats the building without requiring auxiliary heat. At temperatures below the balance point temperature, auxiliary heat must be used with the heat pump to heat the building.

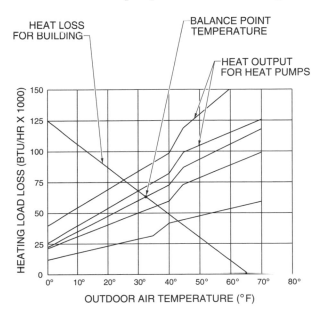

Figure 12-18. At the balance point temperature for the heat pump, the heat output from the heat pump balances the heat loss from the building.

The amount of heat that must be provided by the auxiliary heat source is found on the scale at the left of the graph for any outdoor temperature by determining the difference between the two lines at that temperature (in Btu per hour).

Review Questions

1. How does a heat pump system operate differently than an air conditioning system?

2. Define *heat sink*. How is a heat sink used with a heat pump system?

3. What two substances are used commonly as heat sinks?

4. List and describe the four classifications of heat pump systems.

5. In what types of buildings are air-to-air heat pump systems used?

6. In what types of buildings are water-to-water heat pump systems used?

7. What components are found on a heat pump system that are not normally found on an air conditioning system?

8. What is the primary function of the reversing valve in a heat pump system?

9. Define and describe *solenoid-operated pilot valve*.

10. Why can a capillary tube be used as an expansion device regardless of the direction of refrigerant flow?

11. Why is the orifice at one end of a refrigerant control valve larger than the orifice at the other end of the valve?

12. Why is bypass piping required with thermostatic expansion valves on a heat pump system?

13. Why is auxiliary heat required when a heat pump is used to heat a building?

14. Can a heat pump be used with auxiliary heat other than electric resistance heat? Explain.

15. Describe a heat pump thermostat. How does it control the heat pump during both the cooling and the heating modes?

16. What three main parts of a heat pump system does a heat pump thermostat control?

17. Name the functions of each bulb when a four-bulb, two-stage heat and two-stage cooling thermostat is used for controlling a heat pump.

18. What three components of a heat pump system are controlled directly during a defrost cycle?

19. Describe the operation of the time clock in a defrost control.

20. Describe the function of an initiation sensor in a heat pump system.

21. Describe the function of the termination sensor in a heat pump system.

22. How does refrigerant temperature actuate a defrost cycle?

23. How does refrigerant pressure actuate a defrost cycle?

24. How does the pressure, temperature, and enthalpy of the refrigerant at different points in a heat pump system show how heat is transferred in the system?

Control Systems

Chapter **13**

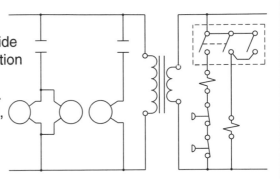

A control system regulates the operation of the various components of a heating and/or cooling system to provide comfort in a building. The components are the combination of devices that may include all or part of the heating, ventilating, and air conditioning equipment in a building. Three types of control systems are electrical, electronic, and pneumatic control systems.

CONTROL SYSTEM FUNCTIONS

A control system links the heating, ventilating, and air conditioning equipment to provide a comfortable climate in a building. Control systems contain sensors and operators that use signals and feedback to control the operation of equipment.

Sensors

A *sensor* is a device that changes size, shape, or resistance due to a change in conditions. Conditions are temperature, humidity, or pressure. A sensor determines existing conditions and compares them with desired conditions. *Existing conditions* are conditions sensed by the sensor. *Desired conditions* are the setpoint conditions. Sensors include thermostats, aquastats, and pressurestats. The sensor sends a signal to the operator through the connecting circuitry to indicate that an action should be taken. Signals are generated by the sensors depending on the kind of system, the operators, and feedback signals. See Figure 13-1.

Operators

An *operator* is a mechanical device that switches the heating, ventilating, and air conditioning equip-

ment ON or OFF. Operators include relays, contactors, solenoids, and primary control systems. A *primary control system* is a combination of operating and safety controls for power burners. For example, in a heating system, the mercury bulb switch in a thermostat (sensor) closes on temperature fall. This sends a signal to a relay (operator) which switches the furnace burner ON.

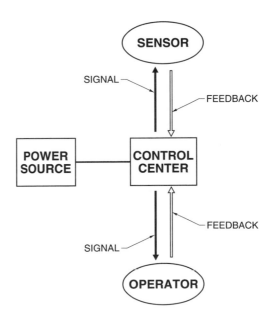

Figure 13-1. A control system links the heating, ventilating, and air conditioning equipment to provide comfort in a building.

Control Signals

A *control signal* is the medium used to communicate between the sensor and operator. AC electricity, DC electricity, and air are used as control signals. An *electrical control system* is a control system that uses AC electricity as a control signal. *AC (alternating current) electricity* is electric current that is continuously changing directions. Electrical control systems are used in residential heating, ventilating, and air conditioning systems. An *electronic control system* is a control system that uses low-voltage DC electricity as a control signal. *DC (direct current) electricity* is electric current that flows in the same direction. Electronic control systems are used in residential and commercial applications that have a high degree of sophistication. A *pneumatic control system* is a control system that uses compressed air pressure as a control signal. Pneumatic control systems are used where control of explosion hazards is required, such as where flammable chemicals are used near electricity.

Feedback

Feedback is the change in a signal. Feedback indicates that the conditions at a control point satisfy setpoint conditions. The *control point* is the point at which the sensor for the control system measures conditions. A heating unit must be turned ON when a building requires heat and shut OFF when the building reaches the setpoint conditions. In a heating unit control system, the control circuit is normally closed on a call for heat and open when no heat is required. The open circuit is feedback that indicates that the setpoint conditions are satisfied.

ELECTRICAL CONTROL SYSTEMS

An electrical control system uses electricity as the control signal. Control devices control electric current in electrical control systems. The connections between the components of the system are electric conductors. Electrical control systems may use any voltage, but low-voltage AC electricity is most common. Low voltage allows the use of lightweight conductors, terminals, and contacts in the control

devices. Low voltage also provides more accurate control of the heating and cooling produced.

Standard electrical symbols identify controls used in control systems. These symbols are used in control system drawings and schematic diagrams. A *schematic diagram* is a diagram that uses lines and symbols to represent the electrical circuits and components in an electrical control system. Schematic diagrams make understanding control systems easier and make the relationships between the components clearer. See Figure 13-2. Different symbols are used for electronic and pneumatic control systems.

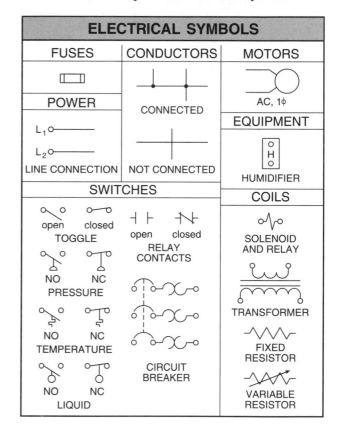

Figure 13-2. Standard electrical symbols are used in schematic diagrams.

Heating Control System

A *heating control system* controls the temperature in a building by cycling heating equipment ON and OFF to maintain the temperature of a building within a few degrees of a setpoint. The sensor in a heating control system is a thermostat. A thermostat senses temperature and sends a signal to an operator. The operator turns the heating unit ON or OFF. A thermostat contains a setpoint adjustor lever or dial that is used to set the desired setpoint temperature. The control point may be indoors, outdoors, or in a controlled medium such as boiler water. A control system may sense conditions in one area or any number of areas.

In a heating control system, when the temperature at the thermostat falls below the setpoint, the thermostat switch closes. When the thermostat switch closes, a signal is sent to an operator, which turns the heating unit ON. The signal is sent to the operator, which indicates that the setpoint conditions are not met. When the temperature at the thermostat rises above the setpoint, the thermostat switch opens, which shuts the signal to the operator OFF. The absence of a signal to the operator is feedback, which indicates that setpoint conditions are met. The operator may be a gas valve solenoid, relay, or primary control.

The basic heating control system consists of power, operating, safety controls, and conductors. Power controls control the flow of electricity to a heating unit. Operating controls cycle equipment ON or OFF. Safety controls monitor the operation of the heating unit.

Power Controls. Power controls are installed in an electrical circuit between the power source and the heating unit. Power controls should be installed per article 424 of the National Electrical Code®. Power controls should be installed by a licensed electrician.

The electrical power circuit to a heating unit contains at least two disconnects. A disconnect is a manual switch that shuts OFF the current to a heating unit. One disconnect is located at the electrical control panel and the other is located at the heating unit. Overcurrent protection devices such as fuses or circuit breakers are placed in the electrical circuit to protect the conductors in an electrical circuit from excessive current flow. See Figure 13-3. If overcurrent conditions occur, the conductor will overheat and become a fire hazard.

Operating Controls. Operating controls on a residential heating unit include the transformer, relays, and gas valve solenoid. A transformer is an electric device that lowers the voltage in an electrical circuit

to the voltage required in the control circuit. Most control circuit transformers are step-down transformers with a secondary voltage of 24 V and are installed at the factory as part of the basic system control package.

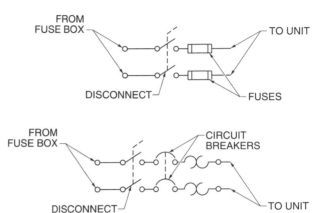

Figure 13-3. In an electrical circuit, fuses and circuit breakers protect the conductors in the circuit from excessive current flow.

On a gas-fired furnace, the primary side of the transformer is connected to the electrical power source. The secondary side of the transformer is connected to blower relay coils. See Figure 13-4.

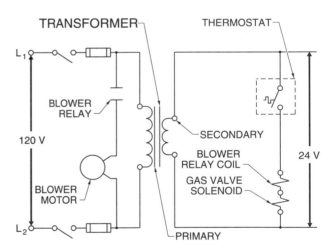

Figure 13-4. The primary side of a transformer is connected to the power source and the loads in the system. The secondary side of the transformer is connected to the thermostat and devices in the system that control the loads.

One low-voltage lead runs from the secondary side of the transformer to the thermostat in the system. The other lead from the secondary side of the transformer runs to devices in the system that control the loads. These devices may be a relay coil, contactor coil, or gas valve solenoid. The circuit is completed by running a conductor from the thermostat to the relay coil, contactor coil, or solenoid.

A relay is a device that controls the flow of electric current in one circuit (load circuit) with the electric current in another circuit (control circuit). Relays use a magnetic coil to open or close one or more sets of contacts. The magnetic coil is connected in the control circuit and the contacts are connected in the load circuit. One side of each set of contacts is fixed and the other side is movable and is controlled by the magnetic coil. The moving contacts in a normally open relay are held open by a spring. When the coil is energized by a signal from a sensor, the contacts close and the load circuit is energized.

Safety Controls. A limit switch is a temperature-actuated electric switch. Limit switches have normally closed contacts that open on temperature increase. Limit switches shut down the burner on a heating unit if the unit becomes overheated. Limit switches sense air or water temperature. A limit switch is wired in the electrical control circuit to the burner control. See Figure 13-5.

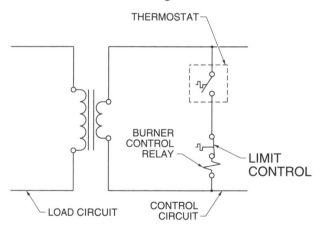

Figure 13-5. Limit controls sense air or water temperature. If a furnace becomes overheated, limit controls shut OFF the burner.

A stack switch contains a bimetal element located inside the flue at the furnace opening. A stack switch senses flue-gas temperature and converts it to mechanical motion that controls a set of contacts. On a call for heat, current flows through a safety switch

heater in the burner control circuit. The safety switch heater is wired to a set of cold contacts (normally closed contacts). The current flow through the cold contacts allows the fuel valve to remain open.

If the temperature of the flue gas does not increase in a reasonable time, the safety switch heater will open the cold contacts. Opening the cold contacts opens the burner control circuit. If ignition does occur, the bimetal element will expand because of the hot flue gas. This action moves a metal rod that closes a set of hot contacts (normally open contacts) in the burner control circuit. See Figure 13-6.

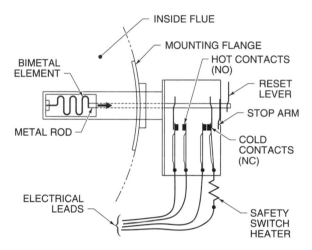

Figure 13-6. A stack switch contains a bimetal element that senses flue-gas temperature and converts the temperature to mechanical motion.

A flame surveillance control is an electronic combustion safety control that consists of a light-sensitive device, which detects flame. The device, which is called a cad cell, is aimed at the combustion chamber to detect light when the flame is ON. Cadmium sulfide is a material in which the resistance to electric current varies as the intensity of the light striking the material varies. When a cad cell is exposed to light, the resistance to the flow of electricity through the cell is low. When the cad cell is in darkness, the resistance through the cell is high. Current flow through the cell is monitored by the control center to determine if the cad cell is detecting light.

The cad cell is located in a burner in direct line-of-sight with the flame from the burner. If the burner fires and a flame is established on a call for heat, the resistance through the cad cell will be low. The low resistance allows the electrical signal to pass to the control center and the unit continues firing.

If firing does not occur, the resistance through the cad cell will be high and prevent the electrical signal from reaching the control center. After a reasonable length of time, the control shuts the unit OFF.

A flame rod is an electronic combustion safety control used on large commercial furnaces. A flame rod uses the flame as an electric conductor. When the unit is firing, a control device sends out a low-voltage electrical signal to a rod located in the flame. When the flame is established, the flame forms the other side of the circuit. See Figure 13-7.

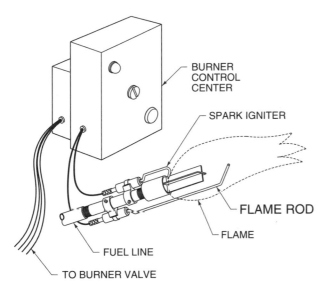

Figure 13-7. A flame rod combustion safety control uses flame as an electric conductor.

On a call for heat, if a flame is established, the electrical safety circuit is made through the flame. If a flame is not established within a reasonable length of time, the electronic devices in the burner control center will prevent the heating control from actuating. The heating control may be a gas valve, oil burner control, or resistance heating relay.

Air Flow Control System

Air flow control is the control of the circulation of air through building spaces and the introduction of ventilation air into a building. Air flow control is achieved by actuating the blower(s) in a furnace for constant air circulation or dampers for the regulation of ventilation air.

Forced-air heating units require air flow through the unit. This air flow is produced by the furnace blower and is used for circulating heated air to building spaces. A blower control is a temperature-actuated electric switch that controls the blower motor in a furnace.

The blower control system is a line voltage control. One conductor from the electrical supply runs to one side of the blower control. A conductor from the other side of the blower control runs to one side of the blower motor. A conductor from the neutral side of the electrical supply runs to the other side of the motor. A blower control on a furnace consists of a bimetal element that operates an electric switch. The electric switch closes when the temperature increases and opens when the temperature decreases. Once the thermostat is satisfied, the blower relay coil is de-energized. This action would normally shut the blower OFF, but the blower control allows the blower to remain ON to cool the heat exchanger. See Figure 13-8.

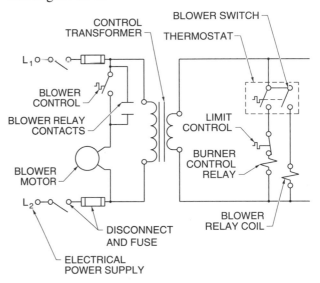

Figure 13-8. In a blower control, the bimetal element is wired from the electrical power circuit to the blower motor.

Temperature Control System

The two types of temperature control systems are digital control systems and modulating control systems. In a digital control system, the sensor either energizes or de-energizes the operator. Digital temperature control systems are used in residential and small commercial applications where close temperature control is not required. A modulating control system regulates the operation of a controlled device in proportion to changes in existing conditions, which are sensed by the sensor. Modulating control systems can provide precise temperature and humidity control.

Digital Control System. A *digital control system* is a two-position system. The sensor in a digital control system contains an electric switch that opens or closes a circuit as the sensor indicates changing conditions. An example is a mercury bulb switch in a thermostat. The switch is wired in with an operator in the system. The operator is an electrically operated device such as a valve or damper operator motor, a relay coil, or a solenoid coil. Digital control is adequate for many operations, but cannot achieve close temperature control. See Figure 13-9.

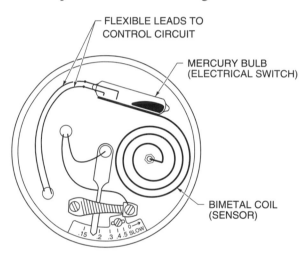

Figure 13-9. The sensor in a digital electrical control system contains an electric switch that opens or closes a circuit as the sensor indicates changing conditions.

Modulating Control System. A *modulating control system* is a control system in which the sensor regulates the operator in proportion to changes in existing conditions. For example, a modulating burner in a furnace is regulated at a percentage of its full capacity in proportion to temperature variations above or below a setpoint.

Electrical modulating control systems use a proportional thermostat that controls a potentiometer. A potentiometer is a coil with a wiper that moves back and forth across the coil. The wiper is controlled by the thermostat current. A potentiometer can be a control device or a feedback device.

The wiper on a control potentiometer is connected to the secondary side of the transformer. The ends of the control potentiometer coil run to the actuator. When the wiper is in the middle of the potentiometer coil, the current is identical at the ends of the potentiometer coil. If the wiper moves to either end of the potentiometer coil, the current at the ends of the potentiometer coil will become unbalanced. A greater current flows in the end of the potentiometer coil closest to the wiper. Because the wiper is controlled by the thermostat current, the current at the ends of the potentiometer coil is proportional to any variation from the setpoint.

The actuator for an electrical modulating control system contains a balancing relay, a bidirectional motor, and a feedback potentiometer. In the balancing relay, a lever arm is attached on a pivot between two balancing relay coils. Each balancing relay coil exerts magnetic force on the lever arm. When the electric current flow through both balancing relay coils is the same, the lever arm remains stationary. If the current flow through one relay coil is greater than the current flow through the other relay coil, the lever arm will tip toward the relay coil with the greater current flow. See Figure 13-10.

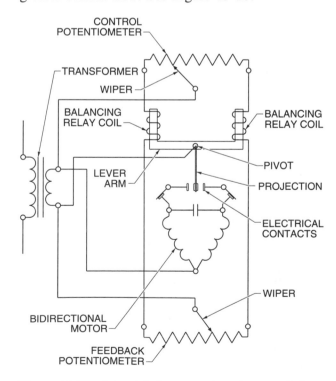

Figure 13-10. A modulating control system controls a device in proportion to changes in the control variable.

A projection extends from the center of the lever arm. One end of the projection has contacts on opposite sides. The contacts are connected to the transformer and are centered between two other contacts. The two other contacts are connected to a bidirectional motor. A *bidirectional motor* is a motor that can turn in either direction. One contact drives the motor in a clockwise direction. The other contact drives the motor in a counterclockwise direction. The bidirectional motor drives dampers or a valve to vary the amount of water or air flow.

The common terminal for the bidirectional motor is connected to the secondary side of the transformer opposite the connection of the center contacts on the balancing relay. This causes the motor to turn clockwise when the lever arm tips in one direction and turn counterclockwise when the lever arm tips in the other direction.

The shaft of the bidirectional motor is connected to a damper or valve and to the wiper on the feedback potentiometer. The wiper on the feedback potentiometer is connected to the transformer opposite the connection of the thermostat potentiometer wiper. The two balancing relay coils are connected to the ends of the feedback potentiometer coil.

As the motor moves a damper or valve, it also moves the wiper across the feedback potentiometer coil. The wiper on the feedback potentiometer coil moves in the opposite direction to the wiper in the control potentiometer coil. The feedback potentiometer wiper moves in the opposite direction just far enough to balance the current flow in the electrical circuits running between the two potentiometers and through the balancing relay coils. When the motor damper blade moves far enough to balance the current flow, the balancing relay contacts are broken. As a result, motor operation is proportionally controlled by the thermostat.

Water Temperature Control. A *water temperature control system* is the group of components that controls the temperature of water in a hydronic heating system. Water temperature control is achieved by controlling the burners in a boiler.

In a hydronic heating system, an aquastat controls the temperature of the water in the boiler. An aquastat is a temperature-actuated electric switch. A bellows that is connected to a remote sensing bulb by a capillary tube actuates the switch. The sensing bulb is charged with a small amount of refrigerant and is located either in a well in the water tank on the boiler or immersed directly in the boiler water.

If the temperature of the water increases, the refrigerant in the bulb will boil off and increase the pressure in the bulb. The increase in pressure pushes the bellows out. If the temperature of the water decreases, the refrigerant in the bulb will condense and decrease the pressure in the bulb. The decrease in pressure moves the bellows in. The movement of the bellows opens or closes the contacts.

If the temperature of the water in the boiler increases above the setpoint of the aquastat, the switch will open. If the temperature of the water falls below the setpoint of the aquastat, the switch will close. The switch is wired into the electrical control circuit that controls the boiler burner or heating elements. See Figure 13-11.

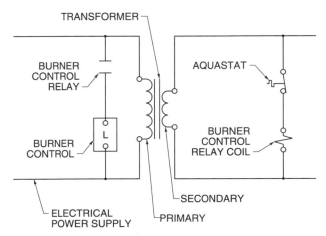

Figure 13-11. In a hydronic heating system, an aquastat controls the temperature of the water in the boiler.

Zone valves are valves used in the piping of a hydronic heating or cooling system to regulate the flow of water. Manual zone valves are set by hand to balance the flow of water in a piping loop. Automatic zone valves contain valve motors that open or close the valve automatically. The valve motor is driven by electricity in an electrical or electronic control system or by air in a pneumatic control system. Zone valve motors are automatically controlled by a zone thermostat.

Zone valve motors are available as ON/OFF or modulating controls. An ON/OFF zone valve motor is a two-position valve. The motor is driven so the valve is completely open or completely closed. A modulating zone valve motor is an infinite position valve. The motor is driven so the valve modulates to the open or closed positions or stays at any intermediate position in response to the control signal it receives.

Humidity Control System

A *humidity control system* controls the amount of moisture in the air to maintain comfort in building spaces. A humidifier controls the level of moisture in the air. When the humidity level falls below a setpoint, the humidifier is activated to add moisture to the air. When the humidity level increases above a setpoint, the outdoor dampers are activated to bring in fresh air or the air conditioning system is activated to dry the air.

The sensor in a humidity control system is a humidistat. A *humidistat* is a device that contains a humidity-sensing element that changes characteristics with changes in humidity. The humidity-sensing element may be hair, manufactured fiber, or a solid-state sensor. On a change in humidity, the humidistat closes a contact, which sends a signal to the coil in the humidity control relay. The coil closes the contact in the humidifying or dehumidifying equipment, which switches ON the equipment. A humidistat is wired into the electrical control circuit of a system. See Figure 13-12.

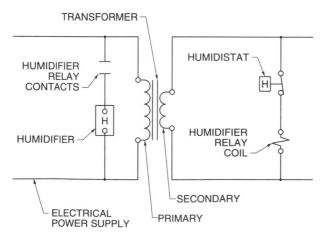

Figure 13-12. A humidity control system controls the amount of moisture in the air to maintain a comfortable climate in building spaces.

Cooling Control System

In an air conditioning system, if the temperature increases above the setpoint, the thermostat switch closes sending a signal to the system operator which turns ON the air conditioning unit. The system operator may be a relay coil or solenoid coil. When the temperature falls below the setpoint, the thermostat switch opens shutting OFF the signal to the system operator.

Power Controls. The electrical power supply for most air conditioning units have two or three conductors in each circuit and each conductor is fused. When circuit breakers are used as overcurrent protection devices, they are wired into the electrical circuits the same way as fuses. One circuit breaker is wired into each line in the circuit. Large air conditioning units use cartridge fuses for protecting electrical circuits.

Operating Controls. Operating controls on a residential air conditioning unit include the transformer, relays, and pressure controls. On an air conditioning unit, the primary side of the transformer is connected to the electrical power source. The secondary side of the transformer is connected to the operators in the unit, which may be a relay, contactor, or magnetic starter. The loads in an air conditioning unit are the evaporator blower motor, condenser blower motor, and the compressor motor.

One low-voltage lead runs from the secondary side of the transformer to the thermostat. The other lead from the secondary side of the transformer runs to devices in the system that control the loads. These may be a relay coil, contactor coil, or magnetic starter coil. The circuit is completed by running a conductor from the thermostat to the relay coil, contactor coil, or magnetic starter coil. The transformer, relays, and magnetic starters are normally installed at the factory as part of the basic control package. The thermostat is field installed. See Figure 13-13.

On small air conditioning units, the thermostat controls the compressor motor which turns ON the motor when the temperature increases above the setpoint. The compressor motor is connected to the condenser blower motor so that the condenser blower runs when the compressor runs. This allows the condenser to reject the heat as soon as the compressor begins to operate. The relay, contactor, or magnetic starter that controls the compressor motor also controls the condenser blower motor so they run simultaneously.

The evaporator blower motor is controlled by a separate relay, contactor, or magnetic starter. The evaporator blower motor relay is energized by a signal from the thermostat. This signal allows the evaporator blower to operate without the compressor or condenser motors being energized. This action allows circulation of the air in the building spaces. Most cooling thermostats are wired internally so the blower terminal is energized when the thermostat calls for cooling or when the blower switch on the thermostat is set to constant operation.

On large air conditioning units, the condenser blower motor may have a separate contactor or magnetic starter, but the control coil on the condenser blower motor is connected to the control coil on the compressor starter. The compressor motor and condenser blower motor still run at the same time.

Safety Controls. High- and low-pressure switches are electric switches installed in the control circuit of an air conditioning unit. The contacts in a pressure switch are opened and closed by a spring-loaded lever arrangement. The lever moves according to the amount of pressure acting on a bellows element.

A *bellows element* is an accordion-like device that converts pressure variation into mechanical movement. A capillary tube connects the bellows element to the section of refrigerant line where the pressure is measured. Pressure is transmitted to the bellows element through the capillary tube. Any change in the refrigerant pressure in the air conditioning unit moves the element. This movement is transmitted to a lever by mechanical linkage. The switch is located in the control circuit that operates the compressor in the air conditioning unit. High pressure in an air conditioning unit can damage the compressor and burn out the compressor motor. Low pressure in an air conditioning unit causes low temperature in the evaporator, which can freeze and damage the evaporator coil. See Figure 13-14.

Some air conditioning units use a pumpdown control system. A *pumpdown control system* is a control system that has a solenoid valve in the liquid line. The solenoid valve shuts OFF refrigerant flow to the evaporator coil when the thermostat is satisfied. The compressor is shut OFF by the low-pressure switch. On a call for cooling, the solenoid valve opens and the pressure switch allows the compressor to start. This control system is used when the evaporator section of the air conditioning unit is in a cold location such as a package unit that is located outdoors. By pumping the refrigerant out of the low-pressure side of the system, the liquid refrigerant in the evaporator will not condense on shut down.

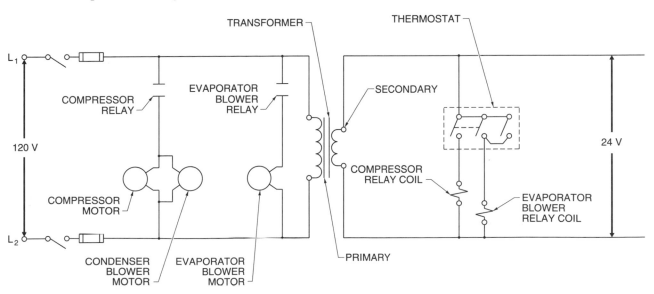

Figure 13-13. In an air conditioning system, the primary side of the transformer is connected to the electrical power source, and the secondary side of the transformer is connected to the control circuits.

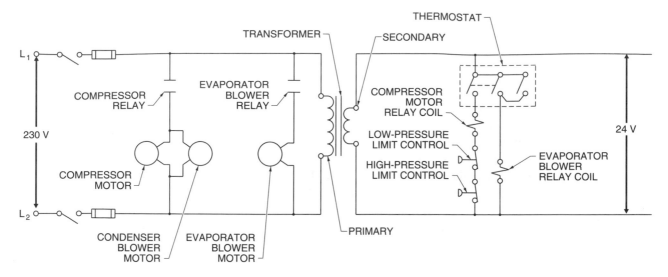

Figure 13-14. High- and low-pressure switches are electric switches located in the control circuit that operates the compressor in an air conditioning unit.

In cooler climates, outdoor air is often cool enough to use for cooling a building. A damper arrangement may be used to bring outdoor air into a building. The outdoor air damper admits outdoor air when the air is cool enough to provide a cooling effect. The outdoor air damper closes when the outdoor air is too warm. An economizer package maintains the desired temperature in the building by controlling the outdoor air dampers and the air conditioning unit.

An *economizer package* is the package of damper controls that brings outdoor air indoors to cool building spaces. An economizer package saves energy by using outdoor air to cool the air in building spaces. The damper arrangement on an economizer package has an outdoor air intake with a damper, a return air damper, and an exhaust damper. The outdoor air intake and the return air damper are linked together so that one opens as the other closes. The exhaust damper is linked to the return air damper so the exhaust damper opens as the return air damper closes. The mixed air controller mixes outdoor and return air to provide comfort in building spaces. If the outdoor air temperature gets too high, the motor on the outdoor air damper will close the outdoor air damper. The motor also locks out the compressor when the outdoor temperature is low enough to cool the building. See Figure 13-15.

An economizer package provides air at a setpoint temperature to building spaces by mixing outdoor

and return air. The air temperature is usually around 55°F to 60°F. If the outdoor air temperature falls below the setpoint temperature, the outdoor dampers will open to admit as much outdoor air as possible. If the outdoor air temperature increases above 80°F, which is the return air temperature, the outdoor air damper will close and the air conditioning equipment will turn ON.

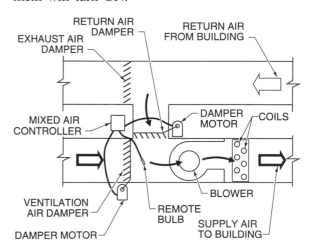

Figure 13-15. An economizer package saves energy by using outdoor air for cooling building spaces.

ELECTRONIC CONTROL SYSTEMS

An electronic control system uses low-voltage DC electric current as the control signal. Small electronic devices and conductors can be used with low-voltage DC current. The control devices are solid-state devices. The low-voltage signal is carried from one device to another by low-voltage conductors. The signal is a variable electric signal. An electronic control system is a modulating control system. Modulation is achieved by variations in the control signal that are proportional to variations in the existing conditions. The existing conditions are sensed by the sensor.

An electronic control system includes sensors, operators, conductors, and a control center that contains solid-state devices. Solid-state devices monitor the low-voltage signals and convert them to AC electric signals. The AC electric signals operate various operators such as relays or contactors.

Sensors used in electronic control systems sense temperature, pressure, air flow, or any condition that must be controlled. Sensing elements within a sensor are solid-state devices such as a thermistor. A thermistor is a temperature-sensing element that changes electrical resistance in response to a temperature change. Diodes, transistors, and other electronic control devices are also used in electronic controls. A *diode* is a solid-state device that allows current flow in only one direction. A *transistor* is a solid-state device that allows current to flow through a primary circuit when a secondary circuit is energized. Combinations of these and other devices are connected on printed circuit boards to achieve the desired control. See Figure 13-16.

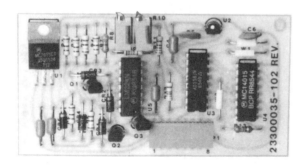

Honeywell Inc.

Figure 13-16. In electronic control systems, combinations of electronic components are connected on printed circuit boards to achieve the desired control functions.

The control center in an electronic control system contains several printed circuit boards containing solid-state devices. AC electrical power is transformed to low-voltage DC power for the system signal. The low-voltage DC signal is sent to the sensor where it is modulated according to the existing conditions. A modulating DC signal is returned to the control center. A *modulating DC signal* is a varying DC signal. The modulating DC signal is identified, compared to other signals, and modified to provide an outgoing signal to the operators in the system. This action is automatic and is performed instantaneously by the solid-state devices in the circuits.

Operators in electronic control systems are similar to operators used in electrical control systems. Special electronic operators have applications specific to an electronic system. The signal from the control center may be either constant or modulating and the operator is selected to respond to the signal. If the signal is a constant-voltage signal, an electrical operator such as a relay, contactor, or solenoid valve may be used. If a system uses a variable electrical signal, a modulating operator may be used.

A self-contained hydraulic motor is a special operator used with electronic control systems. A *self-contained hydraulic motor* is a motor that contains an electronically controlled motor, a hydraulic fluid reservoir, a small pump, and a piston operator. The pump on the motor is actuated by the signal from the control center in the system. Hydraulic pressure developed by the pump moves the piston in its cylinder. The piston is connected by mechanical linkage to the devices it operates.

One advantage of an electronic control system is that the systems are efficient and effective. The control center of an electronic control system can monitor several signals at once, determine actions to be taken in response to the signals, and send the signals to the operators.

The control system selected for a particular application depends on the degree of control sophistication desired. Control systems range from simple ON/OFF (digital) controls for an entire building to very sophisticated systems in which both water and air flow are modulated for individual zones in a building. The level of sophistication required for a given application is determined by the degree of temperature control required. The control system should be designed and installed to achieve that degree of sophistication.

PNEUMATIC CONTROL SYSTEMS

A pneumatic control system uses compressed air as the control signal. Small-diameter tubing carries compressed air between parts of the system. A pneumatic control system is a low-pressure system with a maximum air pressure of about 15 psi. A pneumatic control system is a modulating system. Control is achieved by varying the air pressure in the line between the sensor and the operator.

A pneumatic control system consists of an air source, sensors, and operators. The air source is a compressor and storage tank. Compressed air leaves the air tank and flows through a filter-dryer and a pressure regulator. The pressure regulator is set for the maximum pressure desired in the system. From the regulator, the air is carried to the sensors in the system. The sensors may be thermostats, humidistats, or pressure sensors. See Figure 13-17.

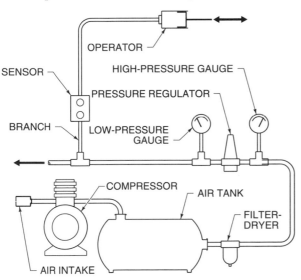

Figure 13-17. A pneumatic control system consists of an air source, sensors, and operators.

Pneumatic sensors modify the air pressure received from the supply line in relation to the existing conditions. For example, a thermostat modifies the pressure in relation to the temperature.

Air with the modified pressure from the sensor is sent to the operator. The operators in a pneumatic system are either bellows, diaphragm, or piston operators. Bellows or diaphragm operators have an element that expands or contracts as control pressure changes. The element is connected mechanically to a valve or other device. A piston operator is a spring-return piston in a cylinder. The piston moves back and forth within the cylinder as control pressure varies. See Figure 13-18.

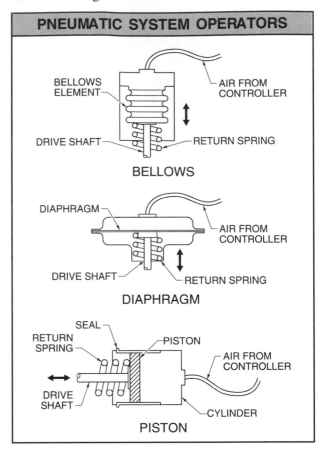

Figure 13-18. The operators in a pneumatic system are bellows, diaphragm, or piston operators.

SENSITIVITY

Sensitivity is the degree of control achieved by a control system compared to the degree of control desired. Control systems may achieve almost any degree of sensitivity. Most people will not notice a temperature difference of 2°F or 3°F above or below a comfortable setpoint. As a result, a control system for a residential application does not require greater sensitivity than 2°F or 3°F above or below a setpoint. A warehouse may not require any greater sensitivity than 10°F above or below a setpoint.

Sensitivity and Systems

The degree of sensitivity achieved by a heating or air conditioning system depends on the type of equipment used for the application. A central heating or air conditioning system that has a single thermostat can only achieve a certain degree of sensitivity because the temperature is sensed only at the location of the thermostat. If a system that contains a single thermostat is used in a large building, temperature control in remote areas may be poor. A large office building that requires close control of conditions in multiple offices must have equipment that can control the conditions in each room. The control system of a large office building must be designed to function with this equipment.

Several combinations of equipment and control systems may be used to control the conditions in an individual room. The equipment and control systems of central heating and air conditioning systems are configured differently to achieve different levels of control.

Cooling-Ventilation Systems. A *cooling-ventilation system* is a basic control system that circulates cool air to all spaces in a building and uses reheat units, which are located in each building space. A *reheat unit* is a heater that reheats air from a central air supply. If the thermostat in the building space does not call for heat, cool air will enter the building space. If the thermostat calls for heat, the reheat unit will heat the cool air that is supplied by the central blower. Cooling-ventilation systems are restricted by many energy codes.

Hot and Cold Duct Systems. A *hot and cold duct system* is a control system that has two duct systems. The duct systems supply warm air and cool air to building spaces. Mixing dampers on the ductwork provides air to each building space. Thermostats control mixing dampers in building spaces.

Variable-volume Duct Systems. A *variable-volume duct system* is a control system that has variable-volume dampers on a primary cooling duct. Variable-volume dampers mix the cool air taken from the duct with return air from the building space

being controlled. The dampers then introduce the mixed air into the building space. Thermostats in each building space control the variable-volume dampers. Heat is provided as needed by reheat units in the dampers.

Hydronic Distribution Systems. A hydronic heating or cooling system often has water-to-air heat pumps located in each building space to provide heating or cooling for the spaces. Zone thermostats control the heat pumps. The heat pumps are connected to a water supply for a zone in the building. One advantage of using heat pumps is that it is possible to heat one zone of a building and cool another zone at the same time. Heat can be transferred from the zone being cooled to a zone that needs heat.

APPLICATIONS

The control system that is chosen for an application depends on the category of the building in which it will be used. Buildings are categorized as residential, commercial, industrial, and special. Establishing definite descriptions defining each category is impossible, and the categories overlap.

A *residence* is any dwelling in which one family resides. Multi-unit dwellings may be included if each unit is designed for one family. A *commercial building* is a building that involves a large number of people. Office buildings, schools, auditoriums, and similar buildings are considered commercial buildings. An *industrial building* is a building in which industrial processes are performed and some heating or cooling processes are used for controlling the climate. An industrial application may be a cross between a commercial application and an industrial application because people and processes are involved. A *special application* is one in which a combination of any of the above applications is to be controlled by one control system. An example would be a manufacturing building with offices that are controlled by the same control system.

Review Questions

1. A control system links which components of a heating, ventilating, or air conditioning system?

2. What is the main purpose of a control system?

3. What are the four elements of a typical control system?

4. Which control operates the heating unit in a hydronic system?

5. What is the purpose of a limit control in a heating control system?

6. Define *signal* as used in a control system.

7. List and describe three control systems.

8. What are the functions of the disconnect and the fuses in the electrical supply lines of an electrical control system?

9. What is the function of the transformer in an electrical control system?

10. What is the function of an operator in a control system?

11. What is the function of combustion safety controls on a heating unit?

12. List and describe two flame surveillance control systems.

13. What devices make an electrical control system proportional instead of digital?

14. What kind of signal is used in a typical electronic control system?

15. What is the function of the balancing relay in a proportional control system?

16. What moves the motor potentiometer wiper blade along the coil in a proportional control system?

17. Define *balancing relay*. Why does a balancing relay tip?

18. How does a blower control on a heating unit control the blower on the furnace?

19. Define *sensor*. Describe the function of a sensor in a control system.

20. What is the function of a pressure switch in an air conditioning control system?

21. Which damper control system can provide free cooling with outdoor air? Explain how it works.

22. An economizer damper control system mixes what two air sources? How?

23. In a pumpdown control system used on an air conditioning unit, what control turns the compressor ON and OFF?

24. What is the function of the control potentiometer in a proportional control system?

25. Define *thermistor, diode,* and *transistor.* Explain the primary function of each.

Heating and Cooling Loads

Chapter **14**

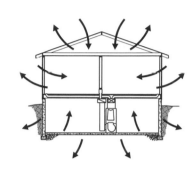

Heating and cooling loads are the heat lost or heat gained by a building. The loads occur because of a difference between the indoor temperature and outdoor temperature, infiltration or ventilation in the building, and internal loads. When the air is cold outdoors, heat must be added to a building to offset the building heating load to maintain a comfortable temperature indoors. When the air is warm outdoors, the air indoors must be cooled to offset the building cooling load to maintain a comfortable temperature indoors.

HEATING AND COOLING LOADS

Heating load is the amount of heat lost by a building. *Cooling load* is the amount of heat gained by a building. Heating and cooling loads occur because of a difference between the indoor temperature and outdoor temperature, infiltration or ventilation in the building, and internal loads. Internal loads are important when calculating cooling loads. Heating and cooling loads are adjusted for internal loads, duct losses, and any other losses or gains not directly related to temperature difference.

Heating and cooling loads are calculated either manually or by a computer. The size of the heating and cooling equipment required in a building depends on the heating and cooling loads of the building. Heating and cooling loads are also necessary for designing a distribution system. Heating and cooling loads are found by using variables and factors. Variables and factors are arranged in a format that allows easy mathematical calculations.

Variables

Variables are data that are unique to a building. Variables relate to the specific location of the building and the specifications of the particular building. Variables include design temperature difference, ar-

eas of exposed building surfaces, infiltration or ventilation rates, people, lights, and appliances.

Design Temperature. *Design temperature* is the temperature of the air at a predetermined set of conditions. *Indoor design temperature* is the temperature selected for the inside of a building. The indoor design temperature used for buildings occupied by people is between 70°F and 75°F. *Outdoor design temperature* is the expected outdoor temperature that a heating or cooling load must balance.

The outdoor design temperature for an area is found on tables of outdoor design temperatures. *Outdoor design temperature tables* are tables of data developed from records of temperatures that have occurred in an area over many years. Outdoor design temperature tables are available from many sources. Outdoor design temperature tables include winter dry bulb temperature, summer dry bulb and wet bulb temperatures, daily range, and latitude degree. See Figure 14-1.

Winter dry bulb temperature is the coldest temperature expected to occur in an area while disregarding the lowest temperatures that occur in from 1% to 2½% of the total hours in the three coldest months of the year. The coldest temperatures are disregarded because they occur for a short period of time during the night, when some temperature swing is acceptable. *Temperature swing* is the difference between the setpoint temperature and the actual temperature. Temperature swing slightly lowers the temperature in a building. A heating load for winter conditions is calculated at the winter dry bulb temperature.

OUTDOOR DESIGN TEMPERATURE					
State and City	Lat.*	Winter	Summer		
		DB§	DB§	DR#	WB§
ALABAMA					
Alexander City	32	18	96	21	79
Auburn	32	18	96	21	79
Birmingham	33	17	96	21	78
Mobile	30	25	95	18	80
Montgomery	32	22	96	21	79
Talladega	33	18	97	21	79
Tuscaloosa	33	20	98	22	79
ALASKA					
Anchorage	61	–23	71	15	60
Fairbanks	64	–51	82	24	64
Juneau	58	– 4	74	15	61
Kodiak	57	10	69	10	60
Nome	64	–31	66	10	58

* in degrees   •*Air Conditioning Contractors of America*
§ in °F
# daily range in °F

Figure 14-1. Tables of outdoor design temperatures are developed based on temperature records over a period of many years.

Summer dry bulb temperature is the warmest dry bulb temperature expected to occur in an area while disregarding the highest temperature that occurs in from 1% to 5% of the total hours in the three hottest months of the year. *Summer wet bulb temperature* is the wet bulb temperature that occurs concurrently with the summer dry bulb temperature. Wet bulb temperature is considered in cooling loads because of the effect of humidity on comfort at higher temperatures and because of the effect of humidity on cooling equipment capacity. A cooling load is calculated for summer conditions at the summer dry bulb temperature. *Design temperature difference* is the difference between the desired indoor temperature and the outdoor temperature for a particular season.

Area of Exposed Surfaces. To calculate heat loss and gain, the area of all exposed surfaces must be found. *Exposed surfaces* are building surfaces that are exposed to outdoor temperatures. Areas of the exposed surfaces of a building are calculated using building dimensions. The area of each wall, roof, or other exposed surface is calculated separately. For cooling loads, the areas should also be separated according to their exposure. *Exposure* is the geographic direction a wall faces. Exposure is important because of the solar gain on the surfaces. The solar gain on a surface depends on the exposure of the exterior walls, windows, and doors. Factors for solar gain, which are included in heating and cooling loads, are given for the different exposures. The area of each flat surface is found by multiplying the length of the surface by the height.

Gross wall area is the total area of a wall including windows, doors, and other openings. Gross wall area is calculated separately for each exposed wall of a building. Gross wall area is found by applying the formula:

$$A_g = w \times h$$

where

A_g = gross wall area (in sq ft)

w = width (in ft)

h = height (in ft)

Net wall area is the area of a wall after the area of windows, doors, and other openings have been subtracted. Net wall area is used for calculating heating and cooling loads. Net wall area is found by applying the formula:

$$A_n = A_g - A_o$$

where

A_n = Net wall area (in sq ft)

A_g = gross wall area (in sq ft)

A_o = area of opening (in sq ft)

Example: Finding Net Wall Area

The south wall of a building is 30′ wide and 8′ high. The wall has a door 3′ wide and 7′ high. Find the net wall area.

1. Find gross wall area.

$$A_g = w \times h$$

$$A_g = 30 \times 8$$

$$A_g = \textbf{240 sq ft}$$

2. Find the area of the opening.

$$A_o = w \times h$$

$$A_o = 3 \times 7$$

$$A_o = \textbf{21 sq ft}$$

3. Find net wall area.

$$A_n = A_g - A_o$$

$$A_n = 240 - 21$$

$$A_n = \textbf{219 sq ft}$$

The area of each window on each exposure of the building is calculated separately. The area of each door is also calculated separately. The values are categorized by exposure. The total area of windows and doors in each exposure are used for calculating window and door gain or loss and net wall area.

The area of the ceiling is calculated using inside dimensions. Overhangs that do not affect the transmission of heat are not included. For sloping ceilings such as cathedral ceilings, the exposed area, not the area of the horizontal plane, is used for calculating the area of the ceiling. See Figure 14-2.

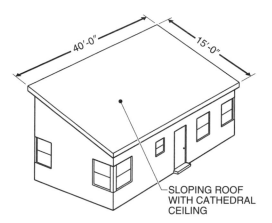

Figure 14-2. For sloping ceilings, the exposed area is used for calculating the area of the ceiling.

The area of a floor that is exposed to a space that is not conditioned is used to calculate heat loss or gain. In a building where a concrete slab floor is located on the ground, most of the heat loss from the floor occurs around the edge. Heat loss is calculated as the distance around the perimeter of the exposed edge of the floor. A concrete slab floor has no heat gain.

Number of People. Estimating the number of people occupying a residential building and estimating the number of people occupying a commercial building are done with different methods. When estimating the number of people that occupy a residential building, the living habits of the individuals must be considered. If a family entertains frequently, the number of people used for calculating loads should include family and guests. For family only, the estimate is based on the number of bedrooms in the building. The number is calculated based on two people in the main bedroom and one person for each additional bedroom. To include guests in the number of people, use an estimate of the actual number of guests.

In a commercial building, the estimate of the number of occupants is based on the area (in square feet) of floor space required per person and the activity level of the person. Tables of occupancy estimates show the area required per person for different activities. Care should be taken to ensure that only space that a person can actually occupy is used. Approximately one-half of the floor space is usually occupied by furniture, appliances, or displays.

Number of Lights and Appliances. Any device that produces heat inside a building must be compensated for by a cooling load. Electric lights produce heat as well as light. Appliances are powered by electricity or fuel. In either case, much of the energy that operates the appliance is converted to heat. The number of lights and appliances must be calculated to find the amount of heat given off by the devices.

Factors

Factors are numerical values that represent the heat produced or transferred under some specific condition. Factors are found in tables that have been prepared for typical applications. Factors include heat loss or gain from conduction, infiltration, ventilation, people, electricity used for lights, electricity or fuel used for operating appliances, and solar energy. Heat gain from people, lights, appliances, and solar energy normally only apply to cooling loads.

Conduction. According to the second law of thermodynamics, when the temperature on two sides of a material is different, heat flows from the warmer side to the cooler side of the material. The outside surfaces of a building, such as walls, windows, doors, floors, and ceilings are composed of materials across which temperature differences exist. When the temperature of the air outside a building is cooler than the temperature inside, heat flows from the indoors to the outdoors. When the temperature of the air inside a building is warmer, heat flows from the outdoors to the indoors. A *building component* is a main part of a building structure such as the exterior walls. See Figure 14-3.

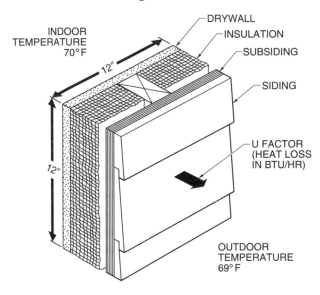

Figure 14-3. Conduction factors identify the amount of heat that flows through a building component because of a temperature difference.

A *conduction factor* (U) is a factor that represents the amount of heat that flows through a building component because of a temperature difference. Conduction factors are expressed in Btu per hour per square foot of material per degree Fahrenheit temperature difference through the material. Factors that identify heat carried by air are based on the sensible heat of air, the amount of air moved, and the difference between the indoor temperature and outdoor temperature.

Heat transfer rates have been calculated for most building components used in the construction of residential and commercial buildings. Heat conduction factors are found in tables of conduction factors through various building materials.

To simplify calculations, conduction factors and either outdoor design temperatures or design temperature differences are combined in a table of heat transfer factors. A *heat transfer factor* is a conduction factor multiplied by a design temperature difference. A table of heat transfer factors is used for calculating cooling or heating loads for residential or small commercial buildings. See Appendix.

Infiltration. *Infiltration* is the process that occurs when outdoor air leaks into a building. *Infiltration air* is air that flows into a building when exterior doors are open or when air leaks in through cracks around doors, windows, or other openings.

Factors for calculating heat loss and gain from the exchange of air at different temperatures is found in tables of infiltration factors. See Figure 14-4. When a table of infiltration factors is used for calculating heat loss or gain for air, the volumetric flow rate (cfm) of air involved is multiplied by the infiltration factor.

The amount of heat transferred by infiltration is represented by factors derived from the specific heat of air. Most factors are expressed in Btu per hour per cubic foot of air per minute and the difference between the indoor and outdoor temperature. Some infiltration tables are based on the heat loss per square foot of exposed building surface.

INFILTRATION FACTORS						
Outdoor design temperature*	85	90	95	100	105	110
Infiltration factors§	0.7	1.1	1.5	1.9	2.2	2.6

* in °F
§ in Btu/hr/sq ft
Note: Infiltration factors are for an indoor design temperature of 75°F.

•*Air Conditioning Contractors of America*

Figure 14-4. Tables of infiltration factors have factors for calculating heat loss and gain from air that seeps into a building space.

To find the volumetric flow rate of air involved, the volume of the building (in cu ft) or the volume of the building spaces involved in the calculation is multiplied by an air change factor. An *air change factor* is a value that represents the number of times per hour that the air in the building is completely replaced by outdoor air. Air change factors are found in tables for selected buildings. See Figure 14-5.

NUMBER OF AIR CHANGES PER HOUR			
Building Type	**Construction Type**		
	Loose	**Medium**	**Tight**
Residential	1.5	1.0	.5
Commercial	3.0	1.5	.5
Industrial	5.0	3.0	1.0

•*Air Conditioning Contractors of America*

Note: Construction type refers to type fit of doors and windows.

Figure 14-5. The number of air changes per hour in a building depends on the type of building and the construction of the building.

Ventilation. *Ventilation* is the process that occurs when outdoor air is brought into a building. *Ventilation air* is air that is brought into a building to keep building air fresh. The air must be heated if it is cooler than the indoor air or cooled if it is warmer than the indoor air.

The volumetric flow rate of ventilation air required for a building is based on volumetric flow rate of air required per person inside the building. The flow rate of air required per person is found in the Uniform Building Code (UBC), or in local adaptations to the UBC. To find the required volumetric flow rate of ventilation air, the volumetric flow rate per person is multiplied by the number of people that occupy the building. The total volume of ventilation air required for a building is found by applying the formula:

$$Q_t = p \times Q_p$$

where

Q_t = total volumetric flow rate of ventilation air required (in cfm)

p = number of people

Q_p = required volumetric flow rate of air per person (in cfm)

Example: Finding Required Total Volumetric Flow Rate — Ventilation Air

A building is occupied by 70 people. The Uniform Building Code calls for 30 cfm of ventilation air per person. Find the required total volumetric flow rate of ventilation air.

$$Q_t = p \times Q_p$$

$$Q_t = 70 \times 30$$

$$Q_t = \textbf{2100 cfm}$$

To find the heat loss from the total volumetric flow rate of ventilation air, the flow rate is multiplied by 1.08 and the difference between the indoor and outdoor temperature. The heat loss from ventilation air is found by applying the formula:

$$H_l = Q_t \times 1.08 \times \Delta T$$

where

H_l = heat loss from ventilation air (in Btu/hr)

Q_t = total volumetric flow rate of ventilation air required (in cfm)

1.08 = constant

ΔT = temperature difference (in °F)

Example: Finding Heat Loss — Ventilation Air

A building is occupied by 35 people. The Uniform Building Code calls for 30 cfm of ventilation air per person. The difference between the indoor and outdoor temperature is 40°F. Find the heat loss from ventilation air.

1. Find the total volumetric flow rate of ventilation air required.

$$Q_t = p \times Q_p$$

$$Q_t = 35 \times 30$$

$$Q_t = \textbf{1050 cfm}$$

2. Find the heat loss from ventilation air.

$$H_l = Q_t \times 1.08 \times \Delta T$$

$$H_l = 1050 \times 1.08 \times 40$$

$$H_l = \textbf{45,360 Btu/hr}$$

The heat gain for a cooling load is calculated with the same procedure, except the humidity in the building must be considered. The most accurate way to calculate heat gain from ventilation air is by use of a psychrometric chart.

People. The human body produces heat and moisture. Heat is produced during digestion and is given off from the skin. Moisture is expelled from the body with every breath and during perspiration. The amount of heat given off varies with the amount of activity. A vigorous activity produces more heat than a passive activity. A table of heat gain from occupants is used to identify the heat produced by people when engaged in various activities. See Figure 14-6.

Lights. Heat gain from lights is a factor in a cooling load calculation. Electricity can be converted directly to thermal energy, which is heat. The factor for converting electricity to thermal energy is 3.41 Btu/W (Btu per watt). To calculate the heat gain from lights, the total wattage of the lights is multiplied by 3.41. If fluorescent lights are used, the conversion factor from electrical energy to thermal energy will be 4.25 Btu/W. Fluorescent lights have a larger factor because larger quantities of heat are given off by the ballast in a fluorescent light. A *ballast* is a combination electric device that contains a transformer and an igniter, which generate heat.

Appliances. Heat gain from appliances is calculated by multiplying the wattage of the appliance by 3.41. Calculating heat gain for appliances that use fuel instead of electricity is based on the rate of fuel used and the heat output for the fuel. Heat gain values are also found on tables of heat gain from appliances. See Figure 14-7.

Solar Gain. *Solar gain* is heat gain caused by radiant energy from the sun that strikes opaque objects. Sunlight, which shines through the windows and strikes the exterior of a building, adds heat to the building. This heat is solar gain. Solar gain is only used when calculating cooling loads. A cooling system must compensate for extra heat. In some commercial buildings, solar gain has a positive effect on heating loads.

Equivalent temperature difference is the design temperature difference that is adjusted for solar gain. Equivalent temperature differences are used when calculating heat gain through walls or ceilings of buildings. Equivalent temperature difference is used only for calculating cooling loads. Equivalent temperature differences are used in most cooling load calculations for commercial buildings.

When the sun shines on an exterior surface of a building or opaque objects inside a building, the solar energy becomes thermal energy. The thermal energy, which is heat energy, is then absorbed into the building. For the cooling load for a commercial building, the heat from the solar energy is added to the cooling load of the building because the system must compensate for the additional heat.

HEAT GAIN FROM OCCUPANTS				
Degree of Activity	Typical Application	Total Heat*	Sensible Heat*	Latent Heat*
Seated, at rest	Theater, grade school classroom	330	225	105
Seated, very light work	Office, hotel, apartment, high school classroom	400	245	155
Moderately active office work	Office, hotel, apartment, college classroom	450	250	200
Standing, light work	Drug store, bank	500	250	250
Sedentary work	Restaurant	550	275	275
Light bench work	Factory	750	275	475
Moderate dancing	Dance hall	850	305	545
Walking, moderately heavy work	Factory	1000	375	625
Bowling, heavy work	Bowling alley, factory	1450	580	870

* in Btu/hr   •*Air Conditioning Contractors of America*

Note: The above values are based on 75°F room dry bulb temperature. For 80°F room dry bulb temperature, the total heat gain remains the same.

Figure 14-6. Heat gain from people is calculated by multiplying a factor from a table for heat gain from occupants.

HEAT GAIN FROM APPLIANCES								
	Electric				Gas			
Type of Appliance	Without Hood			Hood	Without Hood			Hood
	Sensible*	Latent*	Total*	All Sensible*	Sensible*	Latent*	Total*	All Sensible*
Broiler-griddle 31″ x 20″ x 18″					11,700	6300	18,000	3600
Coffee brewer/warmer					1750	750	2500	500
per burner	770	230	1000	340				
per warmer	230	70	300	90				
Coffee urn								
3 gal.	2550	850	3400	1000	3500	1500	5000	1000
5 gal.	3850	1250	5100	1600	5250	2250	7500	1500
8 gal.	5200	1600	6800	2100	7000	3000	10,000	2000
Deep fat fryer								
15 lb fat	2800	6600	9400	3000	7500	7500	15,000	3000
21 lb fat	4100	9600	13,700	4300				
Dry food warmer per sq ft top	320	80	400	130	560	140	700	140
Griddle, frying per sq ft top	3000	1600	4600	1500	4900	2600	7500	1500
Hot plate					5300	3600	8900	2800
Short order stove per burner					3200	1800	5000	1000
Toaster								
Continuous								
360 slices per hour	1960	1740	3700	1200	3600	2400	6000	1200
720 slices per hour	2700	2400	5100	1600	600	4000	10,000	2000
Pop-up (4 slice)	2230	1970	4200	1300				
Waffle iron 18″ x 20″ x 13″	1680	1120	2800	900				
Hair dryer								
Blower type	2300	400	2700					
Lab burners								
Bunsen					1680	420	2100	
Meeker	60				3360	840	4200	
Neon sign per foot of tube			60					
Sterilizer	650	1200	1850					
Vending machines								
Hot drink			1200					
Cold drink			625					

* Heat in Btu/hr created in 1 hour. •*Air Conditioning Contractors of America*

Figure 14-7. Tables for heat gain from appliances have factors that are included when calculating loads for commercial buildings.

Using an equivalent temperature difference is one way to include this increase in the cooling load. Tables of equivalent temperature differences show the equivalent temperature difference for sunlit walls and roofs. See Figure 14-8.

Tables of glass heat transfer factors include the Btu per hour gain per square foot of glass surface for typical windows. These tables also provide data for modifying the factor value for different types of glass such as regular, double-glazed, and heat-absorbing glass. See Figure 14-9.

Solar gain on walls and ceilings is combined with conduction gain in some factor tables. In other ta-

bles, solar gain is included with an equivalent temperature difference for the heating and cooling loads. When using a particular set of tables for calculating heat gain, the method for including solar gain through the building surfaces must be determined.

LOAD FORMS

Load forms are documents that are used by design technicians for arranging the heating and cooling load variables and factors. Load forms simplify the calculation of heating and cooling loads. Two basic forms are prepared forms and columnar forms.

EQUIVALENT TEMPERATURE DIFFERENCES*								
Design Temperature§	85	90	95		100		105	110
Daily Temperature Range	L or M	L or M	M	H	M	H	H	H
Walls and Doors								
Wood frame walls and doors	13.6	18.6	23.6	18.6	28.6	23.6	28.6	33.6
Solid masonry, block, or brick walls	6.3	11.3	16.3	11.3	21.3	16.3	21.3	26.3
Partitions	5.0	10.0	15.0	10.0	20.0	15.0	20.0	25.0
Ceilings and Roofs								
Dark exterior	34.0	39.0	44.0	39.0	49.0	44.0	49.0	54.0
Light exterior	26.0	31.0	36.0	31.0	41.0	36.0	41.0	46.0
Floors								
Over unconditioned, vented, or open space	5.0	10.0	15.0	10.0	20.0	15.0	20.0	25.0
Over conditioned space or on or below grade	0	0	0	0	0	0	0	0

* Equivalent temperature differences are for an indoor design temperature of 75°F. •*Air Conditioning Contractors of America*
§ in °F

Figure 14-8. Equivalent temperature difference is the design temperature difference adjusted for solar gain.

Prepared Forms

A *prepared form* is a preprinted form consisting of columns and rows that identify required information. The proper values for the variables and factors for an application are inserted in the spaces. The calculations are performed by working across the lines on the form.

Many different prepared forms are available. A prepared form has space at the top for the job name, date of the calculations, name of the person who prepared the form, and design temperatures. See Figure 14-10. The left-hand column has spaces for factors such as building components or other elements of the loads. The other columns are for information relating to individual rooms or zones in a building. Headings across the top of the form identify each room or zone column. Factors are entered in the factors column for each building component.

Prepared forms have multiple columns that are used to calculate loads on individual rooms or zones. Heat loss or gain is calculated for each room or zone so the hydronic piping or forced-air ductwork system can provide the correct amount of heat to each room or zone.

COOLING HEAT TRANSFER FACTORS — WINDOWS AND DOORS*									
Exposure	Single Glass Temperature Diff.			Double Glass Temperature Diff.			Triple Glass Temperature Diff.		
	15°	20°	25°	15°	20°	25°	15°	20°	25°
N	18	22	26	14	16	18	11	12	13
NE & NW	37	41	46	31	33	35	26	27	28
E & W	52	56	60	44	46	48	38	39	40
SE & SW	45	49	53	39	41	43	33	34	35
S	28	32	36	23	25	27	19	20	21
Doors									
Wood	8.6	10.9	13.2	8.6	10.9	13.2	8.6	10.9	13.2
Metal	3.5	4.5	5.4	3.5	4.5	5.4	3.5	4.5	5.4

* Inside shading by venetian blinds or draperies. •*Air Conditioning Contractors of America*

Figure 14-9. Tables of glass heat transfer factors (cooling) contain the Btu per hour gain per square foot of glass surface for different types of windows.

1	Name of Room				1			4			5			Building
2	Running Feet of Exposed Wall													Component
3	Room Dimensions													Subtotals
4	Ceiling Height	Exposure												
	Types of		Factors		Area	Btu/hr	Btu/hr	Area	Btu/hr	Btu/hr	Area	Btu/hr	Btu/hr	Btu/hr	Btu/hr
	Exposure		H	C		H	C		H	C		H	C	H	C

Contractor: _____   Date: _____

Name of Job: _____   By: _____

Address: _____

Winter: Indoor Design Temp. ____ Outdoor Design Temp. _____ Design Temperature Difference _____

Summer: Outdoor Design Temp. ____ Indoor Design Temp. _____ Design Temperature Difference _____

•*Air Conditioning Contractors of America*

Figure 14-10. A prepared form has headings for variables and factors used when calculating heating and cooling loads.

Small rooms or adjoining rooms without partitions may be combined on the form. A room or zone is identified at the top of each column. The area of each building component in the room or zone is entered in a space in the column. Factors are entered in the factors column. The areas in each column are multiplied by the factors on the same line. Factors are arranged to be entered once regardless of the number of rooms or zones.

Columnar Forms

A *columnar form* is a blank table that is divided into columns and rows by vertical and horizontal lines. Spaces at the top of a columnar form identify the job with the job name, date of the calculations, name of the person who prepared the form, and design temperatures.

A columnar form should be arranged with the procedure used for a prepared form. Headings are entered at the top of the columns and on the left-hand column on the form to identify information that is entered in the spaces. When a columnar form is used for calculating heating and cooling loads, the columns are used for room or zone data and the rows are used for variables and factors. Heating and cooling loads are calculated on two separate sections of the form. The data relating to rooms or zones is placed in individual columns. Factors are placed in the rows. See Figure 14-11.

A columnar form organizes the required data for heating and cooling loads so that the mathematical calculations are easy to perform. The load calculation data is readily available and the columns are easily added for sizing components of the distribution system and for balancing a system. A design technician using a columnar form for calculating heating or cooling loads should be familiar with the method of calculating loads with a prepared form and the load calculation process.

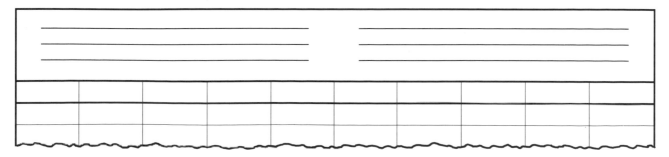

Figure 14-11. A columnar form is a blank grid. A design technician fills in headings for the columns and lines to fit each application.

Review Questions

1. List three elements that may be involved in calculating heating and cooling loads.

2. What two methods are used for calculating heating and cooling loads?

3. What two categories identify data used for calculating heating and cooling loads?

4. Define *variables* in relation to heating and cooling load data.

5. List and describe six variables used when calculating heating and cooling loads.

6. Why are the areas of exposed building surfaces important in calculating heating and cooling loads?

7. Define *exposure* in relation to calculating heating and cooling loads.

8. Explain how gross wall area is calculated differently than net wall area.

9. Why is net wall area of a building important?

10. List and explain the two ways window and door areas are used in calculating heating and cooling loads.

11. Explain how the number of people in a building is used when calculating heating and cooling loads.

12. Define *factor* in relation to heating and cooling load calculations.

13. Define *conduction factor* in relation to heat loss and heat gain.

14. What is the difference between a heat conduction factor and a heat transfer factor?

15. Define *infiltration.*

16. How are air change factors used in heating and cooling load calculations?

17. What is the difference between infiltration and ventilation?

18. How are design temperature difference figures adjusted to find the equivalent temperature difference?

19. What is the difference between a prepared load calculation form and a columnar load calculation form?

20. Explain the advantage of using a columnar form instead of a prepared form.

Load Calculations   Chapter

Heating and cooling load calculations determine the amount of heat or cooling that is required to maintain a constant temperature in a building. Heating and cooling loads are calculated by a conventional method or with a computer.

    Computer programs rapidly calculate heating and cooling loads. If a computer program is used, however, the conventional method should also be understood. The conventional method helps a design technician understand the load calculation procedure.

LOAD CALCULATIONS — CONVENTIONAL METHOD

Conventional load calculations are performed manually. Two load forms are the prepared form and columnar form. The information for the cooling load is entered on the form first because more data is required for calculating cooling loads than for heating loads. The heating load information is taken from the cooling load information.

Once a load form is selected, the information about the job is entered in the spaces at the top of the form. This information includes the name of the contractor, name and address of job, date, and the name of the person who will prepare the form. See Figure 15-1.

Variables and factors are entered on a load calculation form and are used to calculate heat losses and gains for a specific building. Additional losses or gains that occur through the ductwork or piping are added to the building losses and gains to determine the total heating and cooling loads.

Variables

Variables for calculating heating and cooling loads include design temperature differences, area of building components, data for calculating infiltration or ventilation loss or gain, number of people occupying the building, and number of lights and appliances in the building. For most cooling applications, solar gain must also be considered. Solar gain through windows is usually calculated directly. Solar gain through walls and the roof is either included in the factors used for heat gain through walls and roof or is included as an equivalent temperature difference.

Design Temperature. Design temperature is the temperature of the air at a predetermined set of conditions. Indoor design temperature is the temperature selected for the inside of a building. A winter indoor design temperature of 70°F is usually selected to provide comfort for a normally clothed person during winter. A summer indoor design temperature of 75°F is usually selected to provide comfort for a normally clothed person during summer. Outdoor design temperatures are the coldest winter temperatures expected to occur in an area during 97.5% of the winter, and the warmest summer temperatures expected in an area during 95% of the summer.

Outdoor design temperatures are found in tables of outdoor design temperatures. Tables of outdoor design temperatures contain specific temperatures for many different areas. The outdoor design temperature for a building is found on a table by finding

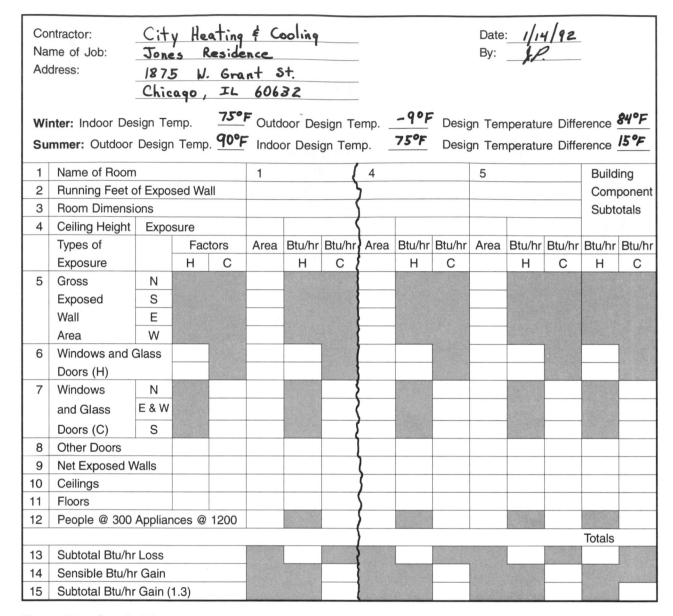

Figure 15-1. Specific information must be entered at the top of a load form.

the area where the building is located. See Figure 15-2. See Appendix.

Design temperature difference is the difference between the indoor design temperature and the outdoor design temperature. Winter design temperature difference is found by subtracting the winter outdoor design temperature from the indoor design temperature. This data is entered on the form in the spaces at the top of the form. Winter design temperature difference is found by applying the formula:

$$\Delta T_W = T_i - T_o$$

where

ΔT_W = winter design temperature difference (in °F)

T_i = winter indoor design temperature (in °F)

T_o = winter outdoor design temperature (in °F)

Summer design temperature difference is found by subtracting the summer indoor design temperature from the summer outdoor design temperature for the area in which the building is located. This data is entered on the form in the corresponding spaces. Summer design temperature difference is found by applying the formula:

$$\Delta T_S = T_o - T_i$$

where

ΔT_S = summer design temperature difference (in °F)

T_o = summer outdoor design temperature (in °F)

T_i = summer indoor design temperature (in °F)

Example: Finding Design Temperature Difference

Find the winter and summer design temperature difference for a residence located in Chicago, Illinois. Winter design temperature difference and summer design temperature difference are found by applying the procedure:

1. Enter 75°F as the winter and summer indoor design temperatures on the load form.

2. On a table of outdoor design temperatures, locate the winter dry bulb and summer dry bulb temperatures for Chicago, Illinois. Enter the temperatures on the load form.

3. Find the winter design temperature difference.

$$\Delta T_W = T_i - T_o$$

$$\Delta T_W = 75 - (-9)$$

$$\Delta T_W = \mathbf{84°F}$$

OUTDOOR DESIGN TEMPERATURE					
		Winter	Summer		
State and City	Lat.*	DB§	DB§	DR#	WB§
ILLINOIS					
Aurora	41	− 6	93	20	79
Champaign	40	− 3	95	21	78
Chicago	41	− 9	90	15	79
Galesburg	40	− 7	93	22	78
Joliet	41	− 5	93	20	78
Kankakee	41	− 4	93	21	78
Macomb	40	− 5	95	22	79
WASHINGTON					
Aberdeen	46	25	80	16	65
Olympia	46	16	87	32	67
Seattle-Tacoma	47	17	84	22	66
Spokane	47	− 6	93	28	65
Walla Walla	46	0	97	27	69

* in degrees   •*Air Conditioning Contractors of America*
§ in °F
\# daily range in °F

Figure 15-2. Tables of outdoor design temperatures contain design temperatures for many different areas.

4. Find summer design temperature difference.

$$\Delta T_S = T_o - T_i$$

$$\Delta T_S = 90 - 75$$

$$\Delta T_S = \mathbf{15°F}$$

Building Data. *Building data* is the name of each room, running feet of exposed wall, dimensions, ceiling height, and exposure of each room. Building data is entered at the top of the load calculation form. For a building in the planning or construction stage, the building data is found on the plans and specifications for the building. *Plans* are drawings of a building that show dimensions, construction materials, location, and arrangement of the spaces within the building. *Specifications* are written supplements to plans that describe the materials used for a building. Plans and specifications should be available from the building contractor.

When plans and specifications are not available for a building that is already built, the floor plan of the building is sketched, and notes about the materials and construction methods are taken. The building data for the load form is taken from the sketches and notes and entered in the appropriate spaces on the form. See Figure 15-3.

Dimensions for walls, ceilings, and floors may be rounded to the nearest 6″. Dimensions for windows and doors may be rounded to the nearest 3″. Dimensions for rooms or zones should be taken from the outside of the building to the center of partitions between adjacent rooms or zones. Hallways, closets, stairways, and other spaces in the building not part of a room or zone are included in the adjacent rooms or zones. The dining room and kitchen of the Chicago residence are combined because they are not divided by a partition. The areas of the hallways are combined with the living room and bath.

Gross wall area is the total area of a wall including windows, doors, and other openings. Gross wall area is calculated separately for each exposed wall of the building. The gross exposed wall area for each room is entered on the load form according to the exposure of the wall. Exposure is the geographic direction a wall faces. Gross wall dimensions are taken from the outdoor dimensions of the building, which are found on sketches or plans of the building.

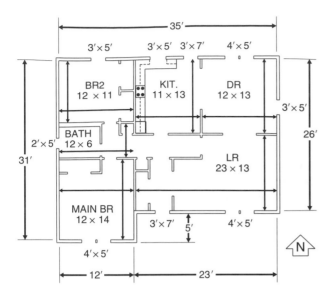

NOTES:
LOCATION – CHICAGO, ILLINOIS
WINDOWS – DOUBLE GLASS WITH STORM SASH
DOORS – WOOD WITH STORM DOOR
WALLS – WOOD FRAME WITH R-11 INSULATION
CEILING – R-44 INSULATION UNDER
                UNCONDITIONED SPACE
FLOOR – WOOD OVER VENTED SPACE WITH R-11
            INSULATION

CHICAGO RESIDENCE

Figure 15-3. Information for load calculations is taken from sketches and notes describing a specific building.

The information for calculating the area of windows and glass doors is found on sketches or plans of the building. For heating load calculations, the area of windows and glass doors is calculated and entered on the form. For cooling load calculations, the area of windows and glass doors is calculated and entered on the form according to exposure. The area of doors that are not glass is also entered on the form.

The net area of exposed walls is found by subtracting the area of windows and doors from the gross exposed wall area for each room. The area of the ceiling and floor is found and entered on the form. The area of the ceiling and floor is found by multiplying the dimensions of the rooms.

Three different methods are used to calculate the rate of infiltration of air into a building. In the first method, an infiltration factor is included with either the wall factors or the window factors taken from heat transfer factor tables. When calculating loads, factor tables should be checked to ensure that infiltration factors are included. For the Chicago residence, the infiltration factors are included in the heat transfer factors table.

The second method is the crackage method. In the crackage method, the actual length of crackage in a building is added up. *Crackage* is the opening around windows, doors, or other openings in the building such as between the foundation and framework of a building. A factor for calculating the flow rate or actual loss or gain for each foot of crackage is found in some calculation factor tables.

The third method is the air change method. The air change method is based on the assumption that infiltrated air completely changes the air in a building. In the air change method, the volume of the building is found by multiplying the length times the width times the ceiling height of the building. The volume of the building is multiplied by a factor related to the number of times air change occurs per hour.

The number of people that occupy a building must also be determined. For a residence, the number of people that occupy a building is determined by the number of bedrooms in the building. The Chicago residence has two bedrooms. The main bedroom represents two people and the other bedroom represents one person. The number of people in the residence for load calculations is three. Enter this figure on the load form in the rooms that the people occupy during the day when cooling is required. This information is entered in the living room and kitchen columns of the building on the load calculation form.

The lighting load is usually not calculated for a residential building because the lights are not on during the day when the cooling load occurs. For residential applications, 1200 Btu/hr of heat gain is used for appliances and cooking processes. This amount is added to the kitchen gain for the cooling load. See Figure 15-4.

Factors

Factors used in calculating heating and cooling loads include conduction factors, solar gain, infiltration factors, and factors for heat produced by people, lights, and appliances. Factors are taken from tables that have been developed for heating and cooling load calculations.

Heat transfer factors are used for calculating heat flow through individual components of a building. The heat transfer is due to a temperature difference between the two sides of a material. Tables of heat transfer factors are divided into separate sections based on building components.

Heating load factors are selected from a table of heat transfer factors. See Appendix. Factors for heat transfer through doors, windows and glass doors, walls, ceilings, and floors are entered on the load form according to the type of building material and design temperature difference.

Cooling load factors for walls, ceilings, and floors are selected from a table of cooling heat transfer factors and are entered on the load form. Cooling load factors for heat transfer through windows and doors are selected from a table of cooling heat transfer factors for doors and windows according to the exposure of the doors or windows and the difference between the indoor and outdoor temperatures. See Appendix. Infiltration factors are included with the wall, window, and door factors. Some factor tables include infiltration. Other tables require that infiltration be calculated separately. See Figure 15-5.

Contractor: City Heating & Cooling												Date: 1/14/92			
Name of Job: Jones Residence												By: L.P.			
Address: 1875 W. Grant St.															
Chicago, IL 60632															

Winter: Indoor Design Temp. ____ Outdoor Design Temp. ____ Design Temperature Difference ____
Summer: Outdoor Design Temp. ____ Indoor Design Temp. ____ Design Temperature Difference ____

				1	LR		4	BR2		5	KIT. / DINE		Building
1	Name of Room												Component
2	Running Feet of Exposed Wall				36			23			36		Subtotals
3	Room Dimensions				23' x 13'			12' x 11'			23' x 13'		
4	Ceiling Height	Exposure			8'	S-E	8'	N-W		8'	N-E		
	Type of		Factors		Area	Btu/hr	Btu/hr	Area	Btu/hr	Btu/hr	Area	Btu/hr	Btu/hr	Btu/hr	Btu/hr
	Exposure		H	C		H	C		H	C		H	C	H	C
5	Gross	N						96			184				
	Exposed	S			184										
	Wall	E			104						104				
	Area	W						88							
6	Windows and Glass Doors (H)				20			15			50				
7	Windows	N						15			35				
	and Glass	E & W									15				
	Doors (C)	S			20										
8	Other Doors				21						21				
9	Net Exposed Walls				247			169			217				
10	Ceilings				299			132			299				
11	Floors				299			132			299				
12	People @ 300 Appliances @ 1200				2 P	600					P + K	1500			

Figure 15-4. Variables used in calculating heating and cooling loads include design temperature differences, area of building components, data related to infiltration, number of people occupying the building, and number of lights and appliances in a building.

1	Name of Room			
2	Running Feet of Exposed Wall			
3	Room Dimensions			
4	Ceiling Height	Exposure		
	Type of		Factors	
	Exposure		H	C
5	Gross	N		
	Exposed	S		
	Wall	E		
	Area	W		
6	Windows and Glass		75	
	Doors (H)			
7	Windows	N		14
	and Glass	E & W		44
	Doors (C)	S		23
8	Other Doors		205	8.6
9	Net Exposed Walls		6	1.7
10	Ceilings		2	.9
11	Floors		6	.8
12	People @ 300 Appliances @ 1200			

Figure 15-5. Heat transfer factors are used for calculating heat flow through individual components of a building due to temperature differences.

When using a table of infiltration factors that is based on crackage, the total amount of crackage (in feet) is multiplied by the appropriate factor. When using a table of infiltration factors that is based on flow rate of air, multiply the infiltration air flow rate, a factor of 1.08, and the winter design temperature difference to find heat loss or gain. The flow rate of infiltration air is the cooling load due to infiltration, but the actual cooling load is found on a psychrometric chart. The psychrometric chart gives a cooling load that includes the latent heat load for moisture in the air. Latent heat load for moisture in the air does not have to be considered in the heating load, but it must be considered in the cooling load. To find the heat loss for infiltration apply the formula:

$$H = Q \times \Delta T \times 1.08$$

where

H = heat transferred (in Btu/hr)

Q = flow rate (in cfm)

ΔT = temperature difference (in °F)

1.08 = constant

Example: Finding Heat Loss — Infiltration

Outdoor air at a dry bulb temperature of 45°F and a flow rate of 150 cfm is introduced into the heating system of a building for ventilation. The dry bulb temperature of the indoor air is 75°F. Find the heat loss for the air.

$$H = Q \times \Delta T \times 1.08$$

$$H = 150 \times (75 - 45) \times 1.08$$

$$H = 150 \times 30 \times 1.08$$

$$H = \textbf{4860 Btu/hr}$$

The heat gain from people is not usually calculated in a heating load. Heat gain from people is usually disregarded in residential load calculations because relatively few people are included. For residential cooling applications, heat gain from people is 300 Btu/hr of sensible heat per person with a latent heat allowance added. This value is multiplied by the number of people and the resulting amount is entered on the load form in the corresponding spaces. The heat gain factor from appliances is 1200 Btu/hr and is added to the kitchen gains.

CALCULATIONS

After the variables and factors have been entered on a load form, calculations must be performed. The variables in the columns are multiplied by the factors in the factor column. The heating losses must be kept separate from the cooling gains. The results are the heating and cooling loads for each room or zone broken down into building components. Building components are the main part of a building structure such as the exterior walls, windows, doors, ceiling, and floor.

Subtotals

To find the subtotals for each room or zone, all of the losses or gains for the building components in a column are added together. The sum in the heating loss column is the heat loss for the room or zone. The sum in the cooling gain column is the sensible heat gain for each room or zone.

For a residence, the latent heat gain is 30% of the sensible heat gain. The total heat gain for each room or zone is the sensible heat gain plus the latent heat gain. To simplify the mathematical process of adding 30% to each sensible load, the load is multiplied by a factor of 1.3. *Note:* The component gains are for sensible heat gain only. To find the total component gain, multiply the sensible heat gains by 1.3. See Figure 15-6.

Totals

Total heat loss or gain for a building is the sum of all the room or zone subtotals. The sum is shown on the far right bottom of the load form. The total of all room or zone subtotals must equal the total of all building component subtotals.

ADJUSTMENTS

To obtain the actual heating and cooling loads for a building, adjustments have to be made to the totals. Adjustments for duct loss or gain are made to the totals. Adjustments for the actual heating or refrigeration effect of HVAC equipment are also made to the totals.

Ductwork or Piping

Heat lost or gained through the ductwork or piping that distributes air or water from the heating and/or cooling system is part of the total load for the building. If any part of the ductwork or piping system runs through an unconditioned space, heat will be lost in the ductwork or piping. This loss is added to the total heating load of the building.

Heat gain through ductwork or piping is usually not added in the calculation. Heat gain through ductwork or piping is normally not significant because of the lower temperature difference between the air or water and the material around it. Ductwork and piping in cooling systems should always be insulated to reduce heat gain. The load adjustment for loss through ductwork or piping is made before the heating equipment is selected.

Load adjustment for ductwork or piping loss is found by using a table of duct heat loss multipliers that approximates the loss through the surfaces of the ductwork. A table of duct loss multiplies provides factors for finding ductwork heat loss.

A table of duct heat loss multipliers shows the increase in loss or gain as a percentage of the calculated loss or gain. See Appendix. To use the table, select the location and insulation of the ductwork and the winter outdoor design temperature. The ductwork heat loss multiplier value is multiplied by the building heating load. The result is an adjusted total that considers heat loss due to ductwork. Loss due to ductwork is found by applying the formula:

$$L_{DW} = TL \times DL$$

where

L_{DW} = loss due to ductwork (in Btu/hr)

TL = total loss (in Btu/hr)

DL = ductwork heat loss multiplier

Example: Finding Adjusted Total Due to Duct Heat Loss

A building has a total heat loss of 57,194 Btu/hr. The ductwork in the building runs the full length of the building in an unheated crawlspace. The ductwork is insulated with R-4 insulation. The outdoor design temperature is below 15°F. Find the adjusted heat loss due to ductwork.

On a table of duct loss multipliers, the duct loss multiplier is 1.10. See Appendix.

$$L_{DW} = TL \times DL$$

$$L_{DW} = 57,194 \times 1.10$$

$$L_{DW} = \textbf{62,913.4 Btu/hr}$$

Equipment Capacity

Heating and cooling equipment must have enough output capacity to satisfy the heating and cooling load on a building. The distribution system for a job must be sized for the equipment output capacity and not for the calculated load.

Since furnaces and air conditioners are manufactured in a limited selection of sizes, the equipment chosen for a job is usually oversized for the loads. Regardless of the capacity, the equipment is designed by the manufacturer to operate efficiently with a given supply of air or water circulating through it. If the heating and/or cooling equipment is oversized, the components of the distribution system must be oversized to fit the equipment.

Contractor: City Heating & Cooling
Name of Job: Jones Residence
Address: 1875 W. Grant St. Chicago, IL 60632

Winter: Indoor Design Temp. 75°F **Outdoor Design Temp.** −9°F **Design Temperature Difference** 84°F
Summer: Outdoor Design Temp. 90°F **Indoor Design Temp.** 75°F **Design Temperature Difference** 15°F

Date 1/14/92 **By:** L.P.

	Factors H	Factors C	1 LR Area	LR Btu/hr H	LR Btu/hr C	2 MAIN BR Area	BR H	BR C	3 BATH Area	BATH H	BATH C	4 BR2 Area	BR2 H	BR2 C	5 KIT./DR Area	KIT H	KIT C	Bldg. Comp. Subtot. H	Subtot. C
1 Name of Room			LR			MAIN BR			BATH			BR2			KIT./DR				
2 Running Feet of Exposed Wall			36			31			6			23			36				
3 Room Dimensions			23'×13'			12'×14'			12'×6'			12'×11'			23'×13'				
4 Ceiling Height — Exposure			8' S-E			8' W-S-E			8' W			8' N-W			8' N-E				
5 Gross Exposed Wall N/S/E/W (Area)			184 / 104			96 / 112			48			96 / 88			184 / 104				
6 Windows and Glass / Doors (H)	75		20	1500	20	20	1500	10	10	750	15	15	1125	15	50	3750	50	8625	1125
7 Windows and Glass / Doors (C) N		14												88			104	920	
E&W		44			460			460			440			440			3750	700	1100
S		23 / 20			112			48			10			96				490	660
8 Other Doors			20	1482	249										17	1302	369	2784	
9 Net Exposed Walls	21	.9	205	4305	181		1368	228		228	65		1014	119	205	4305	181	11220	362
10 Ceilings	2	.9	299	598	269	168	336	151	72	144	65	132	264	119	299	598	269	1940	873
11 Floors	6	.8	299	1794	239	168	1008	134	72	432	58	132	792	106	299	1794	239	5820	776
12 People @ 300 Appliances @ 1200			2 P		600	2 P									P + K		1500		2100
13 Subtotal Btu/hr Loss				9679			4212			1554			3195			11,749		30,389	
14 Sensible Btu/hr Gain					1998			1133			628			722			3708		8189
15 Subtotal Btu/hr Gain (1.3)					2597			1473			816			939			4820		10,646

Figure 15-6. Total heat loss or gain for a building is the sum of all the room or zone subtotals.

To oversize the system components, an equipment ratio factor is used. The equipment ratio factor is found by dividing the equipment capacity by the calculated load. The ratio is then multiplied by the heating and cooling load for each room to adjust for the output capacity of the equipment. This allows the distribution system to be designed for the output capacity of the equipment. The sheet containing the equipment adjustment ratios becomes part of the permanent record of the calculations for the job. Heating equipment ratio is found by applying the formula:

$$R_{HE} = \frac{O_F}{TL}$$

where

R_{HE} = heating equipment ratio

O_F = furnace output capacity (in Btu/hr)

TL = total load (in Btu/hr)

Example: Finding Heating Equipment Ratio

The total heat load for a building is 62,913 Btu/hr. The furnace with an output closest to and above 62,913 Btu/hr is a 75,000 Btu/hr model. Find the heating equipment ratio.

$$R_{HE} = \frac{O_F}{TL}$$

$$R_{HE} = \frac{75,000}{62,913.4}$$

$$R_{HE} = \mathbf{1.19}$$

After the air conditioning equipment has been selected for a job, the actual refrigeration effect of the equipment is compared with the calculated heat gain. The cooling capacity of the air conditioner is divided by the total calculated heat gain to find the cooling equipment ratio. The room loads are multiplied by this ratio to find the air conditioning equipment adjustment factor. Cooling equipment ratio is found by applying the formula:

$$R_{CE} = \frac{O_{AC}}{TG}$$

where

R_{CE} = cooling equipment ratio

O_{AC} = air conditioner output capacity (in Btu/hr)

TG = total heat gain (in Btu/hr)

Example: Finding Cooling Equipment Ratio

The total heat gain for a building is 17,559 Btu/hr. The air conditioner with an output closest to and above 17,559 Btu/hr is a 24,000 Btu/hr model. Find the cooling equipment ratio.

$$R_{CE} = \frac{O_{AC}}{TG}$$

$$R_{CE} = \frac{24,000}{17,559}$$

$$R_{CE} = \mathbf{1.37}$$

Example: Finding Heating and Cooling Load — Conventional Method

A sketch of a residence contains the dimensions required for calculating the heating and cooling load of the building. See Figure 15-7. Notes on the sketch describe construction features that affect the loads. The residence is located in Seattle, Washington. The dining room and kitchen are combined because they are not divided by a partition. The baths and utility room are combined because they are so small.

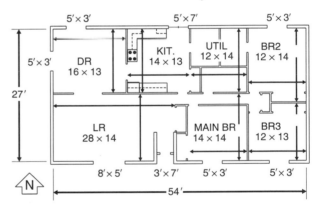

NOTES:

LOCATION – SEATTLE, WASHINGTON

WINDOWS – HORIZONTAL SLIDE, DOUBLE GLASS

DOORS – WOOD WITH STORM DOOR IN LIVING ROOM
DOUBLE SLIDING GLASS DOOR IN KITCHEN

WALLS – WOOD FRAME WITH SHEATHING AND
SIDING WITH R-11 INSULATION

CEILING – UNDER UNCONDITIONED SPACE WITH
R-38 INSULATION

FLOOR – WOOD OVER VENTED SPACE WITH R-11
INSULATION

SEATTLE RESIDENCE

Figure 15-7. A sketch of a residence shows the dimensions that are needed for heating and cooling load calculations. Notes on the sketch describe construction features that affect the loads.

1. Enter job data at the top of the load form.
2. Enter indoor design temperatures, outdoor design temperatures, and design temperature differences on the load form.

   The outdoor design temperature is taken from an outdoor design temperature table under the listing for Seattle, Washington. The winter design temperature is 17°F db and the summer design temperature is 84°F db. See Appendix.
3. Enter names of rooms, running feet of exposed wall, room dimensions, ceiling height, and the exposure of the individual rooms in the appropriate areas on the load form. See Figure 15-8.
4. Find the gross exposed wall area for each room.
5. Find the area of windows and glass doors. Enter the area on the form for the heating load. Enter the area on the form according to exposure for the cooling load. Calculate area of other doors and enter on the form.
6. Find net exposed wall area for the individual rooms by subtracting the area of the windows from the gross exposed wall area. Enter the net exposed wall area on the form.
7. Find area of ceilings and floors. Enter the area on the form.
8. Determine the number of people occupying the building based on the number of bedrooms. A heat gain factor of 300 Btu/hr per person is added to the dining room and living room.
9. Determine the heat gain from appliances. A heat gain factor of 1200 Btu/hr is added to the kitchen of the residence.

Contractor:	Best Heating & Cooling										Date: 1/6/92			
Name of Job:	Patrick Residence										By: G. Z.			
Address:	1541 N. Johnson													
	Seattle, WA 98031													

Winter: Indoor Design Temp. **75°F** Outdoor Design Temp. **17°F** Design Temperature Difference **58°F**

Summer: Outdoor Design Temp. **84°F** Indoor Design Temp. **75°F** Design Temperature Difference **9°F**

					1	LR		5	BR3		6	MAIN BR		Building
1	Name of Room													
2	Running Feet of Exposed Wall				42			25			14			Component
3	Room Dimensions				28' x 14'			12' x 13'			14' x 14'			Subtotals
4	Ceiling Height	Exposure			8'	S-W		8'	E-S		8'	S		
	Types of		Factors		Area	Btu/hr	Btu/hr	Area	Btu/hr	Btu/hr	Area	Btu/hr	Btu/hr	Btu/hr	Btu/hr
	Exposure		H	C		H	C		H	C		H	C	H	C
5	Gross	N													
	Exposed	S			224			96			112				
	Wall	E						104							
	Area	W			112										
6	Windows and Glass Doors (H)				40			15			15				
7	Windows and Glass Doors (C)	N													
		E & W													
		S			40			15			15				
8	Other Doors				21										
9	Net Exposed Walls				275			185			97				
10	Ceilings				392			156			196				
11	Floors				392			156			196				
12	People @ 300 Appliances @ 1200				2 P		600								

Figure 15-8. The areas of each of the rooms, windows and glass doors, other doors, net exposed wall area, ceilings, and floors in the building are entered on the load form.

10. Find factors for the heating load from a table of heat transfer factors. Find factors for cooling load from table of cooling heat transfer factors. See Appendix. See Figure 15-9.

1	Name of Room				
2	Running Feet of Exposed Wall				
3	Room Dimensions				
4	Ceiling Height		Exposure		
	Type of			Factors	
	Exposure			H	C
5	Gross	N			
	Exposed	S			
	Wall	E			
	Area	W			
6	Windows and Glass Doors (H)			60	
7	Windows and Glass Doors (C)	N			14
		E & W			44
		S			23
8	Other Doors			145	8.6
9	Net Exposed Walls			4	1.7
10	Ceilings			2	1
11	Floors			4	.8
12	People @ 300 Appliances @ 1200				

Figure 15-9. The factors for the heating load are found on a table of heat transfer factors. The factors for the cooling load are found on a table of cooling heat transfer factors.

11. Find heating and cooling load for each room by multiplying factors by variables for each room.
12. Find building component subtotals by adding the heating and cooling loads for all rooms in the building. See Figure 15-10.
13. Find the individual room loss or gain subtotals by adding the building component loss or gains for each room.
14. Make adjustments for ductwork and piping.
15. Find equipment adjustment ratios.

The required heating capacity is 25,293 Btu/hr. The heating unit with the nearest output is 30,000 Btu/hr. The heating adjustment ratio is 1.19. This value is multiplied by the subtotal heat loss for each room for designing the heating distribution system.

The cooling capacity required is 14,026 Btu/hr. The air conditioning unit with the nearest output is 24,000 Btu/hr. The cooling equipment ratio is 1.71. This value is multiplied by the subtotal heat gain

for each room for designing the air conditioning distribution system.

LOAD CALCULATIONS — COMPUTER-AIDED

Computer-aided load calculations are heating and cooling load calculations that are performed by a computer. A load calculation program is used to calculate heating and cooling loads for a building.

Hardware

Hardware consists of all physical units of a computer system. Hardware includes the computer, monitor, keyboard, disk drive, and printer. Many different kinds of computers are available. The computer system most commonly used to run load calculation programs is the personal computer (PC). Personal computers contain input devices, a central processing unit (CPU), and output devices. See Figure 15-11.

An *input device* is a device that allows an operator to enter information into the computer system. The most common input device is the keyboard. The operator strikes keys on the keyboard that are translated into corresponding codes. The computer system responds to the codes, which appear on the screen as characters.

The *central processing unit* (*CPU*) is the control center of the computer system. The CPU receives information through the input devices. The CPU also stores, manipulates, and controls the output of data. The CPU is controlled by a computer program. A computer program is a series of commands that controls the operation of the computer system.

The CPU stores information on disk drives and internal memory. A *disk drive* is a device that stores data on and retrieves data from a disk. One or more disk drives are part of a computer system. Some load calculation programs require two or more disks and disk drives for operation.

Output devices are devices that allow the CPU to output information. Output devices include monitors and printers. A *monitor* is a device that is similar to a television screen. A monitor displays information on the screen. A *printer* is a device that prints information on paper.

Project Information

Contractor:	Best Heating & Cooling
Name of Job:	Patrick Residence
Address:	1541 N. Johnson, Seattle, WA. 98031
Winter: Indoor Design Temp.	75°F
Summer: Outdoor Design Temp.	84°F
Outdoor Design Temp.	17°F
Indoor Design Temp.	75°F
Date	1/6/92
By:	[signature]
Design Temperature Difference (Winter)	58°F
Design Temperature Difference (Summer)	9°F

Heat Loss / Heat Gain Worksheet

	1 LR			2 DR/KIT			3 UTIL			4 BR2			5 BR3			6 MAIN BR			Totals	
	Area	Btu/hr H	Btu/hr C	Area	Btu/hr H	Btu/hr C	Area	Btu/hr H	Btu/hr C	Area	Btu/hr H	Btu/hr C	Area	Btu/hr H	Btu/hr C	Area	Btu/hr H	Btu/hr C	H	C
1 Name of Room	LR			DR/KIT			UTIL			BR2			BR3			MAIN BR				
2 Running Feet of Exposed Wall	42			43			12			26			25			14				
3 Room Dimensions	28'x14'			30'x13'			12'x13'			12'x14'			12'x13'			14'x14'				
4 Ceiling Height / Exposure	8' S–W			8' N–W			8' N			8' N–E			8' E–S			8' S				
6 Windows and Glass	60	3045	181	65	3900		15	900	15	15	900	15	15	900	15		900		9000	
8 Other Doors / Doors (C)	40	920	104	50	700	660											345			
9 Net Exposed Walls	275	1100	468	279	1116	474	96	384	163	193	772	328	185	740	315	97	388	165	4500	2403
10 Ceilings	392	784	392	390	780	390	156	312	156	168	336	168	156	312	156	196	392	196	2916	1458
11 Floors	392	1568	314	390	1560	312	156	624	125	168	672	134	156	624	125	196	784	157	5832	1167
12 People @ 300 / Appliances @ 1200			600 (2 P)			1800 (2A+k)														2400
13 Subtotal Btu/hr Loss	8897			7356			1320			2680			2576			2464			25,243	
14 Sensible Btu/hr Gain			2875			4336			934			840			941			863		10,789
15 Subtotal Btu/hr Gain (1.3)			3738			5637			1214			1092			1223			1122		14,026

(Building Component Subtotals shown in the Totals column.)

Figure 15-10. Total heat loss or gain for a building is the sum of all the room or zone subtotals.

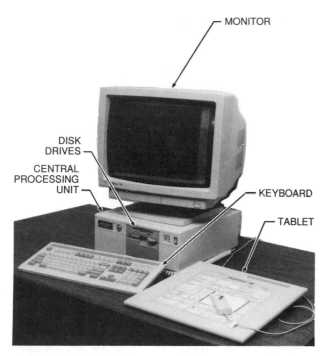

Figure 15-11. Personal computers consist of an input device, a central processing unit, and an output device.

Software

Software is the program that runs the computer system. The software distributes information to the computer system and directs various functions. A *load calculation software program* is a series of commands that requests data from the operator and manipulates the data to determine the heating and cooling loads. The process is similar to a conventional load calculation, except that the computer does the calculations. The data about a building is entered and recorded on a disk. The disk is converted to a format that is used for the calculations of the loads by the computer system.

A load calculation program requests the same variables and factors that are used for a conventional load calculation. The speed of the computer greatly reduces the amount of time required to perform the calculations. A person who operates load calculation programs should be familiar with conventional load calculation procedures.

Load calculation programs are available from engineering and technical associations, manufacturers of heating and air conditioning equipment, and other companies. An instruction manual is included in each program package. The instruction manual describes preparations that must be made to run the program and how to run the program.

When using a load calculation program, the program is run on a computer system and data is input when requested. The computer performs the required calculations. Most load calculation programs are written so the loads are calculated on rooms or zones automatically and the totals are obtained from the room or zone subtotals.

Most load calculation program instructions include a job information sheet. A job information sheet provides spaces for all information that is required for the calculations. The information includes the job name, date, design temperatures, dimensions, building materials, and other information. A job information sheet should be completed before running the program. Data from the sheet is then input to the computer.

When all of the information requested by the program has been input, the computer calculates the loads. When the calculations are complete, the results are shown on the monitor. Some computer load calculation programs print the results automatically.

Running a Program

To run a load calculation program, the software disk containing the program is inserted in a disk drive. The disk drive loads the program from the disk into the memory of the CPU. As the program runs, the computer requests the factors and variables. The information that is input depends on the program.

In most load calculation programs, questions that ask for input appear on the monitor. The input is data that is required for the calculation. The questions usually begin by requesting general information such as the job, geographical location, and design temperatures. The next series of questions request the sizes of the building, rooms, or zones. Some programs have factors stored on the disk. Questions relating to factors are about construction methods and materials. Programs that do not have factors on the disk ask for many factors. To accurately calculate a load, the factors are obtained from charts and tables of factors.

As a load calculation program runs, the operator inputs responses to questions that appear on the monitor. Explanations about procedure appear on the monitor, and many programs have options that give help when requested. A program instruction manual should answer any questions that the operator may have about the program. An instruction sheet that shows step-by-step procedures for running a program is usually included with a program.

After loads have been calculated, the results are shown on the monitor. The display includes a breakdown that shows subtotals for the components and for individual rooms or zones. Some programs also display information relating to the job variables, such as areas by exposure, infiltration data, and people and equipment input data. This information is used for checking the accuracy of the loads. A printout of the loads is made at this point. The printout may be used as a worksheet for checking the loads and is a permanent file copy.

The results of most load calculation programs are saved on a disk. Saving the job data is important when results need to be checked or modified. Several small jobs may be saved on one disk. Large jobs should be saved on individual disks.

Analysis

After load calculations for a building have been run, the results of the calculations should be checked for accuracy. Accuracy is important because the loads are used to design the entire heating or cooling system of a building. Computer programs are checked before they are sold to ensure correct operation, but errors can occur if information is input incorrectly. By comparing the results of the computer load calculations with estimated loads, accuracy is assured.

Total Loads. An analysis of computed loads usually starts with a check of the total loads. The total loads from the computer program can be compared with a standard for the type and location of the building or with a load calculated with the conventional method for a part of the building. Comparisons are made by preparing a sheet showing the results from the computer calculations and the results from a standard or conventional calculation method. See Figure 15-12.

Components		Heat Gain			Heat Loss		
		Conventional	Computer		Conventional	Computer	
Walls		1371	1128		3652	4361	
Windows	Includes solar gain and infiltration	7155	6021	Includes solar gain & infiltration	10,500	6191	
Doors		158	756		3045	2923	
Ceiling		1054	874		2108	3376	
Floor		528	611		4216	4725	
Infiltration			840			3240	
People & Appl.		2400	1200 + 871				
Subtotals		12,666	12,301				
		× 1.3	× 1.3				
Totals		16,466	15,991		23,521	24,816	

Figure 15-12. Comparisons for checking computer-aided load calculations are made by preparing a worksheet that shows the results from the computer calculations and the results from a standard or a conventional calculation method.

Because different factors are used for each load calculation method, there is usually some difference between computer-aided load figures and conventional load figures. The totals of a computer-aided load calculation should be within 5% of the totals taken from a reliable standard. If the totals are not within 5%, the calculations should be checked further to find the reason for the difference.

Subtotals. Most commercial computer programs are written so a printout of the results shows subtotals of losses and gains for the rooms or zones in a building. These subtotals should be checked against a reliable standard to see if any errors appear. A positive error in one component can be offset by a negative error in another. If there is a noticeable difference between the computer subtotals and the standard, the component factors and variables used to get the subtotal should be checked. The best way to compare subtotals is by listing them on a comparison sheet.

Component Totals. If the computer printout shows component losses and gains, the component totals can be checked to see if any noticeable differences appear. A quick check of the component loads may indicate a higher or lower load on one component in a particular part of the building.

Variables and Factors. Make sure all data on the job information sheet is correct, and verify the data on the job information sheet against the data input in the computer. The computer program calculates component subtotals the same way they are calculated in a conventional program. Variables are multiplied by factors and the results are added.

Some load calculation programs include heat loss for ductwork and some do not. For many programs, adjustments must be made to the calculated loss data for the actual capacity of the equipment. This step must be done for a conventional calculation.

Review Questions

1. List two ways that heating and cooling load calculations can be performed.

2. List two different load forms that are used for calculating heating and cooling loads.

3. Why should a cooling load form be filled in first when preparing forms for calculating loads?

4. List at least five of the six variables that are considered in a load calculation.

5. Which variable is not included in heating loads, but must be considered when calculating a cooling load?

6. Define *design temperature.*

7. Where are design temperatures found?

8. How is information that is needed for calculating heating and cooling loads obtained when plans and specifications for a building are not available?

9. Explain the difference between gross wall area and net wall area.

10. How are hallways, closets, and stairways included in load calculations?

11. List and describe three ways that infiltration loads are calculated.

12. What does air change mean in relation to infiltration loads?

13. How many people would be included in the calculations for the heating and cooling loads for a three-bedroom residence?

14. A heat transfer factor combines a conduction factor with what other element?

15. Why is the heating load for people not usually included in a residential heating load?

16. Where are factors included on a heating and cooling load form?

17. What is a building component?

18. How is sensible heat gain used to account for latent heat gain?

19. List and explain two adjustments that may have to be made to a heating or cooling load after the loads are calculated.

20. Why is the heating or cooling equipment usually oversized when it is chosen for a job?

21. Why do the heating and cooling loads have to be adjusted for the equipment size before a system is designed?

22. Describe how an equipment ratio is found.

23. What is a computer-aided load calculation?

24. List and describe five components of a computer system.

25. What program is used for calculating heating and cooling loads?

26. How can errors occur when running a computer program?

Equipment Selection

Chapter 16

When designing a heating system or an air conditioning system for a building, selecting the proper heating and air conditioning equipment is important. A properly selected heating or air conditioning system heats or cools a building with maximum efficiency. To design an efficient system, the kind of equipment used, size and construction of the building, availability and cost of fuels and electricity, and size and construction of the distribution systems must be considered.

FORCED-AIR HEATING SYSTEMS

A forced-air heating system consists of a furnace, blower, ductwork, registers and grills, and controls. Air is heated in a furnace and distributed to building spaces. See Figure 16-1.

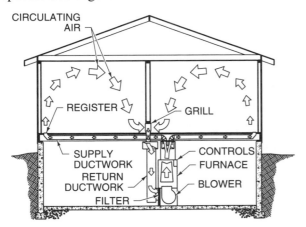

Figure 16-1. In a forced-air heating system, air is heated in a furnace and distributed to building spaces through ductwork.

Furnaces

The furnace is the central component in a forced-air heating system. The fuel or energy used for producing heat, direction of air flow, and output capacity vary among different furnaces. Design technicians should be familiar with standard furnace sizes that are available from furnace manufacturers. A *design technician* is a person who has the knowledge and skill to plan heating and/or air conditioning systems. A design technician selects the proper equipment for meeting the specifications of a heating or air conditioning system.

Catalogs and engineering specifications sheets from manufacturers are the best source of information about heating and air conditioning equipment. Catalogs and engineering specifications sheets contain descriptions of and specifications for furnaces. See Figure 16-2.

Fuel and Electricity. Furnace selection is based on the availability and cost of fuel or electricity. A natural gas-burning furnace is often chosen because natural gas is available in most areas. Using natural gas is convenient because the fuel can be piped directly to a furnace. Natural gas is available in most areas, is a clean-burning fuel, and has a reasonable cost compared to other fuels.

If natural gas is not available, either LP gas or fuel oil may be used. LP gas-burning furnaces are similar to natural gas-burning furnaces, and many LP gas-burning furnaces can be converted to natural gas-burning furnaces. The cost of LP gas is usually higher than other fuels.

	UPFLOW GAS FURNACES						
Model Number	Dimensions*			Input Rating§	Output Rating§	Shipping Weight#	AFUE**
	L	W	H				
005-AX-10	28.5	14	46	40,000	30,000	100	75
010-AX-10	28.5	14	46	44,000	35,000	105	79
050-AX-6	28.5	18	46	50,000	38,000	120	76
070-AX-5	28.5	20	46	75,000	58,000	120	77
100-AX-3	28.5	20	46	100,000	75,000	166	75
200-AX-3	28.5	20	46	125,000	97,000	178	78
300-AX-6	28.5	24	46	132,000	105,000	205	80
400-AX-0	28.5	24	46	150,000	127,000	255	85

* in inches
§ in Btu/hr
# in lb
** Annual fuel utilization efficiency

Figure 16-2. Catalogs and engineering specifications sheets contain descriptions of and specifications for furnaces.

LP gas is transported in tanks to rural areas where it is used for residential heating. LP gas-burning furnaces are also used for standby in commercial or industrial applications.

Fuel oil-burning furnaces are used for residential, commercial, and industrial heating. Fuel oil is usually delivered by truck or railroad tank car and is stored in tanks. Fuel oil is also used in rural areas where there are no underground gas lines.

Electric furnaces are often used because electricity is available in most areas. The combustion process is not required in electric furnaces, and the equipment is fairly simple. Heat produced by electricity produces no air contamination at the point of use. Electricity is used for heating in residences and in commercial and industrial applications.

Coal is used primarily for industrial heating applications. Coal is used to produce electricity for electric furnaces. Transporting and storing coal is costly unless the coal is used in large quantities.

Air Flow Direction. Three categories of furnaces are based on the direction of air flow out of the furnace. Upflow, downflow, and horizontal are the three categories of furnaces. Each of the categories is used for a different application. See Figure 16-3.

In an upflow furnace, heated air flows upward to pass through and out of the furnace. Upflow furnaces are used where the supply ductwork is located above the furnace. In a residence with a basement, an upflow furnace is located in the basement and the supply ductwork for the upper floors is located in the ceiling of the basement. An upflow furnace can also be located on the main floor of a building where the ductwork is located along the ceiling of the same floor.

In a horizontal furnace, heated air flows horizontally to pass through and out of the furnace. Return air enters horizontally on one end of the furnace and supply air exits horizontally on the other end. Horizontal furnaces are installed where the furnace and ductwork are located in an attic or crawlspace of a building. The ductwork of a horizontal furnace is located on the same level as the furnace.

In a downflow furnace, heated air flows downward to pass through and out of the furnace. Downflow furnaces are used where the ductwork is located below the furnace. An example is a furnace that is located on one floor of a building and the supply ducts are in the ceiling of the floor below.

Furnaces are often designated as residential, commercial, or special purpose. The difference between residential and commercial furnaces is the output capacity. Furnaces with output capacities of up to 300,000 Btu/hr are used for residential applications. Furnaces with output capacities higher than 300,000 Btu/hr are used for commercial applications.

Special purpose heating units include unit heaters, duct heaters, space heaters, and direct-fired heaters.

Depending on the output of the heater, unit heaters and duct heaters may be used for either residential or commercial applications. Unit heaters and duct heaters are found more often in commercial or industrial applications, where temperature control is not precise. Direct-fired heaters are used almost exclusively for industrial applications because the heaters require 100% outdoor air. A special purpose heating unit that is selected should have the best characteristics for a particular application.

Residential or small commercial buildings have heating loads up to 300,000 Btu/hr. Larger commercial or industrial buildings have heating loads that are greater than 300,000 Btu/hr. A commercial building would require a commercial-size furnace. If the building has a flat roof, there will be little room in the ceiling space for ductwork. If there is space for ductwork between roof trusses and in hallway ceilings, a heating system that contains one or more furnaces for that particular type of construction will be required.

Heating Capacity. The heating capacity (output rating) of a furnace is the heat output of the furnace at design conditions. *Design conditions* are the conditions of the air at which the heating or air conditioning equipment provides comfort in building spaces. Catalogs and specifications sheets for furnaces contain input and output ratings for furnaces.

Input rating is the amount of heat produced per unit of fuel as fuel burns. A combustion furnace should burn fuel as efficiently as possible. The heat exchanger in a furnace should transfer as much heat as possible from the hot products of combustion to the distribution system.

Output rating is the amount of heat that a furnace produces in 1 hour. The input rating and output rating of a furnace are different because of the heat loss due to the heat allowed to flow up the flue to create a draft. In a furnace with a conventional burner and heat exchanger, 20% of the heat is lost to provide a draft up the flue. A burner relies on heat from the burner to provide draft. A conventional heat exchanger should extract 80% of the heat from the products of combustion. The output rating of a furnace with a conventional heat exchanger is 80% of the input rating. See Figure 16-4.

High-efficiency furnaces with power burners and condensing heat exchangers have higher output efficiencies than conventional furnaces. The output rating of a high-efficiency furnace may be as high

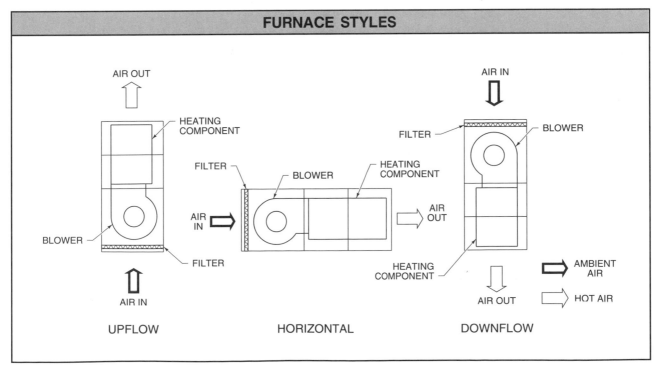

Figure 16-3. Upflow, downflow, and horizontal furnaces are used for different applications.

as 97% of the input rating. The output rating of a furnace in a building should always be equal to or higher than the adjusted heat loss of the building.

Electric furnaces have no difference between input and output ratings because combustion is not involved, and draft is not necessary. Some efficiency loss occurs in electric furnaces due to energy lost in the resistance heating elements.

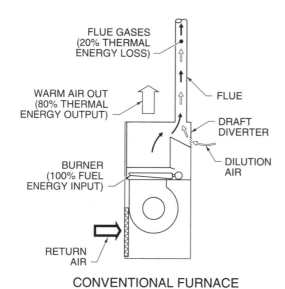

Figure 16-4. Conventional combustion furnaces have an output capacity loss because some heat flows up the flue to create a draft. A conventional furnace loses 20% of its heating capacity for draft.

Ductwork

Selecting a ductwork system for a forced-air heating system depends on the location of the ductwork. In a building that has no room for ductwork in the building frame, the ductwork may be placed under the floor. The ductwork in a multistory building may be installed in a combination ceiling/floor structure between floors. In a building that covers a large area, the ductwork may be located in the ceiling spaces or on the roof.

Registers and Grills

The location of registers in a building depends on the supply air pattern desired in each space and the location of the supply ductwork. If the supply ductwork is under the floor, the floor or the low sidewall will be the best place for the register. If the ductwork

is overhead, the ceiling or high sidewall will be the best place for the register.

Grills are located where a connection to the return air ductwork can be made. Air return efficiency in all parts of the building is also considered.

Blowers

The blower in a forced-air heating system is usually part of the furnace. In a built-up heating system, the blower should move the required quantity of air using a small amount of energy. If the blower is located at a point where noise should be low, the noise level of the blower should also be considered.

Control System

The use of a building determines what kind of control system works best for the building. If the only requirement for a building space is to keep the temperature above the freezing point, the control system will not be sophisticated. One thermostat located centrally in the building may be the only control required. If close control of the temperature in individual spaces in a building is required, such as in a range of a few degrees, a control system that provides that level of sophistication will be required.

Heating Air Flow

In a forced-air heating system, the furnace heats air and the blower circulates the air through a building. The *air flow rate* is the volumetric flow rate of air circulated by a blower. Air flow rate is related to the output rating of the furnace and the temperature increase of the air as it flows through the heat exchanger. The air flow rate from a blower is found by applying the formula:

$$Q = \frac{OR}{\Delta T \times 1.08}$$

where

Q = air flow rate from blower (in cfm)

OR = output rating of furnace (in Btu/hr)

ΔT = temperature increase of the air (in °F)

1.08 = constant

Example: Finding Air Flow Rate — Furnace

A furnace has an output rating of 240,000 Btu/hr and a temperature increase of 88.6°F. Find the air flow rate from the blower.

$$Q = \frac{OR}{\Delta T \times 1.08}$$

$$Q = \frac{240,000}{88.6 \times 1.08}$$

$$Q = \frac{24,000}{95.7}$$

Q = 2508 cfm

Forced-air heating equipment operates most efficiently with a certain temperature increase through the heat exchanger. Temperature increase through a heat exchanger is the difference between the temperature of the air entering the heat exchanger and the temperature of the air leaving the heat exchanger.

Combustion furnaces usually operate with a temperature increase of about 80°F to 100°F. Electric furnaces usually operate with a temperature increase of around 50°F or less. Heat pumps without auxiliary heat operate with a temperature increase of about 20°F. Heat pumps with auxiliary heat operate with a temperature increase of about 50°F or less.

After selecting a furnace that provides the amount of heat required to heat a building, the air flow rate must be found. Air flow rate is the amount of air in cubic feet per minute that meets the heating requirements of a building. The air flow rate required for heating a building is a function of the furnace output rating and the temperature increase of the air through the heat exchanger.

For spaces occupied by people, air velocity must not be above 40 fpm. Air that strikes a person at a rate above 40 fpm feels uncomfortable because of the cooling effect of moving air. For a typical application, the air flow rate in a building space should be less than 1 cfm per square foot of floor space. This flow rate can be increased if the area is not occupied by people. The air flow rate required to heat a building space without causing drafts is a function of the air flow rate and area of the building. The proper air flow rate per unit area of building space is found by applying the formula:

$$Q_A = \frac{Q}{A}$$

where

Q_A = air flow rate per unit area (in cfm/sq ft)

Q = air flow rate (in cfm)

A = area of building (in sq ft)

Example: Finding Air Flow Rate per Unit Area

A room has an area of 1200 sq ft. A furnace blower circulates 1000 cfm of air. Find the proper air flow rate for heating the room.

$$Q_A = \frac{Q}{A}$$

$$Q_A = \frac{1000}{1200}$$

Q_A = .83 cfm/sq ft

The flow rate of this heating system is acceptable because the flow rate is less than 1 cfm/sq ft. The heating system will not create a draft.

FORCED AIR CONDITIONING SYSTEMS

In a forced air conditioning system, air is the evaporating medium. Forced air conditioning systems contain ductwork, registers and grills, blowers, and controls that are similar to the equipment used in forced-air heating systems. A forced air conditioning system contains an air conditioner that cools air. The air is distributed to building spaces.

Air Conditioners

The central component in a forced air conditioning system is the air conditioner. An air conditioner produces a refrigeration effect, which causes heat transfer from air to a cooler substance. Air conditioners that use mechanical compression for a refrigeration effect are commonly used. Mechanical compression air conditioning systems use electricity as a source of energy.

Some air conditioners use absorption refrigeration to provide the refrigeration effect. An absorption refrigeration system uses electricity to operate motors and controls but may use gas or oil to operate the generator in the system. When some form of waste heat can be used for operating the system, an absorption refrigeration system may be better than a mechanical compression refrigeration system. Waste heat is steam or hot water that is discharged after a manufacturing process.

The arrangement of an air conditioning system depends on the construction of the building. Air conditioners may be arranged as package systems and split systems. A package air conditioning system consists of a cabinet that holds all operating components. A split system consists of different cabinets that hold the operating components.

Package air conditioning systems are used where ductwork to building spaces can be kept relatively short and where individual control of building spaces is needed. Package systems are often located on the roof of a building or building space.

For example, package air conditioning units are used in large offices or retail stores that have large spaces with high internal heat gains. The supply ductwork system in these applications is short, and multiple units cover the heat gain of the space.

Split systems are used when the evaporator is located in one part of the air distribution system, and the condenser and compressor must be located in another part of the air distribution system. Split systems are used where air-cooled condensers are placed outdoors and the evaporator is located in ductwork inside the building. Many large air conditioning systems are delivered from a factory as components, which are assembled at the job site. These systems are usually split systems. In a split air conditioning system, air-cooled condensers are often located on the roof of a building and blower coil units are located in the building spaces.

Cooling Capacity. The cooling capacity of an air conditioner is the refrigeration effect produced by an air conditioner. The cooling capacity of an air conditioner must meet the cooling load of the building spaces. When selecting an air conditioner for an application, the cooling capacity and the air flow rate must be considered.

The cooling capacity of an air conditioner must be as great or greater than the calculated cooling load of the building. Air conditioners produce a certain refrigeration effect. The output rating (cooling capacity) of air conditioning equipment is expressed in Btu per hour or in nominal tons of cooling. One ton of cooling is equal to 12,000 Btu/hr. Nominal size is the closest rounded-off size. Nominal size is found by applying the formula:

$$N = \frac{cc}{12,000}$$

N = nominal size (in tons)

cc = cooling capacity (in Btu/hr)

12,000 = constant

Example: Finding Nominal Size

An air conditioning system has a cooling capacity of 49,000 Btu/hr. Find the nominal size of the air conditioning system.

$$N = \frac{cc}{12,000}$$

$$N = \frac{49,000}{12,000}$$

$N =$ **4 tons**

Nominal sizes of refrigeration equipment range from one-half ton increments in smaller sizes to multiple ton increments in larger sizes. Equipment catalogs show the different types and sizes of air conditioning systems and include engineering specifications. See Figure 16-5.

Package air conditioners and components of built-up systems are rated at standard conditions per the Air Conditioning and Refrigeration Institute. The *Air Conditioning and Refrigeration Institute* (*ARI*) is an association of manufacturers of refrigeration and air conditioning equipment and allied products. An important function of this association is establishing product standards for air conditioning and refrigeration equipment. ARI standard conditions for different types of equipment are described in ARI bulletins. The conditions for standard equipment are described in Standard 210-81 for Unitary Air-Conditioning Equipment.

If a temperature swing above the desired temperature is acceptable, some adjustment may be made when selecting an air conditioner. Temperature swing is a variation in the actual indoor temperature compared to the setpoint temperature. Because the hottest temperatures and highest humidity occur for only a few hours of a year, some temperature swing may be acceptable.

AIR CONDITIONING SYSTEMS							
Model Number	**Dimensions***			**Nominal Tons**	**Nominal CFM**	**Operating Weight§**	**EER**
	L	**W**	**H**				
K150-Q	15	6	4	10	8000	1516	9.3
K250-Q	15	6	4	15	12,000	2200	9.2
K500-Q	16	6	4	20	16,000	3027	9.0
K1000-Q	16	6	5	25	20,000	3976	8.7
K2400-P	16	6	5	30	24,000	4621	8.6
K3200-P	16	6	5	40	36,000	5307	8.2

* in ft
§ in lb

Figure 16-5. Air conditioning systems are rated in nominal sizes that are listed in catalogs and engineering specifications sheets.

If temperature swing is acceptable, a table of temperature swing factors is used. A table of temperature swing factors gives factors that are multiplied by the heat gain for a building. The factors are based on the outdoor design temperature and the acceptable temperature swing. When selecting an air conditioner using a temperature swing factor table, a smaller air conditioner may be used to offset the same heat gain to the building. See Figure 16-6.

Cooling Air Flow

When designing an air conditioning system, the air flow rate across the evaporator coil and the air flow rate circulated through building spaces must be considered. The air flow rate across the evaporator coil of an air conditioner is important for proper operation. The air flow rate circulated through building spaces is important for providing comfort.

Most air conditioners operate with about 400 cfm of air across the evaporator coil per ton of cooling. This air flow rate provides a temperature drop of around 20°F across the evaporator coil. A 20°F temperature drop is usually used for cooling calculations. The air flow rate to circulate across the evaporator coil of an air conditioner is found by applying the formula:

$$Q = N \times 400$$

where

Q = air flow rate across evaporator (in cfm)

N = nominal size (in tons of cooling)

400 = constant

Example: Finding Air Flow Rate — Air Conditioner

A building has a cooling load of 35,000 Btu/hr. The air conditioner for the building has a cooling capacity of 36,000 Btu/hr. Find the air flow rate to be circulated across the evaporator coil.

1. Find the nominal size of the air conditioner.

$$N = \frac{cc}{12,000}$$

$$N = \frac{36,000}{12,000}$$

$N = \textbf{3 tons}$

2. Find the air flow rate across the evaporator coil.

$$Q = N \times 400$$

$$Q = 3 \times 400$$

$Q = \textbf{1200 cfm}$

Because sensible heat and latent heat are involved in cooling processes, the actual performance of an air conditioner should be checked with a psychrometric chart to ensure that the refrigeration effect produced can cool the building spaces. An air conditioner must remove the heat from the air and the heat from the moisture in the air to produce a temperature decrease. The condition of the air as it leaves an evaporator coil is found and the actual performance of an air conditioner is checked by applying the procedure:

1. Use a psychrometric chart to find the point that represents the condition of the air as it enters the coil. This point is located at the intersection of the dry bulb temperature and relative humidity lines on the chart.

TEMPERATURE SWING FACTORS

Outdoor Design Temperature*	Desired Indoor Temperature Swing*														
	6					4½					3				
AIR-COOLED UNITS															
85-90	0.07					0.85					1.00				
95	0.75					0.90					1.05				
100	0.80					0.95					1.10				
105	0.85					1.00					1.15				
110	0.90					1.05					1.20				
WATER-COOLED UNITS															
Outdoor Design Temperature*	Leaving Water Temperature*														
	90	95	100	105	110	90	95	100	105	110	90	95	100	105	110
85-90	0.74	0.77	0.81	0.83	0.86	0.88	0.93	0.97	1.00	1.03	1.03	1.08	1.12	1.16	1.20
95	0.78	0.82	0.85	0.88	0.91	0.93	0.97	1.01	1.05	1.08	1.07	1.12	1.17	1.21	1.25
100	0.82	0.87	0.91	0.94	0.97	0.97	1.02	1.06	1.09	1.12	1.10	1.16	1.20	1.24	1.28
105	0.86	0.90	0.94	0.97	1.00	1.00	1.05	1.10	1.14	1.17	1.15	1.20	1.25	1.30	1.34
110	0.89	0.94	0.98	1.01	1.04	1.04	1.09	1.14	1.18	1.22	1.18	1.24	1.30	1.34	1.38

* in °F •Air Conditioning Contractors of America

Figure 16-6. If temperature swing is considered when selecting an air conditioner, a table showing adjustments for various temperature swings is used.

If the air entering the evaporator coil of an air conditioner has a dry bulb temperature of 75°F and 50% relative humidity, the condition of the air will be located at the intersection of the 75°F dry bulb temperature line and the 50% relative humidity lines on a psychrometric chart. See Figure 16-7.

2. The specific volume of the air at this point is found by drawing a line parallel with the specific volume lines on the chart and approximating the specific volume. The specific volume of the air at this point is approximately 13.7 cu ft/lb.

3. The enthalpy of the air at this point is found by drawing a line parallel with the enthalpy lines that extend to the enthalpy scale on the left side of the chart. The enthalpy of the air at this point is 27.4 Btu/lb.

4. The mass flow rate of the air at this point is found by dividing the air flow rate by the specific volume of the air and multiplying by 60. Multiplying by 60 converts mass flow rate from pounds per minute to pounds per hour. Mass flow rate of the air is found by applying the formula:

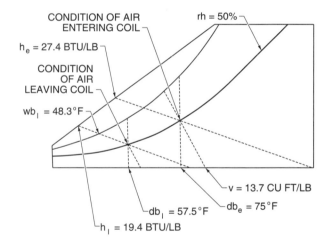

Figure 16-7. A psychrometric chart is used to find the temperature decrease across an evaporator coil in an air conditioner.

$$w = \frac{Q}{v} \times 60$$

where

w = mass flow rate (in lb/hr)

Q = air flow rate (in cfm)

v = specific volume of air (in cu ft/lb)

60 = constant

The blower in the air conditioner circulates 1400 cfm of air and the specific volume is 13.7 cu ft/lb. Find the mass flow rate of the air.

$$w = \frac{Q}{v} \times 60$$

$$w = \frac{1400}{13.7} \times 60$$

$$w = 102.2 \times 60$$

$$w = \textbf{6132 lb/hr}$$

5. The refrigeration effect of the coil is found by dividing the cooling capacity of the air conditioner by the mass flow rate of the air. Refrigeration effect is found by applying the formula:

$$RE = \frac{cc}{w}$$

where

RE = refrigeration effect (in Btu/lb)

cc = cooling capacity (in Btu/hr)

w = mass flow rate (in lb/hr)

The air conditioner has a cooling capacity of 49,000 Btu/hr and the mass flow rate is 6132 lb/hr. Find the refrigeration effect.

$$RE = \frac{cc}{w}$$

$$RE = \frac{49,000}{6132}$$

$$RE = \textbf{8.0 Btu/lb}$$

6. The enthalpy of the air as it leaves the evaporator coil is found by subtracting the refrigeration effect of the evaporator coil from the enthalpy of the air as it enters the evaporator coil. The enthalpy of the air as it leaves the evaporator coil is found by applying the formula:

$$h_1 = h_e - RE$$

where

h_1 = enthalpy of refrigerant leaving the evaporator coil (in Btu/lb)

h_e = enthalpy of refrigerant entering the evaporator coil (in Btu/lb)

RE = refrigeration effect (in Btu/lb)

The refrigeration effect of the air conditioner is 8.0 Btu/lb. The refrigerant enters the evaporator coil

with an enthalpy of 27.4 Btu/lb. Find the enthalpy of the refrigerant as it leaves the evaporator coil.

$$h_1 = h_e - RE$$

$$h_1 = 27.4 - 8.0$$

$$h_1 = \textbf{19.4 Btu/lb}$$

7. On the psychrometric chart, use the relative humidity and enthalpy of the air as it leaves the evaporator to locate the points that represent the dry bulb and wet bulb temperatures of the air leaving the evaporator coil.

The air leaving the evaporator coil is at 50% relative humidity and the enthalpy of the air as it leaves the evaporator coil is 19.4 Btu/lb. The condition of the air leaving the coil is located at the intersection of the 19.4 Btu/lb enthalpy line and the 50% relative humidity line.

The dry bulb temperature of the air at this point is found by following the dry bulb temperature lines to the dry bulb temperature scale on the bottom of the chart. The dry bulb temperature of the air leaving the coil is 57.5°F.

The wet bulb temperature of the air at this point is found by following the wet bulb temperature lines to the wet bulb temperature scale on the curve of the chart. The wet bulb temperature of the air leaving the coil is 48.3°F.

8. Find the temperature decrease.

$$\Delta T = T_1 - T_2$$

$$\Delta T = 75 - 57.5$$

$$\Delta T = \textbf{17.5°F}$$

This temperature decrease is less than 20°F, which is usually used for air conditioning systems. This example illustrates that additional cooling is required because of the latent heat in the air.

Example: Finding Temperature Decrease — Air Conditioner

Air enters the evaporator coil of an air conditioner at a dry bulb temperature of 80°F and 50% relative humidity. The air conditioner has a cooling capacity of 36,000 Btu/hr, and the air flow rate is 1200 cfm across the evaporator coil. Find the dry bulb and wet bulb temperatures of the air leaving the coil when the relative humidity of the air leaving the coil is 50%.

1. Locate the intersection of the 80°F db temperature line and the 50% relative humidity line on a psychrometric chart. See Figure 16-8.

2. Find the specific volume for this point. Specific volume is 13.85 cu ft/lb.

3. Find enthalpy for this point. Enthalpy is 31.8 Btu/lb.

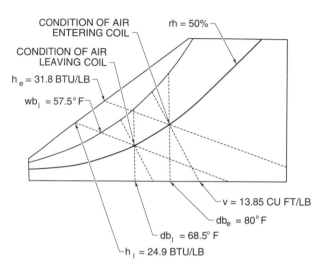

Figure 16-8. The actual performance of an air conditioner is checked by using a psychrometric chart to find the temperature decrease.

4. Find mass flow rate.

$$w = \frac{Q}{v} \times 60$$

$$w = \frac{1200}{13.85} \times 60$$

$$w = 86.6 \times 60$$

$$w = \textbf{5196 lb/hr}$$

5. Find refrigeration effect.

$$RE = \frac{cc}{w}$$

$$RE = \frac{36,000}{5196}$$

$$RE = \textbf{6.9 Btu/lb}$$

6. Find enthalpy of the air as it leaves the evaporator coil.

$$h_l = h_e - RE$$

$$h_l = 31.8 - 6.9$$

$$h_l = \textbf{24.9 Btu/lb}$$

7. Find dry bulb and wet bulb temperature of the air leaving the evaporator coil at 50% relative humidity.

$$db = \textbf{68.5°F}$$

$$wb = \textbf{57.5°F}$$

8. Find the temperature difference.

$$\Delta T = T_1 - T_2$$

$$\Delta T = 80 - 68.5$$

$$\Delta T = \textbf{11.5°F}$$

The temperature decrease is 11.5°F. This is less than 20°F, which is usually used for air conditioning systems. This example illustrates how additional cooling is required to cool air a given temperature because of the latent heat in the air.

HYDRONIC HEATING SYSTEMS

A hydronic heating system is a heating system in which water, steam, or other fluid is the heating medium. Water is used most often in hydronic heating systems because it has a relatively high specific weight and specific heat.

Hydronic heating systems are used where the heating equipment is located far from building spaces that require heat. Hydronic heating systems are also used when several buildings are heated by a central heating plant. Campus school buildings are an example. Another example is a multistory building where the heating equipment is located on one floor and hot water is distributed to terminal devices located on other floors.

A hydronic heating system consists of a boiler, piping, terminal devices, circulating pump(s), and controls. The central component in a hydronic heating system is a boiler. See Figure 16-9.

Boilers

A boiler is a pressure vessel that produces hot water in a hydronic heating system. Heat is produced in a boiler by burners or electric heating elements. When selecting a boiler, a design technician should consider the size of the boiler, the amount of heat required, and the availability and cost of fuels and electricity. Boilers are categorized according to boiler working pressures and temperatures, method of heat production, and heating capacity.

Pressures and Temperatures. Most boilers used for hydronic heating systems are low-pressure boilers. Low-pressure boilers operate at working pressures up to 15 psi steam or 160 psi water. Low-pressure steam boilers are used primarily for heating warehouses, factories, schools, and apartments. Low-pressure hot water boilers operate with water temperatures up to 250°F. Low-pressure hot water boilers are used for residential heating.

High-pressure boilers operate at working pressures above 15 psi steam or above 160 psi water. High-pressure steam boilers are used for processing operations in industry and for generating electricity. High-pressure hot water boilers operate with water temperatures of 250°F or greater and are used primarily for heating systems in which more than one building is to be heated. An example is a school campus where several buildings are heated by one boiler. In this application, the water is piped from the boiler to the various buildings.

Fuel and Electricity. Boilers burn fuel or use electricity to produce heat. Combustion boilers burn fuel such as natural gas, fuel oil, or coal to produce heat. Natural gas- and fuel oil-fired boilers are used primarily in residential and commercial buildings. The availability and convenience of using a fuel determines which fuel is used for an application. Coal-fired boilers are usually used in large commercial or industrial applications where the coal is used in large quantities. These larger applications have facilities that can supply and handle large amounts of coal. Electric boilers use heat produced by resistance heating elements. Electric boilers are often used in residential and commercial applications. The electrical wiring in a building is a convenient source of energy for the boiler.

Many hydronic heating systems can use solar energy, reclaim heat from exhaust air, or transfer heat from one part of a building to another. In a hydronic solar heating system, water is heated directly by the sun. Collectors, which are placed on the roof or other outside surface of a building, efficiently change radiant energy to thermal energy. The thermal energy is transferred to the water. The heated water is then stored in a tank and used for heating the building.

A heat reclaim system collects heat from heated air or water that is exhausted from a building. A hot water heat reclaim coil is placed in the exhaust piping. The waste heat is transferred to the water inside the coil. The heated water is then stored and used as a heat source.

In some applications, water-to-air heat pumps are used to heat and cool a building. Water from a heat pump is heated and transferred through the piping

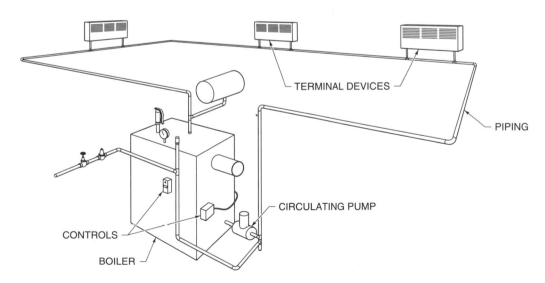

Figure 16-9. In a hydronic heating system, water is heated in a boiler and circulated through piping to terminal devices located in building spaces.

system to a part of the building that requires heat. The water is then used as a heat source for other heat pumps. In this application, the entire distribution system must be designed to use water for both cooling and heating. The amount of heat that must be supplied by a boiler for makeup heat is relatively small. Electric boilers often provide the makeup heat required in these systems.

Heating Capacity. Natural gas- and fuel oil-fired boilers have heating capacities from about 100,000 Btu/hr for residential applications to many hundreds of thousands of Btu/hr for commercial applications. Coal-fired boilers are usually used for large industrial applications, which require millions of Btu/hr. Coal-fired boilers can produce electricity and heat.

A boiler that is selected for heating a building should have a heating capacity that is as great or greater than the heating load on the building. Equipment catalogs and engineering specifications sheets are used to select a boiler for an application. See Figure 16-10.

Piping System

A hydronic piping system distributes water from a boiler to terminal devices, where it is used to heat air. The pipes in a hydronic piping system are much smaller than the ductwork used to carry air in a forced-air heating system.

Piping systems are categorized by the arrangement of the piping loop that carries the water. A typical piping system has a supply line that carries water from the boiler to the terminal devices and a return line that carries water from the terminal devices to the boiler. Piping systems used for hydronic heating systems include one-pipe, two-pipe, three-pipe, and four-pipe systems. Each piping system consists of loops of pipe that include a supply and return line.

One-pipe piping systems are economical to install and give good temperature control in small buildings. One-pipe piping systems are often used in residential and small commercial buildings. Two-pipe piping systems are used in medium and large residential and commercial buildings. Two-pipe piping systems provide better temperature control than one-pipe piping systems, but installing a two-pipe system is more expensive.

Three-pipe piping systems provide good temperature control. Because of the complexity of the system, installing a three-pipe piping system is more expensive than a two-pipe system. A three-pipe piping system is used when different parts of a building require heating and cooling simultaneously. A hot and cold piping system with mixing dampers for zone control is an example of an application for a three-pipe piping system.

Four-pipe piping systems are expensive to install but provide excellent control of air temperature. Four-pipe piping systems are also more economical

HOT WATER BOILERS						
Model Number	Dimensions*			Heating Input§	Heating Capacity§	Shipping Weight#
	L	W	H			
STANDING PILOT — NATURAL GAS						
HWNG020	5	3	4	40,000	34,000	170
HWNG050	6	3	4	60,000	51,000	220
HWNG100	6	3	5	127,000	107,950	300
HWNG150	6	3	6	149,500	127,500	375
STANDING PILOT – LP GAS						
HWLP031	5	4	4	36,000	32,400	165
HWLP063	5	4	4	50,500	42,500	210
HWLP126	5	4	6	125,500	106,675	345
HWLP177	5	4	6	175,500	149,175	405

* in ft
§ in Btu/hr
# in lb

Figure 16-10. Boiler specifications sheets show the size, heating capacity, and weight of boilers.

to operate than three-pipe systems. A four-pipe piping system is used where precise temperature control is required in separate rooms and zones. In this case, the low operating costs would be more important than the high initial cost.

To select the piping system for an application, the building construction must be considered. The piping for a hydronic heating system is considered as part of the building because it is installed within the building frame. The degree of sophistication desired for temperature control should also be considered. Smaller, simpler piping systems have less temperature control than the larger, more complicated piping systems.

Terminal Devices. Terminal devices transfer heat from the hot water in a hydronic heating system to air in building spaces. Terminal devices for hydronic heating systems are combinations of water coils and other components. The water coils act as heat exchangers and the other components control the flow of water and air through the terminal device. The terminal devices used in hydronic heating systems can be categorized as radiators, duct coils, convectors, blower coils, or unit ventilators.

Radiators are heat-distributing devices that consist of coils through which hot water passes. Radiators are only used for heating applications. A radiator is heated as hot water flows through the coils or tubes. Heat radiates from the hot surface of the radiator. Because radiators are unattractive and hard to conceal, they are used mainly in warehouses or factory buildings. A hydronic heating system using radiators is relatively inexpensive to install, but temperature control is marginal.

Duct coils are water coils that are installed in an air duct. Air from a blower, which may be remotely located, is blown over the hot coils and is heated. Duct coils are used when a boiler provides heat to many separate zones in a building. Each zone is supplied with duct coils and a control system. Duct coils are used in applications where zone control is desired. Separate ductwork runs to each zone and duct coils provide temperature control in each zone. The cost of the distribution system is higher when multiple coils are used, but multiple coils give better temperature control.

Convectors are water coils contained in a cabinet. Air flows through the cabinet because of natural convection. Convectors should only be used for heating purposes. Convection requires a large temperature difference to transfer heat. The temperature drop of the air through the coil is not adequate to cause natural convection in a cooling application. Convectors are used in applications that do not require close temperature control.

A unit heater is a forced convection heater that has a propeller fan, hot water coils, and controls in one cabinet. Unit heaters can stand alone or can be installed in a duct system. In a duct system, each unit heater has separate ducts, which act as separate systems within a building. Unit heaters are often used for space heating in warehouses or manufacturing areas where appearance is not important.

A cabinet heater is a forced convection heater that has a blower, hot water coils, filter, and controls in one cabinet. The blower moves the air and the filter cleans the air. Cabinet heaters are often used for heating offices, schoolrooms, and hallways in commercial buildings. Cabinet heaters provide good temperature control but take up floor space.

Unit ventilators are blower coil units that bring outdoor air into building spaces and recirculate indoor air. Unit ventilators are often used in buildings that have a large number of people and that need constant ventilation such as schoolrooms.

The amount of temperature control required in a building determines what kind of terminal device should be used. Radiators and convectors do not give the flexible temperature control of blower coils and unit ventilators.

Circulating Pump

A circulating pump moves water in a hydronic heating system from the boiler through the supply piping and terminal devices and returns it to the boiler. Selecting a circulating pump is a part of designing a hydronic heating system. The arrangement of zones in a building should be considered when selecting equipment. Each control zone, which is identified by a separate piping loop, should have a circulating pump. A circulating pump is chosen for an application based on the location of the pump in the building, the location in the piping loop, and the capacity of the pump.

A circulating pump is always located on the piping loop of a hydronic heating system. Another circulating pump is usually located on the return water piping. A circulating pump should be located near the boiler or chiller. The pump must deliver the required water flow rate against the resistance to flow in the piping system. If the pump is too small, it will not move enough water. If the pump is too large, it will not deliver the water economically.

Control System

The control system for a hydronic heating system consists of boiler controls and terminal device controls. Boiler controls regulate the temperature of the supply water by regulating the boiler burner(s). Terminal device controls regulate the blower in a terminal device, which controls the temperature of the air in the building spaces.

The degree of control desired for an application determines the complexity of the control system for the application. Control systems range from simple ON/OFF manual controls to very sophisticated systems in which both water and air flow are controlled separately for individual zones in a building. The degree of control for an application must be determined before a control system can be designed. Control systems vary in levels of sophistication and can be designed to achieve any level of control.

Heating Water Flow

In a hydronic heating system, water is heated by the burners in the boiler. The circulating pump circulates the hot water through the piping system to terminal devices. The heating capacity of the boiler must be as large or larger than the heating load of the building. The water flow rate in a hydronic heating system must match the heating output capacity of the boiler for a given temperature drop through the system. The water flow rate required to heat a building is found by applying the formula:

$$Q = \frac{OR}{\Delta T \times 500}$$

where

Q = water flow rate (in gpm)

OR = output rating (in Btu/hr)

ΔT = temperature increase of the water (in °F)

500 = constant

Example: Finding Water Flow Rate — Boiler

A building with a heating load of 127,000 Btu/hr requires a boiler that provides at least 127,000 Btu/hr of heat. The temperature decrease should be 20°F. Find the required water flow rate.

$$Q = \frac{OR}{\Delta T \times 500}$$

$$Q = \frac{127,000}{20 \times 500}$$

$$Q = \frac{127,000}{10,000}$$

$$Q = \textbf{12.7 gpm}$$

HYDRONIC AIR CONDITIONING SYSTEMS

A hydronic air conditioning system is a system in which water is chilled and circulated to terminal devices, where it cools air. The cooled air is used for cooling building spaces. The piping system, terminal devices, circulating pump, and control system in a hydronic air conditioning system are similar to those in a hydronic heating system. The chiller is equivalent to the boiler in a hydronic heating system.

Chillers

A chiller is the component in a hydronic air conditioning system that chills water. Mechanical compression chillers and absorption chillers are available. Mechanical compression chillers are used in buildings of all sizes because the chillers are compact and can be adapted to all conditions. Absorption chillers are usually used on very large commercial or industrial applications. The high installation costs of an absorption chiller, due to the extra piping required, makes a mechanical compression chiller more appropriate for small systems. For larger applications, the installation costs of an absorption chiller is comparable to the costs for a mechanical compression chiller.

When selecting a chiller for an application, the cooling capacity and physical characteristics of the chiller should be considered. Chillers used for typical applications are water-cooled package chillers. Water cools the condenser in a hydronic air conditioning system. Water-cooled package chillers are usually located in an equipment room or on the roof of a building. Some air-cooled chillers can be located outdoors. In either arrangement, a piping system runs from the chiller to the terminal devices inside the building. See Figure 16-11.

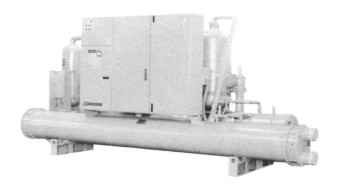

Dunham-Bush

Figure 16-11. When selecting a chiller for an application, the physical characteristics and cooling capacity should be considered.

Chillers have a wide range of cooling capacities. Chillers are available with the cooling capacity for almost any application. Equipment catalogs and engineering specifications sheets should be used for selecting equipment for an application.

Cooling Capacity. The cooling capacity of a chiller is a function of the components of the chiller. Chillers produce a given cooling capacity. Mechanical compression chillers have cooling capacities from 5 tons of cooling to hundreds of tons of cooling. The best source of information on chiller sizes is engineering or equipment specifications sheets. See Figure 16-12.

Absorption refrigeration chillers are built-up systems that consist of components from different manufacturers. The capacity of an absorption refrigeration chiller is a function of the capacity of the components. Information about these built-up systems is found on equipment data sheets for the components as combined in the total package.

RECIPROCATING CHILLERS					
Model Number	Cooling Capacity*	Dimensions§			Shipping Weight#
		L	W	H	
XTAC012	14	6	2	4	2110
XTAC034	29	12	3	4	4920
XTAC056	36	12	3	4	5730
XTAC077	52	12	3	6	6470
XTWC015	16	6	2	4	3750
XTWC025	24	12	3	4	4330
YTAC127	93	14	4	6	7230
YTAC152	107	14	4	6	7860
YTWC075	68	12	3	4	6250
YTWC125	86	14	4	6	6960
YTWC176	135	14	4	6	8540

* in tons of cooling
§ in ft
# in lb

Figure 16-12. Catalogs and engineering specifications sheets are used when selecting equipment for an application.

Cooling Water Flow

In a hydronic air conditioning system, the chiller cools water and the circulating pump circulates the water through the piping system to the terminal devices in a building. The chiller must be able to provide enough cool water to offset the heat gain of the building at a given temperature increase through the piping system. The amount of water circulated through the piping system is related to the cooling load on the building and the temperature increase of the water as it flows through the terminal devices in the system.

Example: Finding Water Flow Rate — Chiller

A chiller has a cooling capacity of 140,000 Btu/hr. The system has a temperature increase of 19.4°F. Find the water flow rate.

$$Q = \frac{cc}{\Delta T \times 500}$$

$$Q = \frac{140{,}000}{19.4 \times 500}$$

$$Q = \frac{140{,}000}{9700}$$

$$Q = \textbf{14.43 gpm}$$

Review Questions

1. List and describe the six main components of a forced-air heating system.

2. What are the four types of fuel or energy used to produce heat in furnaces?

3. What is the most commonly used fuel in furnaces? Why?

4. List and describe the three airflow patterns that categorize furnaces.

5. What output capacity of a furnace is considered the upper limit for residential use?

6. What is the difference between the input and output rating of a furnace?

7. What percentage of the input rating of a conventional furnace is required to produce a draft?

8. Explain why the input of energy and output of heat from an electric furnace are nearly the same.

9. What is the most important consideration when locating ductwork in a building?

10. What is the most important consideration when locating supply registers in a building?

11. What two factors determine the quantity of air that must be circulated through the ductwork in a heating system?

12. Name two types of refrigeration systems used in air conditioning equipment.

13. What is the difference between a package unit and a split system?

14. What relationship must exist between the cooling output of an air conditioner and the cooling load on a building?

15. Define *temperature swing*. Explain in terms of system operation.

16. Why is it necessary to use a psychrometric chart to check air conditioning system performance and not heating system performance?

17. List and describe five components of a hydronic heating system.

18. List and describe five different types of terminal devices used in hydronic heating systems.

19. What is the function of the two sets of controls required in a hydronic heating system?

20. List and describe the four main components in a hydronic air conditioning system.

Forced-air System Design

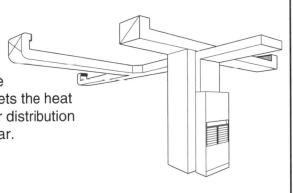

A forced-air system uses air to carry heat. Air is circulated to building spaces through a distribution system from a furnace or an air conditioner. All of the parts of a distribution system must be designed to circulate the required amount of air to building spaces. The air offsets the heat gain or cooling load in each building space. Forced-air distribution systems for heating and cooling applications are similar.

AIR DISTRIBUTION SYSTEMS

Forced-air distribution systems consist of a blower, supply ductwork and registers, and return ductwork and grills. The distribution system circulates warm air from a furnace or cool air from an air conditioner to building spaces. The ductwork consists of supply and return ducts that are made of sheet metal or fiberglass. See Figure 17-1.

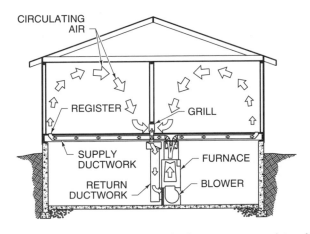

Figure 17-1. A forced-air distribution system consists of a blower, supply and return ductwork, registers, and grills.

Principles of Air Flow

The rate of air flow through a duct is a function of the air velocity, size of the duct, and friction loss.

Air flow is expressed in cubic feet per minute. Duct dimensions are expressed in inches. Friction loss is expressed in inches of water column.

Air Flow Rate. Air flow rate is the volumetric flow rate of the air in the ductwork. Volumetric flow rate of air is the quantity of air moved per unit of time. The total air flow rate required for a forced-air system is a function of the capacity of the heating and cooling equipment. The air flow rate required for a heating system is determined by the temperature difference of the air through the heat exchanger of a furnace. The air flow rate required for an air conditioning system is determined by the temperature difference of the air through the evaporator coil of an air conditioner.

A forced-air distribution system is designed to move air at the air flow rate that provides comfort to each building space. The air flow rate in each building space should be proportional to the heating and/or cooling load in each building space.

The heating and cooling load calculations for a building are used to find the air flow rate for each building space. The heating and cooling loads should include duct loss and duct gain and should be adjusted for the size of equipment. For a combination heating and cooling system, the larger value is used for duct sizing. The air flow rate for a heating load is found by applying the formula:

$$Q = \frac{L_h}{\Delta T \times 1.08}$$

where

Q = air flow rate (in cfm)

L_h = heating load (in Btu/hr)

ΔT = temperature difference (in °F)

1.08 = constant

Example: Finding Air Flow Rate — Heating Load

A room has an adjusted heating load of 13,300 Btu/hr. A temperature difference of 80°F is desired. Find the air flow rate required to provide comfort in the room.

$$Q = \frac{L_h}{\Delta T \times 1.08}$$

$$Q = \frac{13,300}{80 \times 1.08}$$

$$Q = \frac{13,300}{86.4}$$

$$Q = \textbf{153.94 cfm}$$

To find the air flow rate required to cool a room, divide the heat gain (cooling load) for the room by 30. Air conditioners are designed to have 400 cfm of air flowing across the evaporator coil per ton of cooling. With 12,000 Btu/hr per ton of cooling and 400 cfm of air per ton of cooling, 1 cfm of air equals 30 Btu/hr (12,000 ÷ 400 = 30). Dividing the heat gain for a room by 30 gives the air flow rate required to provide comfort in the room. The air flow rate required to cool a room is found by applying the formula:

$$Q = \frac{L_c}{30}$$

where

Q = air flow rate (in cfm)

L_c = cooling load (in Btu/hr)

30 = constant

Example: Finding Air Flow Rate — Cooling Load

A room has an adjusted heat gain of 7135 Btu/hr. Find the air flow rate required to provide comfort in the room.

$$Q = \frac{L_c}{30}$$

$$Q = \frac{7135}{30}$$

$$Q = \textbf{237.83 cfm}$$

Air Velocity. Air velocity is the speed at which air moves from one point to another. Proper circulation in building spaces depends on the correct air velocity. Proper circulation prevents temperature stratification and drafts. Temperature stratification is the variation of air temperature in a building space that occurs when warm air rises to the ceiling and cold air drops to the floor.

When temperature stratification occurs, individuals become uncomfortably warm due to insufficient air circulation. Drafts occur when the air velocity is above 40 fpm. Individuals become uncomfortably cool due to the high rate of evaporation of perspiration from the skin. The air velocity in a duct is a function of the air flow rate and duct size. Air velocity is found by applying the formula:

$$V = \frac{Q}{A}$$

where

V = air velocity (in fpm)

Q = air flow rate (in cfm)

A = area of duct (in sq ft)

Example: Finding Air Velocity

The blower in a forced-air distribution system moves 1200 cfm of air through a 24″ × 12″ duct. Find the air velocity.

$$V = \frac{Q}{A}$$

$$V = \frac{1200}{2}$$

$$V = \textbf{600 fpm}$$

Duct Size. Ducts carry the required air flow rate to building spaces to provide comfort. *Duct size* is the size of a duct, which is expressed in inches of diameter for round ducts and in inches of width and height for rectangular ducts. In dimensions for rectangular ducts, the width should always be given first. For example, a 20″ × 8″ duct is a duct 20″ wide and 8″ high.

Friction Loss. *Friction loss* is the decrease in air pressure due to the friction of the air moving through a duct. Friction loss occurs as air scrubs against the internal surfaces of a duct. Friction loss is due to the turbulence caused by a film of air moving along the surface of the duct. Friction loss is indicated by static pressure drop.

Static Pressure. *Static pressure* is air pressure in a duct measured at right angles to the direction of air flow. Static pressure is the pressure that has a tendency to burst a duct. *Static pressure drop* is the decrease in air pressure caused by friction between the air moving through a duct and the internal surfaces of the duct. See Figure 17-2.

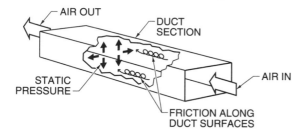

Figure 17-2. Friction loss occurs as air scrubs against the internal surfaces of a duct.

Static pressure is measured with a manometer. A *manometer* is a device that measures the pressures of vapors and gases. The two types of manometers commonly used to measure air pressure in ducts are U-tube and inclined manometers. A *U-tube manometer* is a U-shaped section of glass or plastic tubing that is partially filled with water or mercury.

The liquid in the two legs of a U-tube manometer remains level due to gravity when the pressure on the two legs is equal. If more pressure is exerted on the air in one leg, the liquid in that leg will be pushed down. As the liquid in one leg goes down, the liquid in the other leg rises. The difference in the liquid levels is a measurement of air pressure. See Figure 17-3.

When using a manometer, air pressure is measured in inches of water column or inches of mercury. Water column (WC) is the pressure exerted by a square inch of a column of water. Inches of water column is used to express small pressures above and below atmospheric pressure. One inch of water column equals .036 psi.

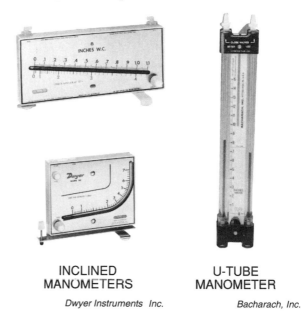

INCLINED MANOMETERS

U-TUBE MANOMETER

Dwyer Instruments Inc.   *Bacharach, Inc.*

Figure 17-3. The difference between the liquid levels in the two legs of a manometer is a measurement of air pressure.

An *inclined manometer* is a U-tube manometer designed so the bottom of the "U" is a long inclined section of glass or plastic tubing. The liquid in an inclined manometer fills part of the inclined tube and part of one leg.

A change in air pressure in one leg of an inclined manometer causes the liquid to move along the inclined tube. The inclined tube is calibrated along the long slope of the incline. A small change in pressure moves the liquid a long distance along the incline. The calibration of an inclined manometer makes it possible to read pressure differentials on an incline manometer more accurately than on a U-tube manometer.

Static pressure drop in a duct reduces the pressure of the air as the air moves along the duct. To measure static pressure in a duct, a small hole is drilled or punched in the side of the duct, and a flexible tube is run from the hole to one leg of a manometer.

The difference between the static pressure readings at two points in a duct is the friction loss that has occurred in the duct between the two points.

Static pressure drop is the difference between the static pressure at the beginning of a duct section and the static pressure at the end of the duct section. Static pressure drop in a duct section is found by dividing the length of the duct section by 100. The length of the duct section is divided by 100 because the design static pressure drop for the distribution system is per 100' of duct. The *design static pressure drop* is the pressure drop per unit length of duct for a given size of duct at a given air flow rate. The result is multiplied by the design static pressure drop. The duct section pressure drop is found by applying the formula:

$$P_d = \frac{L}{100} \times p_d$$

where

P_d = static pressure drop (in in. WC)

L = length of duct section (in ft)

100 = constant

p_d = design static pressure drop (in in. WC)

Example: Finding Duct Static Pressure Drop

The air flow rate in a section of duct 140' long is 1500 cfm. The duct section has a design static pressure drop of .08" WC. The dimensions of the duct are 20" × 12". Find the static pressure drop of the duct section.

$$P_d = \frac{L}{100} \times p_d$$

$$P_d = \frac{140}{100} \times .08$$

$$P_d = 1.4 \times .08$$

$$P_d = \mathbf{0.112'' \ WC}$$

The friction loss in a duct is directly related to the surface area exposed to air flow and the air flow rate. An *equal friction chart* is a chart that shows the relationship between the air flow rate, static pressure drop, duct size, and air velocity. See Figure 17-4.

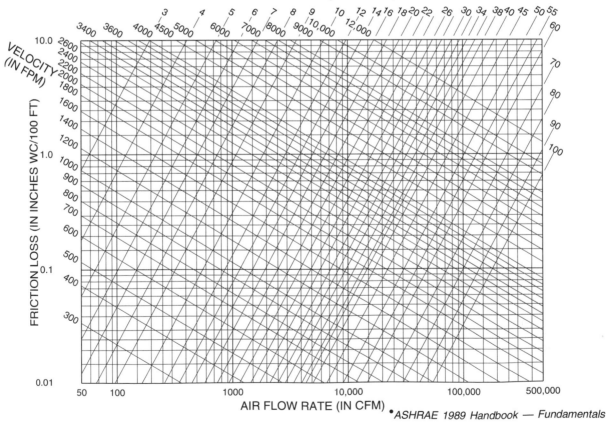

•*ASHRAE 1989 Handbook — Fundamentals*

Figure 17-4. Equal friction charts show the relationship between duct size, flow rate, velocity, and static pressure drop.

An equal friction chart is used to size a duct when two values are given. To find the duct size required to provide an air flow rate at a given static pressure drop, apply the procedure:

1. Locate the air flow rate on the bottom of the equal friction chart and draw vertical line from that point.

2. Locate the static pressure drop on the left side or top of the chart and draw a horizontal line from that point.

3. Locate the size of the duct and the air velocity in the duct at the intersection of the two lines on the graph.

An equal friction chart shows the sizes of round ducts. A conversion table must be used to find the size of square or rectangular ducts that provide the same static pressure drop. See Appendix.

Example: Using Equal Friction Chart

The air flow rate in a section of duct is 1200 cfm. The static pressure drop is .10″ WC. Find the size of the duct using an equal friction chart.

1. Locate 1200 cfm on the scale at the bottom of the chart. Draw a vertical line from this air flow rate. See Figure 17-5.

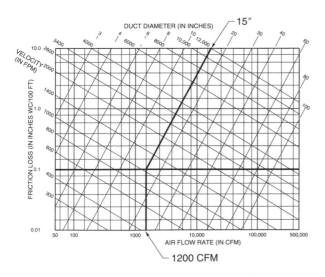

Figure 17-5. The size of a duct is found at the intersection of the air flow rate and the static pressure drop lines on the equal friction chart.

2. Locate .10″ WC on the scale on the left side of the chart. Draw a horizontal line across the chart from this static pressure drop.

3. From the point where the lines intersect, draw a line parallel with the diagonal duct diameter lines to the scale at the top of the chart. The size of a round duct is approximately 15″.

Slide rules (ductulators), digital calculators, and computer programs can also be used for sizing ductwork. Each of these devices shows the same relationships as an equal friction chart.

System Pressure Drop. *System pressure drop* is the total static pressure drop in a forced-air distribution system. Total static pressure drop in a distribution system is the sum of the static pressure drop through all duct sections, fittings, transitions, and accessories such as filters or coils, which add resistance to air flow in the distribution system.

Many forced-air distribution systems have several major duct sections. A *major duct section* is an independent part of a forced-air distribution system through which all or part of the air supply from the blower flows. A major duct section must be considered when system pressure drop is calculated. An *individual duct section* is part of a distribution system between fittings in which the air flow, direction, or velocity changes due to the configuration of the duct. See Figure 17-6.

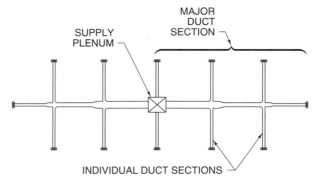

Figure 17-6. A duct section is part of a distribution system between fittings where the air flow, direction, or velocity changes due to the configuration of the duct.

Dynamic Pressure Drop. *Dynamic pressure drop* is the pressure drop in a duct fitting or transition caused by air turbulence as the air flows through the fitting or transition. Turbulence occurs wherever the air flow pattern is disturbed by a change in the air flow rate, direction of air flow, or size of the duct. Turbulence causes a drop in velocity pressure in the duct at the point where turbulence occurs.

Static pressure drop through a fitting or transition is calculated with a table of equivalent lengths or a table of fitting loss coefficients. A *table of equivalent lengths* is a table used to convert dynamic pressure drop to friction loss in feet of duct. A table of equivalent lengths contains the length of duct that gives the same static pressure drop as a particular fitting or transition. See Figure 17-7.

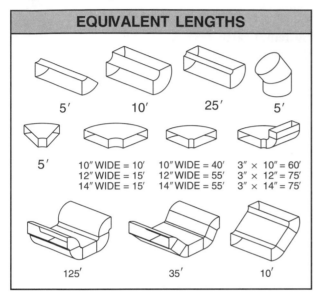

EQUIVALENT LENGTHS

5′   10′   25′   5′

5′   10″ WIDE = 10′   10″ WIDE = 40′   3″ × 10″ = 60′
      12″ WIDE = 15′   12″ WIDE = 55′   3″ × 12″ = 75′
      14″ WIDE = 15′   14″ WIDE = 55′   3″ × 14″ = 75′

125′   35′   10′

Figure 17-7. A table of equivalent lengths shows the length of a duct section that gives the same static pressure drop as a particular fitting or transition.

Velocity pressure is the air pressure in a duct that is measured parallel to the direction of air flow. Measuring velocity pressure without measuring static pressure is difficult. Velocity pressure is usually found by measuring total pressure and static pressure and subtracting the static pressure from the total pressure. The velocity pressure of the air entering a fitting is found by applying the formula:

$$P_v = \left(\frac{V}{4005}\right)^2$$

where

P_v = velocity pressure (in in. WC)

V = air velocity (in fpm)

4005 = constant

Example: Finding Velocity Pressure

A duct fitting handles 1200 cfm of air. The air velocity is 1500 fpm. Find the velocity pressure of the air entering the fitting.

$$P_v = \left(\frac{V}{4005}\right)^2$$

$$P_v = \left(\frac{1500}{4005}\right)^2$$

$$P_v = .3745^2$$

$$P_v = .3745 \times .3745$$

$$P_v = \textbf{.1403″ WC}$$

Fitting loss coefficients (*coefficient C*) are values that represent the ratio between the total static pressure loss through a fitting and the dynamic pressure at the fitting. To find the static pressure drop through a fitting, the coefficient C for the fitting is multiplied by the velocity pressure of the air entering the fitting. The static pressure drop through a fitting using coefficient C is found by applying the formula:

$$P_d = P_v \times C$$

where

P_d = static pressure drop (in in. WC)

P_v = velocity pressure (in in. WC)

C = fitting loss coefficient

Example: Finding Static Pressure Drop Through Fitting

An elbow with a 12″ radius has a velocity pressure of 0.1403″ WC. The fitting is a 18″ × 12″ smooth-radius elbow. Find the pressure drop through the fitting.

1. Find smooth-radius elbow on a fitting loss coefficient table and calculate coefficient C.

2. Multiply velocity pressure by the coefficient C.

$$P_d = P_v \times C$$

$$P_d = .1403 \times .19$$

$$P_d = \textbf{0.0266″ WC}$$

Ductwork Design

A forced-air heating or air conditioning system is designed using a floor plan or sketch of the building. Each floor that has ductwork, registers, or grills requires a floor plan. To design the distribution system, the heating and cooling loads are calculated, and the ductwork is sized.

Layout Drawings. *Layout drawings* are drawings of the floor plan of a building that show walls, partitions, windows, doors, fixtures, and other details that affect the location of the ductwork, registers, or grills. Layout drawings show the heating and cooling loads and required air flow rate for each room or zone. Because forced-air distribution systems are sized for a required air flow rate, the heating and cooling loads for a building must be converted to air flow rate for building space. See Figure 17-8.

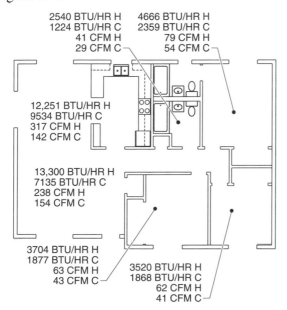

Figure 17-8. Layout drawings show the heating and cooling load for each room or zone.

Layout drawings include several overlays that show the placement of ductwork, registers, and grills on each floor of a building. Tentative locations of the registers and grills are shown on the overlay for each floor of the building.

Registers are located based on the air flow pattern desired out of each register. Grills are located so all of the air supplied by the registers is returned to the system blower. Grills should be located where they are not covered by furniture or fixtures. Air must get back to the blower through the return ductwork. Therefore, grills must be located so they are open at all times. Ceilings or walls are the best locations for grills because floor grills are easily covered with furniture or carpeting and collect dirt.

Grills should not be located close to registers. If grills are located close to registers, air from the reg-

isters will flow directly into the grills. This air flow upsets the distribution pattern of the registers. Grills should be located centrally in the building space. One or more grills may be used in a zone or building space as long as sufficient circulation is achieved. If some building spaces are isolated by partitions or doors, grills should be located in each building space. Grills may be connected to the same return ductwork but must be located so they return air to the system blower from all parts of the building.

The distribution system is drawn on a separate overlay. The distribution system overlay contains the location of the registers and grills, which shows the ducts in relation to the registers and grills. Each register is connected to the supply ductwork, and each grill is connected to the return ductwork. The actual layout of a distribution system depends on the type of system chosen. On the drawings, ductwork must be located where it does not run through beams, columns, stairwells, elevator shafts, or where plumbing or electrical devices may be located. See Figure 17-9.

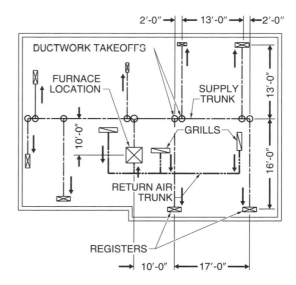

Figure 17-9. An air distribution system overlay shows the ductwork layout in relation to the placement of registers and grills.

After layout drawings and overlays have been prepared, the distribution system is chosen based on the temperature control and sophistication level required for each building space. After the distribution system has been chosen, the registers and grills are chosen based on the heating and cooling needs of each building space.

If the design of part of the distribution system is especially complicated, large-scale plans may be necessary to identify the actual location of each duct section. A ductwork layout drawing should show all fittings and transitions in the distribution system. Fittings and transitions are duct sections that change air flow direction or duct size. Fittings and transitions are shown on a single line drawing by an "X" marked on the duct drawing. A *single line drawing* is a drawing in which the walls and partitions are shown by a single line with no attempt to show actual wall thickness. Fittings and transitions are shown on a double line drawing in plan view as they actually appear. A *double line drawing* is a drawing in which walls and partitions are shown with double lines. Wall thickness is shown at actual scale on the drawing.

Section Identification. Identifying each duct section is necessary for keeping track of the proper size ducts, dampers, and fittings. A *duct section* is a section of ductwork between two fittings where the air flow rate changes. Duct sections occur between takeoffs where air enters or leaves the duct. Duct sections are identified using letters or numbers. The two most common methods of duct section identification are marking the duct sections or marking the fittings at the end of each section.

When letters are used to identify duct sections, the duct section between fittings is identified by the letters on the fittings. The duct section between fittings is identified by the fitting letters. For example, all of the fittings are marked with a different letter. The first fitting, which comes off the furnace or air conditioner, is identified as A. The fitting at the end of the first duct section is identified as B. The duct section between fitting A and B is identified as section A-B. The letters are shown on the distribution system overlay. See Figure 17-10.

Duct Sizing

A duct is sized to allow the required amount of air for proper heating or cooling to flow to each building space. The air flow rate is controlled by the size of the duct, and the duct is sized to carry the air at a given pressure loss.

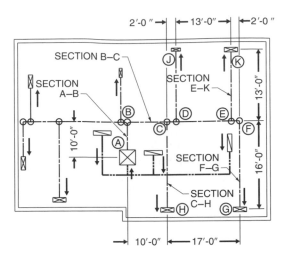

Figure 17-10. A letter identifies each fitting and the duct section between fittings is identified by the fitting letters.

Forced-air distribution systems often consist of major duct sections and individual duct sections. The major duct sections are connected to the supply plenum and individual duct sections run off of the major duct sections. For sizing and calculating pressure drop through a distribution system, each major section can be treated separately.

The most common methods for sizing ductwork are equal friction, velocity reduction, and static regain methods. Small distribution systems used in heating and air conditioning systems as well as small ventilation and exhaust systems are sized by either the equal friction method or velocity reduction method. Large, sophisticated distribution systems are sized using the static regain method. The equal velocity method is used for sizing exhaust systems that carry particulate matter.

In the past, duct sizes were calculated manually or with duct sizing charts, graphs, or ductulators. Today, computer programs are used for sizing ducts. Small jobs are easily sized by conventional methods, but large jobs are more easily sized using a computer duct-sizing program.

The *equal friction method* is duct sizing that considers that the static pressure is approximately equal at each branch takeoff in the distribution system. An equal friction chart is used to size the ductwork for a given static pressure drop using the equal friction method.

To use an equal friction chart for sizing ductwork, the static pressure drop is determined, and the chart is used to find the size of the duct. The static pressure drop can be selected arbitrarily. For larger systems, the static pressure drop is found by selecting an arbitrary air velocity and duct size for the first section of duct off the furnace or air conditioner. This data is used on an equal friction chart to find the static pressure drop. This static pressure drop is used for sizing the other ducts in the distribution system.

Small distribution systems are designed for about .06" WC to .10" WC static pressure drop per 100' of duct. In larger distribution systems, the static pressure drop per 100' of duct may be greater. To size ductwork using the equal friction method, apply the procedure:

1. Prepare layout drawings and overlays that show the heating and cooling load and air flow rate for each building space.

2. Select and locate registers and grills on the first overlay. In some applications, more than one register is required in a room. When more than one register is required in a room, more than one duct branch runs to the room. The air flow rate for the room should be divided between the duct branches supplying the room.

3. Lay out the distribution system on the second overlay.

4. Enter the duct sections in the first column of a columnar form.

5. Enter the air flow rate for each duct section in the second column.

6. Select a design static pressure drop, which is the pressure drop that occurs per 100' of duct in the actual distribution system design.

   The design static pressure drop is determined by selecting an air velocity, picking a duct size for that section of duct, and finding the design static pressure drop on an equal friction chart. The air velocity is for the first duct section off the blower that ensures quiet operation. This design static pressure drop is used for sizing the ductwork.

7. Find the size of each duct section using an equal friction chart. Enter the values in the third column of a columnar form.

8. Convert sizes of round ducts to equivalent rectangular sizes. Enter the values in the fourth column of the columnar form. These values are taken from a round-to-rectangular duct conversion chart. See Appendix.

Example: Duct Sizing — Equal Friction Method

The blower of a furnace used to heat a three bedroom residence moves 800 cfm of air. Size the ductwork using the equal friction method.

1. Prepare layout drawings and overlays that show the heating and cooling load and air flow rate for each building space. See Figure 17-11.

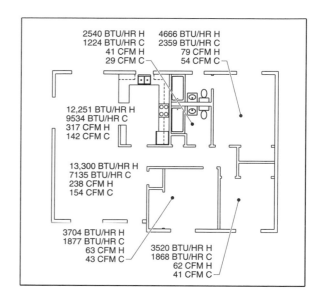

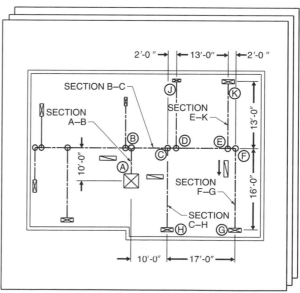

Figure 17-11. Layout drawings show the air flow rate for each building space. Overlays show the location of the ductwork system, registers, and grills.

2. Locate registers and grills on the first overlay.

3. Lay out the ductwork on the second overlay.

4. Enter the identifying letters for the duct sections in the first column of a columnar form.

5. Enter air flow rate for each duct section in the second column. See Figure 17-12.

6. Select a design static pressure drop of .1″ WC for the forced-air distribution system.

7. Find the size for each duct section using an equal friction chart. Enter the values in the third column of the columnar form.

8. For the duct sections that should be rectangular, convert the round duct sizes to equivalent rectangular sizes. Enter the values in the fourth column of the columnar form.

Section	Flow Rate*	Size§	Size#
A – B	800	12.7	20 × 8
B – C	245	8.1	12 × 8
C – D	182	7.3	12 × 8
D – E	141	6.6	12 × 8
E – F	62	4.8	8 × 8
F – G	62	4.8	
C – H	63	4.9	
D – J	41	4.1	
E – K	79	5.3	

3 bedroom residence
.1″ WC

* in cfm
§ in inches diameter
# in inches

Figure 17-12. Duct sections, air flow rate, and duct size are recorded on a columnar form.

Rectangular ducts are used in distribution systems where head room is limited. The width of a rectangular duct can be greater than the height. This allows the bottom of the duct to be higher from the floor of a building space. Rectangular ducts are also used as trunks.

To function most efficiently, the aspect ratio of a rectangular duct should not be more than 3 to 4 times the height. *Aspect ratio* is the ratio between the height and width of a rectangular duct.

When duct sizes are taken from an equal friction chart or ductulator, the actual sizes are recorded on the columnar form. For practical use however, nominal sizes of ducts that are readily available or more economical to manufacture are shown on the drawing. If the actual size of a duct is not the same as the nominal size of an available duct, the next larger nominal size will be used.

The *velocity reduction method* is duct sizing that considers that the air velocity is reduced at each branch takeoff. The air velocity in the first duct section off the furnace or air conditioner and the size of the first duct section are selected arbitrarily. The other duct sections are sized so the air velocity is reduced in each section according to some arbitrary scale. Tables show recommended air velocities in different duct sections. When a distribution system is sized using the velocity reduction method, the design static pressure drop may be different in each duct section. See Figure 17-13.

The *static regain method* is duct sizing that considers that each duct section is sized so that the static pressure increase at each takeoff offsets the friction loss in the preceding duct section. The static regain method is used for duct sizing in long distribution systems with many takeoffs or with registers located on the duct itself.

The static regain method causes approximately the same static pressure at each branch takeoff. The static regain method can result in large ducts with low air velocity at the ends of long runs. The static regain method is used for larger distribution systems and requires many calculations for each duct section and for each fitting.

The *equal velocity method* is duct sizing that considers that each duct section has the same air velocity. The ducts are sized with a ductulator and a velocity scale. The sizing may be done mathematically using the air velocity and the cross-sectional area of the duct. The equal velocity method is more applicable to sizing exhaust systems or material collection systems than it is for sizing distribution ducts for a heating or air conditioning system.

Blower Sizing

When designing a forced-air distribution system, the blower must have the capacity to move air at

RECOMMENDED DUCT AIR VELOCITIES*							
Use		**Residential Buildings**		**Commercial Buildings**		**Industrial Buildings**	
		Recommended	Maximum	Recommended	Maximum	Recommended	Maximum
Total Face Area Velocities	Filters	250	300	300	350	350	350
	Coils	450	500	500	600	600	700
	Intakes	500	800	500	900	500	1200
	Air washers	650	650	650	650	650	650
Net Free Area Velocities	Ducts						
	Main	800	1000	1500	1600	1500	2200
	Branch	600	850	750	1300	900	1800
	Risers	500	800	650	1200	800	1300
	Outlets	1300	1700	1650	2200	2000	2800

* in fpm

Figure 17-13. Recommended duct air velocities are used when sizing ductwork by the velocity reduction method.

the required air flow rate against the resistance in the distribution system. When sizing a blower for a distribution system, the required air flow rate and the total static pressure drop for the system are found on a blower performance table for a blower. The velocity of the blower wheel and the blower horsepower required to turn the blower wheel to produce the required air flow rate are determined from the blower performance table.

The blower in a forced-air distribution system must move the air at the air flow rate required for the system against the resistance to air flow in the system. The air flow rate required is determined when the heating or air conditioning equipment for the system is chosen. Resistance to air flow in a distribution system is due to friction in the ducts and to dynamic pressure losses through fittings and transitions in the ducts. Resistance to air flow occurs in both the supply and return ducts of a distribution system. A blower must be large enough to move the required air flow rate in the major section of supply and return ductwork having the greatest resistance to flow. The total resistance to air flow in a distribution system is calculated as static pressure drop in the system.

If a blower circulates air through more than one major section of supply ductwork and one major

section of return ductwork in a system, the blower will not have to overcome the sum of the static pressure drops in all of the sections. The blower must overcome the static pressure drops in the supply and return sections having the largest static pressure drops.

When sizing a blower for a forced-air distribution system, it is not necessary to calculate the pressure drop through all the sections of a distribution system that have more than one major duct section running to or from the blower. It is necessary to find the supply duct section with the highest static pressure drop and the return duct section with the highest static pressure drop. The two static pressure drop values are added together to find the total static pressure drop against which the blower has to work.

To calculate the total static pressure drop through a major duct section that has been designed by the equal friction method, the actual length of the duct section that has the greatest resistance to air flow is found. The length is taken from the layout drawing. The equivalent length of each fitting or transition in the section is found from a table of equivalent lengths. The equivalent length of each fitting is added to the actual length of duct in the section. The result is the total equivalent length of that section of the distribution system.

The total equivalent length of the duct section is divided by 100′ and multiplied by the design static pressure drop that was used to size the duct. The result is the static pressure loss for that duct section.

To find the total static pressure drop through one section of a forced-air distribution system, start with the columnar form used for sizing the duct. Add a column for the actual length of each duct section and fitting equivalent lengths. See Figure 17-14.

Enter the actual duct length of each section from the overlays. In the next column, enter the equivalent lengths for the fittings in each duct section. Add the lengths of the duct sections, which gives the total actual length of duct in the section. Add the equivalent lengths for the fittings, which gives the total equivalent length for fittings. Add the duct actual length total and the fitting equivalent length total to get the total equivalent length of the major duct section.

Divide the total equivalent length by 100 and multiply by the design static pressure drop to get the static pressure drop through the major duct section. Repeat this step for the section of return ductwork that appears to have the greatest pressure drop. The sum of the pressure drop for the supply ductwork section and the return ductwork section is the total pressure drop for the ductwork in the system.

The static pressure drop for filters, coils, dampers, or other devices that add resistance to air flow in the distribution system should be added to the duct pressure drop. The pressure drop through an accessory is found in specifications sheets. The total static pressure drop for a major duct section is the duct pressure drop plus the pressure drop through accessories in the system.

If a distribution system has been designed by a method other than the equal friction method, the static pressure drop in each duct section between fittings may be different. To find the total static pressure drop in a major duct section, the static pressure drop in each individual duct section must be found. The individual section pressure drops are then added together. The pressure drops through fittings and transitions are calculated by using the equivalent length method or the coefficient C. The result is added to the sum of the section pressure drops. Pressure drop through accessories is added to the duct section and the fitting pressure drop to find the total major section pressure drop.

To find the static pressure drop through an individual duct section, the size of the duct and the air flow rate through the section are used. To find the static pressure drop through an individual duct section apply the procedure:

3 bedroom residence
.1″ WC

Section	Flow Rate*	Size§	Size#	DAL**	FEL§§	TEL##			
A - B	800	12.7	20 × 8	10	55				
B - C	245	8.1	12 × 8	10	30				
C - D	182	7.3	12 × 8	2	25				
D - E	141	6.6	12 × 8	13	25		$\left(\dfrac{355}{100}\right) \times .1 = .355″ \, WC$		
E - F	62	4.8	8 × 8	2	25				
F - G	62	4.8		16	30				
C - H	63	4.9		16	30				
D - J	41	4.1		13	30				
E - K	79	5.3		13	30				
		Total Actual Length		95 +	260 =	355′			

* in cfm
§ in inches diameter
# in inches

** duct actual length in ft
§§ fitting equivalent length in ft
total equivalent length in ft

Figure 17-14. The sum of the lengths of the duct sections and the fitting equivalent lengths is the total actual length of duct in the section.

1. Using an equal friction chart or a ductulator, find the design static pressure drop.

2. Calculate the actual length of the duct section from the duct layout drawing. This is the length of the duct measured from centerline of the fitting at the start of the duct section to the centerline of the fitting at the end of the duct section.

3. Determine the equivalent length for the fittings in the section. Equivalent lengths are taken from an equivalent length table.

4. Add the equivalent lengths for fittings to the actual lengths for the section. The sum is the total equivalent length for the section.

5. Divide the total equivalent length of the section by 100 and multiply by the design static pressure drop for the system.

The result is the static pressure drop for the section of the distribution system. The pressure drop of accessories in the section, which add resistance to air flow, must be added to the duct section pressure drop to determine the total static pressure drop in the section.

The same procedure is followed for each individual duct section through a major section. The sum of the individual section pressure drops is the total static pressure drop for the major section.

Example: Sizing Blower — Velocity Reduction Method

In a forced-air distribution system, the design static pressure drop is .10″ WC per 100′. Size the blower using the velocity reduction method.

1. Enter the length of each individual duct section on a columnar form. See Figure 17-15.

2. Enter the static pressure drop for each duct section on the form.

The static pressure drop in the duct section is found by dividing the length of the duct by 100 and multiplying by the design static pressure drop.

3. Enter a description of the fittings in each duct section on the next column of the form. Describe the fitting so it can be identified for calculating the static pressure loss.

4. List the equivalent lengths for the fittings identified for each duct section on the form.

5. Enter the static pressure drop for the fittings in each section. This is calculated with the method used for the duct sections. The equivalent length for fittings is divided by 100 and multiplied by the system design static pressure drop.

---

3 bedroom residence
.1″ WC

Section	Flow Rate*	Size§	Size#	Length**	SP Drop§§	Fitting	EL##	SP Drop***	Sub-T§§§
A-B	800	12.7	20 x 8	10	.01	plenum T.O.	35	.035	.045
B-C	245	8.1	12 x 8	10	.01	splitter	30	.030	.040
C-D	182	7.3	12 x 8	2	.002	through T.O.	25	.025	.027
D-E	141	6.6	12 x 8	13	.013	through T.O.	25	.025	.038
E-F	62	4.8	8 x 8	2	.002	through T.O.	25	.025	.027
F-G	62	4.8		16	.016	BR T.O. ¢ Boot	30	.030	.046
C-H	63	4.9		16	.016	BR T.O. ¢ Boot	30	.030	.046
D-J	41	4.1		13	.013	BR T.O. ¢ Boot	30	.030	.043
E-K	79	5.3		13	.013	BR T.O. ¢ Boot	30	.030	.043
						Total Pressure Drop			.355

* in cfm
§ in inches diameter
# in inches
** duct length in ft

§§ static pressure drop in inches WC
equivalent length in ft
*** static pressure drop for fitting in inches WC
§§§ subtotal pressure drop in inches WC

Figure 17-15. The static pressure drop in a duct section is found by dividing the length of the duct by 100 and multiplying by the design static pressure drop.

6. Add the static pressure drop for each duct section to the static pressure drop for the fittings in the section. Add across the form. The subtotal static pressure drops for the individual sections are added together to find the total pressure drop for the entire major duct section.

The total pressure drop for the duct section is shown on the bottom right-hand corner of the columnar form. Before this figure is used for sizing a blower, the static pressure drop through the return system and the additional losses through registers, grills, or other accessories in the system should be found and added together.

Register Sizing

Registers and grills are selected and sized to allow the proper amount of air to flow into each building space to offset the heat loss or gain. Registers and grills are located to efficiently distribute the air to the building spaces.

The size of a register is determined by the air flow rate required from the register and the required air velocity in a building space. The air flow rate from a register is a function of the size of the free area of the register. *Free area* of a register is the face area of the register minus the area blocked by the frame or vanes. The air velocity at a register is a function of the free area of the register and the air flow rate. The air velocity at the register determines the distance the air is thrown from the register.

Registers are designed to deliver air in different patterns. Air patterns are defined by throw and spread. Throw is the distance that air travels directly away from a register. Throw is measured to the point where the air velocity drops below 50 fpm. Spread is the distance measured across the envelope of air that spreads out from a register. Spread is the distance on each side of the air envelope where the air velocity drops below 50 fpm. The pattern of the air flow out of the register is controlled by the vanes on the panel.

To size a register, the general type of register is chosen. The air flow rate required from the register, throw and spread characteristics, and the air velocity required to achieve those characteristics are determined. The register that produces these characteristics is found in a catalog. One or more types and sizes of supply registers may be used in a build-

ing space. The registers should be chosen for air pattern and sized to provide complete coverage of the room with conditioned air.

Registers are sized so the face velocity is not high enough to produce noise. Catalogs are marked to indicate the air delivery rate at which the registers produce noise.

Grill Sizing

Grills are sized according to velocity of the air at the grill face. Each grill must return the air from the building spaces at a low face velocity to prevent noise. A face velocity of 500 fpm to 750 fpm is allowable when sizing grills.

To size a grill, the air flow rate through the grill is divided by the face velocity desired (500 fpm to 750 fpm) and is multiplied by a factor of 1.3. A factor of 1.3 allows for 30% of the grill face area to be taken up by vanes and frame. The result is the free area of the face of the grill in feet. Grills are sized in inches. To find the size of the grill in inches, the free grill area in square feet must be converted to square inches. The square inches in a square foot, 144, converts to square inches. Free grill area is found by applying the formula:

$$A_g = \frac{Q}{V} \times 1.3 \times 144$$

A_g = free grill area (in sq in.)

Q = air flow rate (in cfm)

V = air velocity (in fpm)

1.3 = constant

144 = constant

Example: Finding Free Area — Grill

A grill in a forced-air distribution system must handle 1500 cfm of air at a face velocity of 600 fpm. Find the required free area.

$$A_g = \frac{Q}{V} \times 1.3 \times 144$$

$$A_g = \frac{1500}{600} \times 1.3 \times 144$$

$$A_g = 2.5 \times 1.3 \times 144$$

$$A_g = 3.25 \times 144$$

$$A_g = \textbf{468 sq in.}$$

If a square grill is required, the size of the grill is found by taking the square root of the free area of the grill. The size of a square grill is found by applying the formula:

$$G_s = \sqrt{A_g}$$

where

G_s = size of square grill (in inches)

A_g = free grill face area (in sq in.)

Example: Finding Size of Square Grill

A grill has a free area of 468 sq in. A square grill is desired. Find the size of the grill.

$$G_s = \sqrt{A_g}$$

$$G_s = \sqrt{468}$$

$$G_s = \mathbf{21.63''}$$

The nominal grill size is 22″ × 22″.

If a rectangular grill is required, the area in square inches should be divided by the dimension of one side of the grill, and the result is the dimension of the other side of the grill. The size of a rectangular grill is found by applying the formula:

$$w = \frac{A_g}{h}$$

where

w = width (in inches)

A_g = free grill area (in sq in.)

h = height (in inches)

Example: Finding Size of Rectangular Grill

A grill in a distribution system must handle 1000 cfm of air at a face velocity of 500 fpm. The height must be 14″. Find the width of the grill.

1. Find the free area of the grill.

$$A_g = \frac{Q}{V} \times 1.3 \times 144$$

$$A_g = \frac{1000}{500} \times 1.3 \times 144$$

$$A_g = 2 \times 1.3 \times 144$$

$$A_g = 2.6 \times 144$$

$$A_g = \mathbf{374.4 \ sq \ in.}$$

2. Find the width of the grill.

$$w = \frac{A_g}{h}$$

$$w = \frac{374.4}{14}$$

$$w = \mathbf{27''}$$

After the duct, register, and grill sizing for a building is complete, the sizes are shown on the distribution system plans. The size of the duct in each section and the size of all fittings and transitions are shown on the plan.

Review Questions

17

1. List and describe the five main parts of a forced-air distribution system.

2. Why are the heating and cooling loads for a building required when finding the air flow rate?

3. Write the formula for finding the air flow rate when the temperature difference through the heating or cooling equipment and heating or cooling capacity is known.

4. Why is a factor of 30 used when finding the air flow rate for the cooling load for a building space? Explain.

5. The air velocity in a duct is a function of what two factors?

6. What causes friction loss in a duct section?

7. Name four variables of air flow found on an equal friction chart.

8. Explain static pressure drop, system pressure drop, and dynamic pressure drop.

9. Describe the process used to find the velocity pressure of air in a duct section.

10. Describe the first two steps when designing a duct system.

11. What data relative to a duct layout is shown on the layout drawing and overlays?

12. Why is it important that registers and grills not be located too close to each other?

13. What defines the two ends of a duct section?

14. Describe the difference between the three main methods used to size ductwork.

15. Name the two factors that must be considered when determining the design static pressure drop.

16. What is the maximum air velocity recommended for use in the main ducts in a public building?

17. Name the two variables of air flow that are used in sizing a blower for a duct system.

18. Name the two most important factors to consider when sizing registers.

19. What is the acceptable range of air velocity used for sizing grills?

20. Define *aspect ratio*.

Hydronic System Design

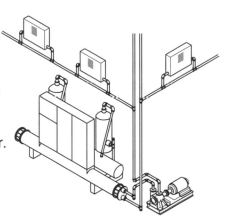

Hydronic systems use water as the heat transfer medium. Heat is transferred from a boiler to the water in a hydronic heating system. The water is circulated to the terminal devices where it heats the air. Heat is transferred to a chiller from the water in the piping of a hydronic air conditioning system. The water is circulated to the terminal devices where it cools the air. The parts of a hydronic system are selected and sized to work together for efficient heat transfer.

WATER DISTRIBUTION SYSTEMS

The distribution system for the water in a hydronic system is the piping system. The piping system distributes water from the boiler or chiller to the terminal devices. The piping system consists of iron or steel pipe or copper tubing, fittings, and valves. The piping system is installed in the frame of a building. The piping system must be designed to distribute the required quantity of water to each building space or zone to offset the heating or cooling loads. See Figure 18-1.

Principles of Water Flow

The rate of water flow through a pipe or an orifice is a function of the pressure exerted on the water and the size of the pipe. Water flow is expressed in gallons per minute. Water pressure is expressed in feet of head or pounds per square inch. Pipe size is indicated by the outside diameter of the pipe and is expressed in inches. *Outside diameter* is the distance from outside edge to outside edge of a pipe or tube. This measurement includes the thickness of the pipe.

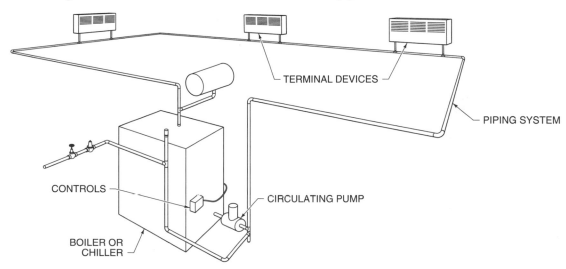

Figure 18-1. In a hydronic system, the piping circulates water from a boiler or chiller to the terminal devices.

Water Flow Rate. *Water flow rate* is the volumetric flow rate of the water as it moves through a given pipe section. Water flow rate is expressed in gallons per minute. Depending on the type of piping system, the flow rate may vary in different pipe sections within a piping loop.

If a piping system is sized for heating and cooling, the larger flow rate is used for sizing the components for the system. Depending on the climate in which the loads are calculated, either the heating or the cooling load may require the larger flow rate.

Commercial water suppliers measure water flow rate with a meter. For design purposes, water flow rate is calculated mathematically. The water flow rate in a hydronic system is based on temperature difference of water at two points in the system and the amount of heat transferred during the change of temperature. The water flow rate is found by applying the formula:

$$Q = \frac{H}{\Delta T \times 500}$$

where

Q = water flow rate (in gpm)

H = heat transferred (in Btu/hr)

ΔT = temperature difference (in °F)

500 = constant

Example: Finding Water Flow Rate — Heating

A heating system has a total adjusted heat loss of 350,000 Btu/hr. A temperature difference of 20°F is used. Find the water flow rate.

$$Q = \frac{H}{\Delta T \times 500}$$

$$Q = \frac{350,000}{20 \times 500}$$

$$Q = \frac{350,000}{10,000}$$

$$Q = \textbf{35 gpm}$$

Water Pressure. Pressure is force exerted per unit of area. Water pressure is expressed in pounds per square inch or in feet of head. *Feet of head* is a unit of measure that expresses the height of a column of water that would be supported by a given pressure. For example, a foot of head is the pressure equal to the weight of a column of water 1′ high. Feet of head can be measured using a special manometer, but it is usually a calculated value.

System Pressure Drop. *Pressure drop* is a drop in water pressure caused by friction between water and the inside surface of a pipe as the water moves through the pipe. *System pressure drop* is the total pressure drop in a piping system. System pressure drop is the difference in pressure between the point where water enters the system and the point where water leaves the system. System pressure drop is the pressure against which the circulating pump must move the water. System pressure drop is expressed in psi or in feet of head.

For water to flow through a pipe, pressure must be exerted on the water at the inlet end of the pipe. The pressure must be great enough to raise the water upward in vertical columns of pipe. This pressure must also overcome the system pressure drop. The sum of the friction head and static head forces is the total head of the system.

Friction head is the effect of friction in a pipe. Friction occurs between the water moving through the pipe and the interior surfaces of the pipe. In a straight pipe section, the friction is a function of the size of the pipe, amount of water flow, and the roughness of the surface of the pipe. Friction head can be calculated mathematically or found on tables that give pipe sizes relative to water flow rate and pressure drop.

Static head is the weight of water in a vertical column above a datum line. A *datum line* is the point at which the pressure would be zero. Pressure exerted by water in a vertical column represented by the weight of the water above the datum line is the static head of the system. Static head is a function of the weight of water and the vertical height of the water in the system above the pump. *Total head* is the sum of friction head and static head. Total head is the system pressure drop.

The pressure drop through a pipe fitting is found by converting the fitting pressure drop to the equivalent number of 90° elbows that would produce the same pressure drop. Tables of elbow equivalents are used. See Figure 18-2.

EQUIVALENT LENGTH OF PIPE FOR 90° ELBOWS*

Velocity§	Pipe Size														
	½	¾	1	1¼	1½	2	2½	3	3½	4	5	6	8	10	12
1	1.2	1.7	2.2	3.0	3.5	4.5	5.4	6.7	7.7	8.6	10.5	12.2	15.4	18.7	22.2
2	1.4	1.9	2.5	3.3	3.9	5.1	6.0	7.5	8.6	9.5	11.7	13.7	17.3	20.8	24.8
3	1.5	2.0	2.7	3.6	4.2	5.4	6.4	8.0	9.2	10.2	12.5	14.6	18.4	22.3	26.5
4	1.5	2.1	2.8	3.7	4.4	5.6	6.7	8.3	9.6	10.6	13.1	15.2	19.2	23.2	27.6
5	1.6	2.2	2.9	3.9	4.5	5.9	7.0	8.7	10.0	11.1	13.6	15.8	19.8	24.2	28.8
6	1.7	2.3	3.0	4.0	4.7	6.0	7.2	8.9	10.3	11.4	14.0	16.3	20.5	24.9	29.6
7	1.7	2.3	3.0	4.1	4.8	6.2	7.4	9.1	10.5	11.7	14.3	16.7	21.0	25.5	30.3
8	1.7	2.4	3.1	4.2	4.9	6.3	7.5	9.3	10.8	11.9	14.6	17.1	21.5	26.1	31.0
9	1.8	2.4	3.2	4.3	5.0	6.4	7.7	9.5	11.0	12.2	14.9	17.4	21.9	26.6	31.6
10	1.8	2.5	3.2	4.3	5.1	6.5	7.8	9.7	11.2	12.4	15.2	17.7	22.2	27.0	32.0

* in ft
§ in fps

•*ASHRAE 1989 Handbook — Fundamentals*

Figure 18-2. Tables of equivalent lengths of straight pipe for 90° elbows converts the pressure drop in piping to pressure drop in 90° elbows.

The elbow equivalent value found on the table is multiplied by the equivalent length of pipe for the 90° elbow based on the size of the pipe and the velocity of the water. To find the equivalent length of pipe that has the same pressure drop as a fitting, apply the procedure:

1. Find the equivalent length of pipe in 90° elbows based on the size of the pipe and the velocity of the water in the pipe on a table of equivalent lengths for 90° elbows.

   If the water flow rate in a 3″ pipe is 6 fps (feet per second), each 90° elbow equals 8.9′ of pipe.

2. Find the number of 90° elbows that is equivalent to the fitting.

   If the fitting is a open globe valve, it will be equivalent to 12 elbows. See Figure 18-3.

3. Calculate the equivalent length of pipe by multiplying the equivalent length of pipe for a 90° elbow by the number of elbows. The equivalent length of pipe is found by applying the formula:

$$L = E \times N$$

where

L = equivalent length of pipe for fitting (in feet)

E = equivalent length of pipe for a 90° elbow (in feet)

N = equivalent number of elbows

If each 90° elbow is equal to 8.9′ of pipe and the fitting is equal to 12 elbows, find the equivalent length of pipe.

ELBOW EQUIVALENTS

Fitting	Iron Pipe	Copper Tubing
Elbow, 90°	1.0	1.0
Elbow, 45°	0.7	0.7
Elbow, 90° long turn	0.5	0.5
Elbow, welded, 90°	0.5	0.5
Reduced coupling	0.4	0.4
Open return bend	1.0	1.0
Angle radiator valve	2.0	3.0
Radiator or convector	3.0	4.0
Boiler or heater	3.0	4.0
Open gate valve	0.5	0.7
Open globe valve	12.0	17.0

•*ASHRAE 1989 Handbook — Fundamentals*

Figure 18-3. Tables of elbow equivalents give fitting pressure drop in number of 90° elbows.

$$L = E \times N$$
$$L = 8.9 \times 12$$
$$L = \textbf{106.8'}$$

Example: Finding Equivalent Length of Pipe

The copper tubing in a hydronic piping system contains an open gate valve. The copper tubing is 1″ in diameter and the velocity of the water in the tubing is 2 fps. Find the equivalent length of pipe that will give the same pressure drop.

1. Find the equivalent length of pipe for 90° elbows.

   On the table for equivalent lengths, the equivalent length of pipe for 90° elbows is 2.5′.

2. Find the number of 90° elbows.

   On the table for elbow equivalents, the gate valve is equal to .7 elbows.

3. Calculate the equivalent length of pipe.

$$L = E \times N$$
$$L = 2.5 \times .7$$
$$L = \textbf{1.75'}$$

Design Static Pressure Drop. *Hydronic design static pressure drop* is the pressure drop per unit length of pipe for a given size of pipe at a given water flow rate. Design static pressure drop is the pressure drop in feet of head per 100′ of pipe or mils per foot of pipe. A *mil* is a unit of measure equal to $\frac{1}{1000}$ of an inch.

Pipe-sizing tables and charts show the size of pipe to use for a given flow rate at design static pressure drops. The charts also show the velocity of the water in feet per minute. Such tables and charts are available for iron, steel, copper, and other kinds of pipe. Sizing tables and charts show the pipe in nominal sizes. Nominal size pipes are the sizes that are normally available from suppliers.

A specific design static pressure drop must be used when using a table or chart for sizing pipe. For most small- to medium-size applications, a design static pressure drop of about 2.5′ of head or 300 mpf (mils per foot) is used. These values give a total pressure drop through a typical piping system within the range of nominal pipe sizes and circulating pumps.

To use the pipe sizing chart, find the 300 mpf row in the friction loss column, which is on the left side of the chart. Follow the row across the chart to the required flow rate. Values for flow rate are listed in the columns under nominal pipe sizes. The heading on the column with the required flow rate is the correct pipe size. The correct pipe size allows the proper water flow at 300 mpf pressure drop. A design pressure drop of 300 mpf of pipe is equivalent to 2.5′ of head. See Figure 18-4.

It is not always possible to find a pipe size that gives the exact flow rate. When this is the case, select a nominal pipe size that provides a flow rate equal to or greater than that needed.

If an arbitrary design pressure drop is not used, the design static pressure drop can be found by applying the procedure:

1. Determine the water flow rate in the first pipe section in the supply side of the system.

2. Choose a velocity that ensures quiet operation for the water in the first pipe section.

3. Select a size of pipe for the first pipe section that provides a reasonable water flow rate for the application.

4. Using the above data, select the static pressure drop from a pipe sizing table. This pressure drop is used as the design static pressure drop for the entire system.

Temperature Difference. Temperature difference is the difference between the initial and final temperature of a material through which heat has been transferred. When designing a piping system, the temperature difference is the difference in temperature of the water at the beginning and end of a heating or cooling process. The temperature difference through each pipe section and terminal device is important because temperature difference represents heat loss or gain. In some piping systems, the temperature difference through each pipe section and terminal device is different.

Total system temperature difference for a heating application is the difference between the temperature of the water entering the piping system at the boiler and the water returning to the boiler. For a cooling application, total system temperature difference is the difference between the temperature of

FLOW OF WATER IN TYPE "L" COPPER TUBING*									
Friction Loss§	Nominal Pipe Size#								Friction Loss**
	3/8	1/2	5/8	3/4	1	1 1/4	1 1/2	2	
100	—	.53	.96	1.44	3.1	5.3	8.5	18.2	.83
125	—	.59	1.05	1.63	3.5	6.0	9.6	20.5	1.04
150	—	.65	1.15	1.79	3.8	6.6	10.6	22.6	1.25
175	—	.71	1.26	1.95	4.2	7.3	11.5	24.6	1.46
200	.41	.76	1.35	2.10	4.5	7.8	12.4	26.5	1.67
225	.44	.81	1.44	2.24	4.7	8.3	13.2	28.3	1.88
250	.46	.86	1.53	2.36	5.0	8.7	14.0	30.0	2.08
275	.48	.91	1.61	2.49	5.3	9.2	14.7	31.5	2.29
300	.51	.95	1.68	2.61	5.6	9.6	15.3	33.0	2.50
325	.53	.99	1.75	2.70	5.8	10.0	16.0	34.5	2.71
350	.56	1.03	1.82	2.82	6.1	10.5	16.7	35.8	2.92
375	.58	1.07	1.90	2.93	6.3	10.8	17.3	37.0	3.13
400	.59	1.11	1.96	3.05	6.5	11.1	18.0	38.3	3.33
425	.61	1.15	2.03	3.15	6.7	11.5	18.5	39.6	3.54
450	.63	1.18	2.10	3.24	6.9	11.9	19.1	40.9	3.75

* in gpm
§ in mpf

\# in inches
** in ft/100′

•*ASHRAE 1989 Handbook — Fundamentals*

Figure 18-4. Pipe sizing charts show the size of pipe to use for a given water flow rate at design pressure drops.

the water entering the piping system at the chiller and the water returning to the chiller. For a typical small- to medium-size application, a temperature difference of about 20°F may be used for both heating and cooling applications. Typically, the temperature of water leaving a boiler is from 160°F to 200°F, and the temperature of water returning to the boiler is from 140°F to 180°F. The temperature of water leaving a chiller is around 45°F and the temperature of the water returning is 65°F.

Piping System Design

A floor plan or sketch of the building is required when designing a piping system for a hydronic system. Each floor that has piping or terminal devices requires a floor plan. The heating and cooling loads must be calculated. When the plans have been made and the loads have been calculated, the pipe sizes can be chosen.

Layout Drawings. Layout drawings are drawings of the floor plan of a building that show walls, partitions, windows, doors, fixtures, and other details that affect the location of the piping and terminal devices. Layout drawings also show the heating and cooling load for each room or zone and the required water flow rate. Piping systems are sized for water flow rates in gallons per minute. Heating and cooling loads for each building must be converted to water flow rate for each section. See Figure 18-5.

Layout drawings contain several overlays, which show the placement of the terminal devices and the piping system. The overlays show the floors in the building that contain terminal devices and the floors where pipes are run. Locations for terminal devices are on the overlay for each floor of the building.

The piping system is drawn on a separate overlay, which contains the location of the terminal devices. With this arrangement, the piping layout can be seen in relation to the terminal devices. On the piping layout, each terminal device is connected to the supply pipe and the return pipe. The actual layout of the piping system depends on the type of piping system chosen.

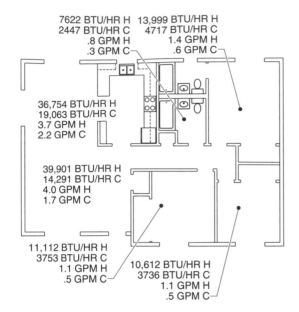

Figure 18-5. Layout drawings are drawings of the floor plan of a building, which contain the heating and cooling load for each room or zone.

On the drawings and in a building, pipes must not run through beams, columns, stairwells, elevator shafts, or areas where plumbing or electrical devices may be located. When the layout drawings and overlays are ready, the piping system is chosen based on the temperature control and sophistication level required in each building space. After the piping system has been chosen, the terminal devices for each building space are chosen based on the heating and cooling requirements for each building space. See Figure 18-6.

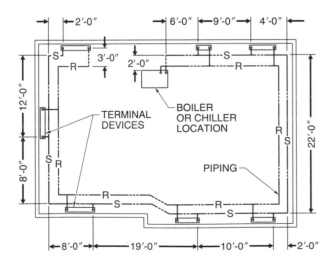

Figure 18-6. A piping system overlay includes the location of piping and terminal devices.

Section Identification. Identifying each pipe section is necessary for keeping track of the proper size pipe, valves, and fittings for each pipe section. A *pipe section* is a length of pipe that runs from one fitting to the next fitting. The length of a pipe section is measured from the outlet of one fitting to the outlet of the previous fitting. Pipe sections are identified by letters or numbers, which are marked on the pipe sections themselves or are marked on the fittings at the ends of each pipe section.

When letters are used to identify pipe sections, the pipe section between fittings is identified by the letters marked on the fittings. For example, all of the fittings are marked with a different letter. The first fitting in a system, which may come off the boiler or chiller, is identified as A. The fitting at the end of this first pipe section, which is at the beginning of the second pipe section, is identified as B. The pipe section between fitting A and B is identified as section A-B. The letters that identify the sections and the fittings are shown on the piping overlay. See Figure 18-7.

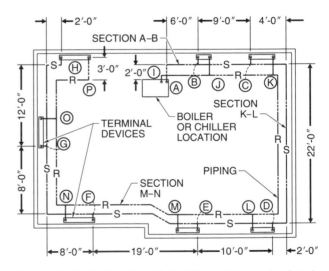

Figure 18-7. Letters identify the fittings at the ends of each pipe section. Pipe sections are identified by letters on the fittings.

Pipe Sizing

Pipe sizing depends on the material of the pipe used. The pipe made from some materials is rougher on the inside than other materials. The roughness of the inside of a pipe affects friction loss. Because one element of pipe sizing is friction loss, the type of pipe used affects the pipe size. Pipes used in hydronic systems are wrought iron, black steel, Schedule 40, or copper tubing. Copper tubing is used for small applications, and iron or steel is used for large applications.

The procedure for sizing a piping system differs depending on the type of piping system used. In a one-pipe series system, the pipe size remains the same throughout the system because the flow rate remains the same. In a one-pipe primary-secondary system, both the supply and return pipe sizes change between each terminal device because the flow rate changes. In a two-pipe direct-return system, the supply pipe and return pipes are usually the same size in each pipe section between terminal devices. The return pipes get progressively smaller as terminal devices are added. In a two-pipe reverse-return system, the supply pipe gets progressively smaller as it goes out from the boiler or chiller, and the return pipe gets progressively larger as it comes back to the boiler or chiller from the terminal device at the end of the piping loop. To size a piping system for a building, apply the procedure:

1. Prepare layout drawings and overlays, which show the heating and cooling load and water flow rate for each building space.

2. Select and locate the required terminal devices on the first overlay.

3. Lay out the piping system on the second overlay.

4. Enter the names of the piping sections in the first column of a columnar form.

5. Enter the water flow rate for each pipe section in the second column.

6. Select a design static pressure drop for the system. This is the pressure drop that occurs per 100′ of pipe in the final pipe design.

   The design static pressure drop should be approximately 300 mils per foot (mpf) or 2.5′ of head. Pipe sized with these pressure drops gives a system average-size pipe within the range of available circulating pumps.

7. Size each pipe section with a pipe sizing chart.

Example: Sizing a Piping System

A small residence requires a hydronic system. The system is a two-pipe reverse-return system made of copper tubing. A 20°F temperature difference is used for the heating flow rate and a 17°F temperature difference is used for the cooling flow rate. Find the pipe sizes for the system.

1. Prepare layout drawings and overlays. Determine the heating and cooling load and water flow rate for each building space. See Figure 18-8.

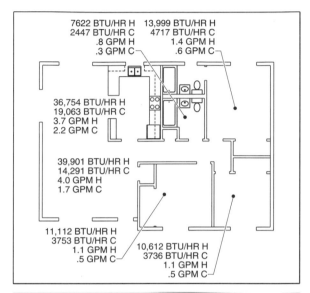

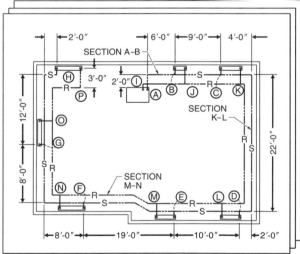

Figure 18-8. Layout drawings and overlays show the heating and cooling load, water flow rate, terminal device layout, and piping system layout.

2. Select and locate the required terminal devices on the first overlay.

3. Lay out the piping system on the second overlay.

4. Enter the names of the piping sections in the first column of a columnar form.

5. Enter the water flow rate for each pipe section in the second column. See Figure 18-9.

6. Select a design static pressure drop.

7. Size each pipe section with a pipe sizing chart.

3 bedroom residence			
300 mpi pressure drop			
Section	Flow Rate*	Size§	
A - B	12.0	1.5	
B - C	11.2	1.5	
C - D	9.8	1.25	
D - E	8.7	1.25	
E - F	7.6	1.25	
F - G	5.6	1	
G - H	3.7	1	
B - J	.8	.5	
C - K	1.4	.625	
D - L	1.1	.625	
E - M	1.1	.625	
F - N	2.0	.75	
G - O	2.0	.75	
H - P	3.7	1	

* in gpm
§ in inches

Figure 18-9. The name of the pipe section, the water flow rate, and the pipe size are entered on a columnar form.

Terminal Devices

The type of terminal device used in a hydronic heating or air conditioning system depends on the degree of temperature control required in the building. Radiators or convectors are used as terminal devices with one-pipe series piping systems. In a one-pipe series piping system, all of the sections of the system, including each terminal device, have the same water flow rate. Each terminal device selected must be sized for that flow rate.

In a one-pipe primary-secondary system, the terminal devices are located on secondary loops that are in parallel with the main system loop. In this type of system, each of the terminal devices can be a different size. The piping to each terminal device must be sized for the flow rate required for the load handled by that terminal device only. The main piping loop is sized for the total load on the main loop. This load usually includes several terminal devices. In a two-pipe system, each terminal device may have a different flow rate.

Selecting Terminal Devices. Heating and cooling capacities of terminal devices are shown on specifications sheets. To choose the correct size, the heating and cooling load on each terminal device must be known. A terminal device specifications sheet that describes the terminal device is also required. The model number of each terminal device is recorded on the overlay that shows the terminal devices. See Figure 18-10.

CAPACITY RATINGS FOR TERMINAL DEVICES IN HYDRONIC SYSTEMS*								
			Hydronic Air Conditioning			Hydronic Heating		
	Flow Rate			45°F Water 80°F db, 67°F wb Air			140°F Water	
Model	Air#	Water**	Pressure Drop§	Total*	Sensible*	Pressure Drop§	50°F Air Total*	70°F Air Total*
Q–100	200	2	1.00	6500	5500	1.4	13,500	10,000
Q–160	250	2.5	1.60	8200	5900	1.6	15,500	12,000
Q–200	275	3	2.00	9500	6400	1.9	18,000	14,000
H–250	300	4.5	2.70	10,000	7600	2.2	21,000	16,000
H–750	400	4	2.80	13,500	10,200	3.9	27,000	21,000
H–1000	600	5	3.90	21,000	16,500	7.1	43,000	34,000

* heat in Btu/hr
§ in ft of water
# in cfm
** in gpm

Figure 18-10. When the type of terminal device has been selected, a model and size is chosen for each room or building space from specifications sheets, which are printed by manufacturers.

Circulating Pump Sizing

There should be one circulating pump on each piping loop in the system. Small piping systems usually have only one pump for the whole system. Larger systems may have several pumps. Circulating pumps are selected and sized to circulate the required amount of water against the total pressure drop in the piping system.

Selecting Circulating Pumps. Graphs of performance data show the capacity of different pumps. The graphs of performance data show the pumping capacity in gallons per minute for different pressure drops. See Figure 18-11.

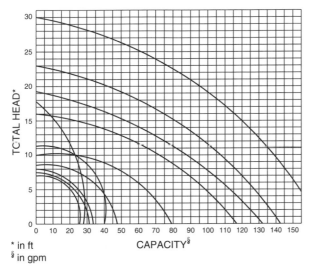

* in ft
§ in gpm

Figure 18-11. Performance data tables show the capacity of circulating pumps.

The total pressure drop in a piping system is the sum of the friction head and the static head in the supply side and return side of a piping system. The friction head in a piping loop is found by multiplying the total equivalent length of the loop by the design pressure drop for the system. The static head is the sum of the lengths of the vertical risers in the loop.

The columnar form is used to find the total head in the system. A column for the actual length of each pipe section is added to the form. The pipe section lengths are taken from the layout drawing of the system. Enter the lengths on the columnar form. The fittings in each pipe section are described in the fifth column of the form. See Figure 18-12.

The equivalent lengths of the fittings in each pipe section are entered in the sixth column on the form. The final column on the form is used to record the equivalent length for each pipe section. The length of each pipe section is found by adding the actual length and equivalent lengths for the fittings horizontally across each row on the sheet. By adding the figures in the last column, the total equivalent length of the system is found. The total equivalent length is used to find the total pressure drop of the system.

The total pressure drop for the system in mils is found by multiplying the total equivalent length by 300. The total pressure drop for the system in feet of head is found by dividing the total equivalent length by 100′ and multiplying by 2.5. The result is the total pressure drop in the pipe and fittings in the system.

The pressure drop through the terminal devices, boiler or chiller, and any other device that would add additional resistance to water flow must be added to the piping pressure drop to find the total system pressure drop. Total system pressure drop is found by applying the formula:

$$PD_t = PD_p + PD_{td} + PD_{hu}$$

where

PD_t = total system pressure drop (in ft of head)

PD_p = piping pressure drop (in ft of head)

PD_{td} = terminal device pressure drop (in ft of head)

PD_{hu} = heating unit pressure drop (in ft of head)

Example: Finding Total System Pressure Drop — Piping

The total system pressure drop through a piping system is 3.5′ of head. The water must flow through six terminal devices with a pressure drop of .65′ of head per device. The water also flows through a boiler with a pressure drop of 1.25′ of head. Find the total pressure drop through the system.

$$PD_t = PD_p + PD_{td} + PD_{hu}$$
$$PD_t = 3.5 + (.65 \times 6) + 1.25$$
$$PD_t = 3.5 + 3.9 + 1.25$$
$$PD_t = \mathbf{8.65′}$$

3 bedroom residence

Section	Flow Rate*	Size§	Length#	Fittings	FEL**	SEL§§			
A-B	12.0	1.5	8	1 Valve 1 El	1.7	9.7			
B-C	11.2	1.5	9	1 Tee	.5	9.5			
C-D	9.8	1.25	28	2 El 1 Tee	2.5	30.5			
D-E	8.7	1.25	10	1 Tee	.5	10.5	138 x 300 = 41,400 mils		
E-F	7.6	1.25	19	2 45°El 1Tee	1.9	20.9			
F-G	5.6	1	16	1 El 1 Tee	1.5	17.5	or		
G-H	3.7	1	14	1 Tee 2 El	2.5	16.5			
							3.45' of head		
B-J	.8	.5	2	1 Tee 1 Valve	1.2	3.2			
C-K	1.4	.625	2	1 Tee 1 Valve	1.2	3.2			
D-L	1.1	.625	2	1 Tee 1 Valve	1.2	3.2			
E-M	1.1	.625	2	1 Tee 1 Valve	1.2	3.2			
F-N	2.0	.75	2	1 Tee 1 Valve	1.2	3.2			
G-O	2.0	.75	2	1 Tee 1 Valve	1.2	3.2			
H-P	3.7	1	2	1 El 1 Valve	1.7	3.7			
				Total Equivalent Length		138			

* in gpm
§ in inches
# in ft

** fitting equivalent length in ft
§§ section equivalent length in ft

Figure 18-12. The length of the pipe sections and the length of the fittings are added to the columnar form.

The size of the circulating pump required for each piping loop is found by using the total flow rate for the piping loop, the total pressure drop in the loop, and a sizing table for the type of pump desired. The circulating pump is selected based on the result.

Fittings and Valves

Pipe fittings and valves control the flow of water. Shutoff valves are located on each side of all major parts of a system, such as the boiler and terminal devices. Balancing valves are located in each branch line off the main supply line and in lines going directly to terminal devices.

Balancing valves are used for balancing water flow through a terminal device to provide the proper heating or cooling effect. Data relating to valves and fittings, especially diverting and balancing valves, should be obtained from manufacturers.

Some piping systems have modulating flow valves, which are located on the supply or return connections to terminal devices. Data and information for using and controlling these valves should be obtained from manufacturers.

Controls

A hydronic control system consists of a control for water temperature in the boiler or chiller and a control for air temperature in the rooms or zones being conditioned. The controls ensure that the setpoint temperature is maintained.

Water Temperature Control. Proper water temperature in a hydronic system is maintained by an aquastat. An aquastat senses water temperature in a boiler or chiller and operates equipment to maintain the setpoint temperature. Aquastats are thermostatically controlled switches, which are wired into electrical control circuits. See Figure 18-13.

An aquastat in a hydronic system is similar to a thermostat in a forced-air system. An aquastat has a setpoint indicator, which is set for the desired water temperature. When the water temperature falls below the setpoint temperature in a heating system, the aquastat will cycle the boiler heating device ON to heat the water. The aquastat also turns the heating device OFF when the water is at the setpoint temperature. The aquastat turns the chiller ON if the water temperature rises above the setpoint temperature in a cooling system and turns the chiller OFF when the aquastat is satisfied.

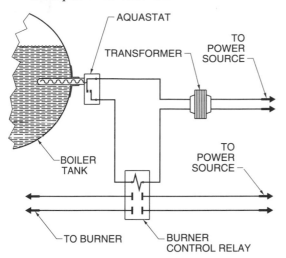

Figure 18-13. Aquastats are thermostatically controlled electric switches.

The control system for any application provides control necessary for the application. Some control systems are more sensitive than others. Outdoor reset is a sensitive control, which automatically resets the boiler water temperature in response to changes in the outdoor temperature. These sensitive controls help a hydronic system offset variations in loads caused by factors that are difficult to include in load calculations. Variable solar loads and rapid changes in outdoor temperature can be accounted for with outdoor reset controls.

Air Temperature Control. Air temperature control may be as simple as a water flow valve, which controls the flow of water to a radiator. This type of simple control maintains a constant air temperature when the water flow rate is constant. This relationship is possible because the temperature of the air leaving a terminal device is a function of the temperature of the water inside the terminal device.

Some control system components are installed in a control panel within a unit. The controls usually include a thermostat, blower speed control, and a ventilation damper control. Other hydronic control systems have components mounted on a wall of the building space.

Sophisticated hydronic control systems have a modulating valve located on the supply or return side of each terminal device. The modulating valve is controlled by a thermostat that monitors room temperature. If room temperature varies, the valve will open or close to allow more or less water to flow through the terminal device. If the system is a heating and cooling system, the valves and thermostat used must be capable of working in both modes. See Figure 18-14.

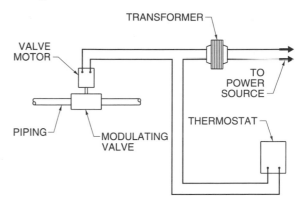

Figure 18-14. A modulating water valve opens or closes to allow more or less water to flow through a terminal device.

In systems using terminal devices with blowers, the control system may control the operation of the blower instead of water flow. A control system that controls water flow and air flow provides excellent control of air temperature. In this type of system, a room thermostat controls a modulating control valve on the water line and a manual or automatic switch controls the blower that circulates the air through ductwork. See Figure 18-15.

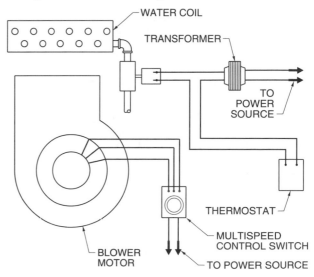

Figure 18-15. Terminal device controls may control blower operation instead of water flow.

Review Questions

## 18

1. Define *hydronic* as related to a heating or air conditioning system.

2. List and describe three parts of a piping system used with a hydronic heating or cooling system.

3. List and describe the two factors used when calculating water flow rate through a pipe.

4. Define *foot of head* as related to pipe sizing.

5. What causes pressure drop in a pipe?

6. How is design static pressure drop used in sizing pipe?

7. Why is the temperature drop through a piping system important?

8. Define *friction head, static head,* and *total head* as related to a hydronic system.

9. Why can water flow rate be different in pipes of the same size when the pipes are made of different material?

10. List the seven steps required for designing a hydronic heating or air conditioning system.

11. What is the difference between a one-pipe series piping system and a one-pipe primary-secondary system?

12. What difference must be considered when selecting terminal devices for a one-pipe series system as compared to a one-pipe primary-secondary system?

13. What two variables are used when selecting a circulating pump for a hydronic heating or air conditioning system?

14. What is the difference between the equivalent lengths of the fittings in a hydronic piping system and the total equivalent length of the system?

15. What are the balancing valves used for in a hydronic piping system?

16. What are shutoff valves used for in a hydronic piping system?

17. What variables are used to find the total pressure drop through a piping system?

18. What two control systems are used in a hydronic heating or air conditioning system?

19. What control device is used for sensing water temperature and for turning a burner ON and OFF in a hydronic heating system?

20. Why are layout drawings important when designing a hydronic system?

Distribution System Balance

Chapter 19

A balanced forced-air or hydronic distribution system provides the required amount of heating or cooling for a building and circulates the required amount of air to each building space. To balance a distribution system, blowers or circulating pumps are adjusted for proper volumetric flow rate. Balancing dampers in the ductwork or valves in the piping system are adjusted for proper medium flow. Dampers in registers or valves in terminal devices are adjusted for proper heating or cooling effect in the building spaces.

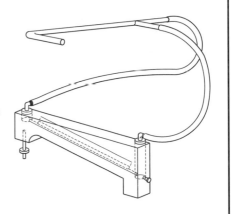

FORCED-AIR DISTRIBUTION SYSTEM BALANCING

A forced-air distribution system is balanced to provide the amount of air that satisfies the heating, cooling, and ventilation requirements of each building space. In a forced-air distribution system, the blower is adjusted to provide the total air flow rate required. The balancing dampers in the main trunks and branches of the ductwork are adjusted to deliver the proper amount of air to each part of the system. The register dampers are adjusted to deliver the proper amount of air from each register to the building space.

Installing a distribution system exactly as it has been designed is almost impossible. Changes in the building design, the location of structural parts of the building, or the location of plumbing or electrical wiring may cause changes in ductwork design. Nominal sizes of registers and grills are usually oversized for operating conditions. These variables affect the operation of a distribution system. To offset the difference between the actual operation of the system and the designed operation, the system should be balanced after all equipment, ductwork, accessories, and controls are installed and operating.

The two forced-air distribution systems used in heating and air conditioning applications are con-

stant air volume (CAV) and variable air volume (VAV) systems. A *CAV system* is an air distribution system in which the air flow rate in each building space remains constant, but the temperature of the supply air is varied by cycling the heating or air conditioning equipment ON and OFF.

A *VAV system* is an air distribution system in which the air flow rate in the building spaces is varied, but the temperature of the supply air remains constant. The air flow rate is varied in a VAV system by mixing dampers, which are located in the branch ductwork or at the registers.

CAV and VAV distribution systems are balanced with the same procedure. The blower is adjusted to provide the total air flow required for a building or building space. Balancing dampers in the distribution systems must be adjusted for the proper air flow to each zone in the building. *Balancing dampers* are air flow dampers located at the branch take-offs from the trunk duct in a distribution system. The dampers in the registers or the variable-volume dampers at the registers must also be adjusted for maximum air flow. *Variable-volume dampers* are dampers that control the air at terminal devices in a VAV system. Both systems should be adjusted for maximum air flow to ensure that all components are large enough to provide proper air flow.

285

Blowers and Air Flow

On small applications, the blower is part of the furnace or air conditioner. See Figure 19-1. On large applications, the blower is a separate component in the system. The blower for a heating or air conditioning system must have enough volumetric capacity for the application. Blowers have a wide range of flow rates at different pressure drops.

An air distribution system is balanced by finding the required air flow rate. The air flow rate of a blower should be checked and adjusted to ensure proper air flow for the application. The required air flow rate is taken from the register or ductwork plan for the building.

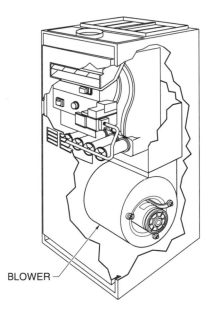

Figure 19-1. The blower is part of the furnace or air conditioner on small applications.

Air flow rate of a blower is a function of blower wheel diameter, blower wheel speed, and system pressure drop. For a blower with a given wheel diameter, the air flow rate is a function of the blower wheel speed and pressure against the blower.

Data sheets that contain blower performance characteristics are used when making the original blower selection and can be used when balancing an air distribution system. The data is presented as a table or as curves on a graph. When selecting a blower, the air flow rate required for the system is plotted against the total pressure drop in the system. The blower table or graph shows which blower should

be used. The total pressure drop in the system is plotted against the blower wheel speed to find the air flow rate from the blower. If more air is needed, the blower speed should be increased. If less air is needed, the blower speed should be reduced. See Figure 19-2.

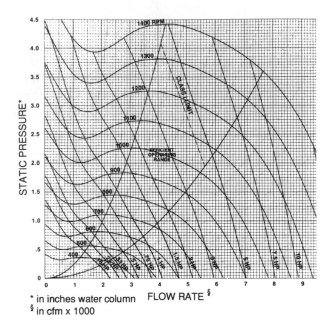

* in inches water column   FLOW RATE §
§ in cfm x 1000

Lau, a Division of Tomkins Industries

Figure 19-2. Blower performance sheets are used for selecting a blower and for balancing a forced-air distribution system.

Air Flow Rate. Air flow rate is the volumetric flow rate of the air in ductwork. Volumetric flow rate of air is the quantity of air moved per unit of time. The air flow rate of a blower is a function of the blower wheel speed. Air flow rate is directly proportional to blower wheel speed. To increase the air flow rate of a blower, the blower wheel speed is increased. To decrease the air flow rate of a blower, the blower wheel speed is decreased.

Because the relationship between air flow rate and blower wheel speed is proportional, when the air flow rate of a blower at one speed is known, the air flow rate at other speeds can be calculated. The air flow rate of a blower is found by applying the formula:

$$Q_F = \frac{Q_I \times N_F}{N_I}$$

where

Q_F = final air flow rate (in cfm)

Q_I = initial air flow rate (in cfm)

N_F = final wheel speed (in rpm)

N_I = initial wheel speed (in rpm)

Example: Finding Air Flow Rate — Blower

A blower produces 1800 cfm at a speed of 978 rpm. Find the air flow rate of the blower at a speed of 1200 rpm.

$$Q_F = \frac{Q_I \times N_F}{N_I}$$

$$Q_F = \frac{1800 \times 1200}{978}$$

$$Q_F = \frac{2{,}160{,}000}{978}$$

$$Q_F = \textbf{2208.6 cfm}$$

Measuring Air Flow Rate. Blower air flow rate is measured with an anemometer or with a pitot tube and a manometer. If the temperature difference of the air through the distribution system is known, the blower air flow rate can be calculated. A pressure drop, the blower wheel speed, and a blower table or graph can also be used to calculate blower air flow rate.

An *anemometer* is a device that measures air velocity or air flow rate. An anemometer consists of a probe, which is inserted into the air stream inside a duct. The velocity or air flow rate of the air is shown on a dial or scale. See Figure 19-3.

When the velocity is read, the velocity and the size of the duct are used to calculate the air flow rate. To calculate the air flow rate of a forced-air system with a anemometer, apply the procedure:

1. Find the area of a cross-section of the duct in square feet. Multiply the dimensions of the duct and divide by 144. Dividing by 144 converts inches to feet.

   If the dimensions of a duct are 24″ × 10″, find the area of a cross-section of the duct.

   $$A = \frac{w \times h}{144}$$

   $$A = \frac{24 \times 10}{144}$$

$$A = \frac{240}{144}$$

$$A = \textbf{1.67 sq ft}$$

2. Find the air flow rate. What is the air flow rate when an anemometer reading is 537 fpm?

   $$Q = A \times V$$

   $$Q = 1.67 \times 537$$

   $$Q = \textbf{896.8 cfm}$$

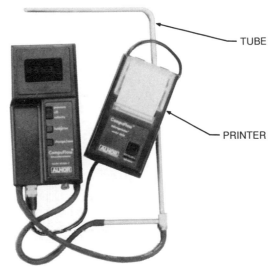

Alnor Instrument Company

Figure 19-3. The probe of an anemometer is inserted into a duct to find the velocity or air flow rate in the duct.

Example: Finding Air Flow Rate — Anemometer

An anemometer in a 24″ × 8″ duct gives a velocity reading of 678 fpm. Find the air flow rate.

1. Find the area of the duct.

   $$A = \frac{w \times h}{144}$$

   $$A = \frac{24 \times 8}{144}$$

   $$A = \frac{192}{144}$$

   $$A = \textbf{1.33 sq ft}$$

2. Find the air flow rate.

   $$Q = A \times V$$

$Q = 1.33 \times 678$

$Q = \mathbf{901.74 \ cfm}$

Blower air flow rate can also be measured with a pitot tube and a manometer. A *pitot tube* is a device that senses static pressure and total pressure in a duct. A pitot tube consists of one tube inside another. The two tubes are arranged with a 90° bend near one end. In the bent end of the tube, the inside tube is open on the end, and the perimeter of the outside tube has a ring of small holes around it. Inside a duct, static pressure is sensed by the outside tube, and total pressure is sensed by the inside tube. These functions occur at the same time. See Figure 19-4.

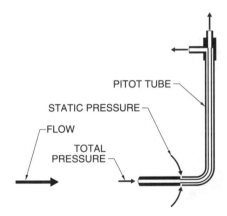

Figure 19-4. When a pitot tube is used to measure air flow rate, total pressure is sensed by the inside tube, and static pressure is sensed by the outside tube.

By connecting a pitot tube to a manometer, the two pressure readings oppose each other. This action indicates the velocity pressure of the blower. Total pressure is the sum of static pressure and velocity pressure. A manometer cannot read velocity pressure directly, but can read total pressure and static pressure. Velocity pressure is found by subtracting static pressure from total pressure. Velocity pressure is converted to velocity and multiplied by the size of the duct to find air flow rate.

Example: Converting Velocity Pressure to Velocity

A velocity pressure reading of .36″ WC is taken with a pitot tube and manometer in a 20″ × 8″ duct. Convert velocity pressure to velocity.

$V = 4005 \times \sqrt{P_v}$

$V = 4005 \times \sqrt{.36}$

$V = 4005 \times .6$

$V = \mathbf{2403 \ fpm}$

When a pitot tube is used to measure air flow rate, the inside tube senses total pressure and the outside tube senses static pressure. The tubes run from inside the duct to an inclined manometer, which is outside the duct. The inside tube is connected to the high end of the manometer and the outside tube is connected to the low end of the manometer. Static pressure from the outside tube is automatically subtracted from the total pressure from the inside tube. See Figure 19-5.

Air flow rate can also be found from the heating capacity and the temperature difference of the air as it flows through a furnace or air conditioner. The temperature difference of the air is found with thermometer readings.

In the temperature difference method, the air flow rate from a blower in a heating system is found by using the output rating of the furnace and the temperature increase of the air through the furnace. Equipment catalogs and data sheets contain output ratings for particular furnaces at standard conditions. Temperature increase is found by measuring the inlet and outlet air temperatures of the furnace.

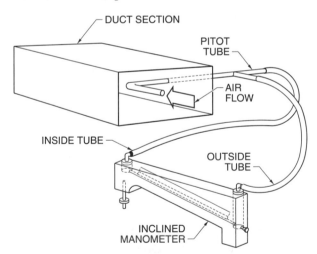

Figure 19-5. A pitot tube is a device that measures static pressure.

Example: Finding Air Flow Rate — Furnace

A furnace has an output rating of 160,000 Btu/hr. The temperature increase of the air through the furnace is 87°F. Find the air flow rate in the furnace.

$$Q = \frac{OR}{\Delta T \times 1.08}$$

$$Q = \frac{160,000}{87 \times 1.08}$$

$$Q = \frac{160,000}{93.96}$$

Q = 1702.85 cfm

Air flow rate through the evaporator coil of an air conditioner can be found if the enthalpy of the air entering and the enthalpy of the air leaving the evaporator coil and the cooling capacity of the evaporator coil are known. The enthalpy values are found on a psychometric chart.

The dry bulb and wet bulb temperatures of the air entering and leaving the air conditioner are measured with a psychrometer. The dry bulb and wet bulb temperature values are used to find the enthalpy of the air on a psychrometric chart. The cooling capacity of the air conditioner is found in an equipment catalog or on a data sheet. The cooling capacity of the air conditioner at existing conditions should be used. The air flow rate for a cooling system is found by applying the formula:

$$Q = \frac{OR}{4.5 \times (h_e - h_l)}$$

where

Q = air flow rate (in cfm)

OR = output rating (in Btu/hr)

4.5 = constant

h_e = enthalpy of air entering evaporator coil (in Btu/lb)

h_l = enthalpy of air leaving evaporator coil (in Btu/lb)

Example: Finding Air Flow Rate — Air Conditioner

An air conditioner has an output rating of 46,000 Btu/hr. The enthalpy of the air entering the air conditioner is 30.9 Btu/lb. The enthalpy of the air leaving the air conditioner is 24.5 Btu/lb. Find the air flow rate.

$$Q = \frac{OR}{4.5 \times (h_e - h_l)}$$

$$Q = \frac{46,000}{4.5 \times (30.9 - 24.5)}$$

$$Q = \frac{46,000}{4.5 \times 6.4}$$

$$Q = \frac{46,000}{28.8}$$

Q = 1597.2 cfm

Air flow rate can also be measured using the pressure drop in the system, the speed of the blower wheel, and a blower performance chart. A manometer is used to measure the pressure drop in the system, and a tachometer is used to measure the speed of the blower wheel. A *tachometer* is a device used for measuring the speed of a shaft or wheel. See Figure 19-6.

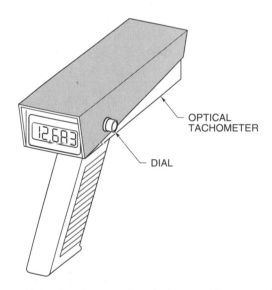

OPTICAL TACHOMETER

DIAL

Figure 19-6. A tachometer is a device used for measuring the speed of a shaft or wheel.

The static pressure drop in a system is found by using a manometer to read the static pressure drop in the supply side and return side of a ductwork system. The readings are taken at the outlet and inlet of the blower. The pressure reading on the return side of the blower is a negative pressure and the pressure on the supply side of the blower is positive pressure. To find the difference between the two readings, the values are added without considering the mathematical signs. Change the negative sign to positive and add. The result is the total static pressure drop. To find the total static pressure drop, apply the formula:

$$P_t = P_r + P_s$$

where

P_t = total pressure (in psi)

P_r = return side pressure (in psi)

P_s = supply side pressure (in psi)

Example: Finding Total Static Pressure Drop

The pressure reading on the return side of the blower is −.1 psi and the pressure on the supply side of the blower is .2 psi. Find the difference between the two pressures.

$$P_t = P_r + P_s$$
$$P_t = (-.1) + .2$$
$$P_t = .1 + .2$$
$$P_t = \textbf{.3 psi}$$

As the blower wheel turns, the speed of the wheel is measured with a tachometer. A *strobe light tachometer* is a tachometer that contains a strobe light, which flashes a beam of light at specified time intervals. The time interval is selected and adjusted with a dial. A spot of white reflective tape is attached to the rim of a motor sheave or blower wheel sheave. As the strobe light flashes, the time interval of the flashes is adjusted until the spot appears to be stationary. At that speed, the flashes are synchronized with the wheel rotation. The speed of the wheel is read from the dial on the strobe light tachometer.

The total static pressure drop and the speed of the blower wheel are necessary for finding the air flow rate for the blower. When these two values are known, the air flow rate for the blower is found on a table or graph for the blower.

Adjusting Air Flow Rate. If the blower air flow rate is lower than the required flow rate for an application, the blower speed should be adjusted to provide the proper flow rate. The flow rate is directly proportional to the blower wheel speed. To change the air flow rate of a blower, the speed of the blower wheel must be changed. Blower wheel speed is changed by changing blower drives.

Blower drives are the connections between the blower motor and the blower wheel. The two basic types of blower drives are direct drive and belt drive. The blower wheel on a direct drive blower is mounted directly on the motor shaft. The blower wheel on a belt drive blower is turned by an arrangement of sheaves and a belt.

On a belt drive blower, the motor sheave is mounted on the motor shaft and the blower sheave is mounted on the blower wheel shaft. A belt or set of belts connects the two sheaves.

To change the air flow rate of a belt drive blower, the blower wheel speed is changed by changing the size of the sheaves. Blower wheel speed is a function of the motor speed and the size of the sheaves. If the blower sheave and the motor sheave on a belt drive system are the same size, the blower wheel speed will be the same as the motor speed. On most blower drives, the motor sheave is smaller than the blower sheave. Blower wheel speed is indirectly proportional to the diameter of the two sheaves.

To adjust a blower for more or less air flow, the size of the motor sheave is changed. On most applications the motor sheave is adjustable. When the air flow rate of a blower is adjusted, the blower speed should also be adjusted for the new flow rate. The blower speed and the size of the motor sheave required for the new air flow rate are found by applying the procedure:

Example: Finding Required Blower Speed

A blower produces 4500 cfm with a 6″ motor sheave and a 10″ blower sheave. The blower is driven by a 1725 rpm motor. A flow rate of 5200 cfm is required from the blower. Find the required blower speed.

1. Find the original blower speed.

$$N_b = \frac{N_m \times PD_m}{PD_b}$$

$$N_b = \frac{1725 \times 6}{10}$$

$$N_b = \frac{10,350}{10}$$

$$N_b = \textbf{1035 rpm}$$

2. Find the new blower speed.

$$N_F = \frac{N_I \times Q_F}{Q_I}$$

$$N_F = \frac{1035 \times 5200}{4500}$$

$$N_F = \frac{5,382,000}{4500}$$

$$N_F = \textbf{1196 rpm}$$

Note: N_F is the new blower speed, which is equal to N_b.

3. Find the blower sheave diameter.

$$PD_b = \frac{N_m \times PD_m}{N_b}$$

$$PD_b = \frac{1725 \times 6}{1196}$$

$$PD_b = \frac{10,350}{1196}$$

$$PD_m = \textbf{8.65''}$$

The blower wheel of a direct drive blower is mounted directly on the motor shaft. The blower wheel turns at the same speed as the motor. To change the speed of the blower wheel in a direct drive blower, the speed of the motor must be changed. Multispeed motors vary the speed of direct drive blowers. Different motor speeds are attained by changing the electrical connections to the motor.

The speeds of a multispeed motor are shown on the specifications sheets for the motor and are usually on the motor nameplate. Because speeds are constant at the different steps of the motor, the speed closest to that required for an air flow rate is used.

Ductwork and Air Flow

The ductwork of an air distribution system is sized to carry the required amount of air through each part of the system. The actual air flow in sections of a new distribution system vary from the design flow rate because of changes in ductwork design and nominal-size ducts. Most distribution systems have balancing dampers in branch takeoffs and at the supply registers. These balancing dampers are adjusted to balance the air flow to the different building spaces.

Measuring Air Flow Rate. The air flow rate in ductwork can be checked with an anemometer or a pitot tube and a manometer. The air flow rate required in each trunk or branch of a distribution system is found in the data used for the system

design. The balancing dampers in the ductwork are adjusted to modify the flow rate in the duct.

Balancing dampers are used to adjust the flow rate in a duct. The balancing damper is opened or closed with a handle that is connected to the damper shaft. The handle extends through one side of the duct. The required air flow rate through ductwork is found in the original design plans. The air flow is balanced by having the balancing damper wide open with the blower set for the proper air flow rate. The balancing damper is slowly closed until the proper flow rate is established.

Registers and Air Flow

The air flow rate at each register is adjusted to provide the design air flow rate from each register. The procedure consists of adjusting the dampers located at the register.

Measuring Air Flow Rate. The air flow rate is measured at each register with a velometer or flow meter. See Figure 19-7. A *velometer* is a device that measures the velocity of air flowing out of a register. To find the air flow rate at a register, the velocity of the air is converted to air flow rate.

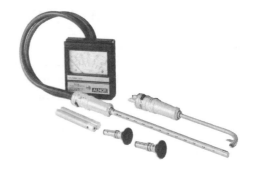

Alnor Instrument Company

Figure 19-7. A velometer is used for measuring the velocity of the air flowing from a register.

Air velocity is measured with a velometer. To measure the air velocity at a large register, several velometer readings are be taken across the surface of the register. The readings are then averaged. After the average velocity of the air leaving a register is found, the free area of the register is used to find the air flow rate. Free area is the area of a register that allows air to pass through. Free area for a register is found in equipment catalogs and specifications sheets.

Example: Finding Register Air Flow Rate

A register with a nominal size of 36″ × 24″ has an area of 4.5 sq ft. Velocity at the register is 475 fpm. Find the register air flow rate.

$$Q = A \times V$$

$$Q = 4.5 \times 475$$

$$Q = \textbf{2137.5 cfm}$$

Adjusting Air Flow. The air flow at each register is regulated by the damper in the register. Air flow rate readings are taken, adjustments are made at the register damper, and the readings are taken again to verify the results. If several registers are located on one duct section, all registers should be adjusted and the air flow rate at each register should be verified before proceeding to the next duct section. Changing the damper setting at one register usually changes the pressure drop in that duct section. The air flow rate changes at other registers on the duct section. Adjusting the register dampers for the proper air flow rate is a matter of trial and error.

HYDRONIC SYSTEM BALANCING

For a hydronic heating or cooling system to work properly, the water flow through the distribution system must be balanced. A balanced water distribution system allows the proper water flow through each terminal device, which provides the required amount of heating or cooling in each building space. The pressure drop in each section of a hydronic distribution system affects the water flow.

Balancing a hydronic distribution system requires adjusting balancing valves in the piping to provide the proper water flow rate in each part of the distribution system. Water flow rate is the volumetric flow rate of water as it moves through a pipe section. Water flow rate is expressed in gallons per minute. The proper water flow rate is required in a hydronic heating or cooling system to provide the desired heating or cooling effect. Balancing, mixing, and diverting valves are used to balance a hydronic distribution system. These valves are located at strategic points in the distribution system.

Balancing a hydronic system is necessary because a hydronic distribution system usually cannot be installed the way it is designed. Actual pressure drops differ from design pressure drops in a distribution system because of variations in construction of the building or relocated piping. The components of a hydronic distribution system are usually larger than necessary because the components are sized for maximum load conditions.

The water flow rate in a hydronic distribution system is balanced by checking water flow rates at one set of conditions and changing the settings of balancing valves in the system. Water flow rates are checked by measuring the temperature difference of the water as it flows through a section of the piping system. Water flow rate is found by checking the pressure drop through a section of the system and using the pipe size and pressure drop.

A hydronic distribution system is balanced by measuring the water flow rate of the circulating pump when all of the balancing valves in the system are open. The balancing valves are adjusted to achieve the water flow rate desired from the pump. The balancing valves are adjusted in each branch in the piping system to ensure an adequate water flow rate in each branch. The water flow rate is adjusted at each terminal device to ensure the required water flow rate through each terminal device. The desired water flow rate is taken from the original design plans.

Circulating Pump and Water Flow

A circulating pump should circulate water through a distribution system at the required water flow rate. The pressure created by the pump should be equal to or greater than the pressure required to overcome the total head of the system. Small distribution systems contain direct drive pumps. The impeller wheel of a direct drive pump turns at the same speed as the electric motor for the pump. See Figure 19-8.

A distribution system that contains a direct drive pump is balanced by changing the pressure or head in the distribution system. The pump has to work against the pressure in the system. The balancing valves in a hydronic system are adjusted to change the pressure in the distribution system.

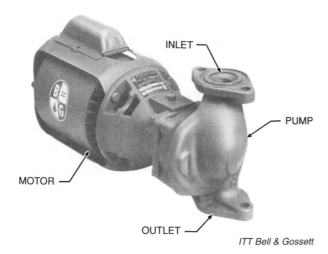

Figure 19-8. Small hydronic distribution systems contain direct drive pumps, which turn at the same speed as the electric motor that drives the pump.

ITT Bell & Gossett

Measuring Water Flow Rate. A hydronic distribution system is balanced by measuring the water flow rate of the pump, which determines if the proper amount of water is being circulated. When measuring the water flow rate of a distribution system, all of the balancing valves in the system are opened to allow full flow. The pump is turned ON and either the temperature difference or the pressure drop through the system is measured. Temperature difference is found by measuring the temperatures of the supply water and return water at the boiler or chiller.

To find the water flow rate in the system by temperature difference, the temperature of the water entering and leaving the boiler or chiller is checked with thermometers located on the supply and return water pipes. The difference between the readings is the temperature decrease or increase in the system.

Example: Finding Water Flow Rate — Temperature Difference

A boiler in a hydronic heating system has an output rating of 150,000 Btu/hr. The temperature difference is 26.4°F. Find the water flow rate.

$$Q = \frac{OR}{\Delta T \times 500}$$

$$Q = \frac{150,000}{26.4 \times 500}$$

$$Q = \frac{150,000}{13,200}$$

$$Q = \textbf{11.36 gpm}$$

Pressure drop is found by measuring the pressure at service ports on the supply and return pipes on each side of the boiler or chiller. See Figure 19-9. Service ports are usually located on fittings on the pipe or on balancing valves or other flow control components used in the system. A service port is a valve port to which a pressure gauge can be connected. The difference between the pressure at the beginning of the piping system and at the end of the piping system is the pressure drop in the system.

To find the water flow rate in a system by pressure drop, the pressure drop through the piping in the system and the pipe size in that part of the system is found. The water flow rate is found on a pump selection chart.

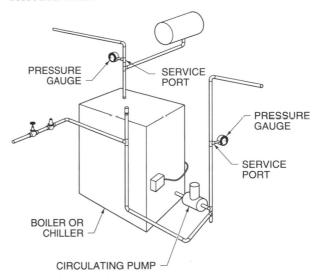

Figure 19-9. Pressure drop is found by measuring the pressure at service ports on the supply and return pipes on a boiler or chiller.

The water flow rate in a distribution system is found by measuring the pressure drop of the system. The pressure drop is found on a pump sizing chart. A horizontal line is drawn across the chart to represent the pressure drop. The size of the pipe at the beginning of the loop is found on the chart. A vertical line is drawn on the table to represent the size of the pipe. The water flow rate is at the point where the two lines intersect. See Figure 19-10. This point represents the water flow rate through the pump and the total water flow rate for the system.

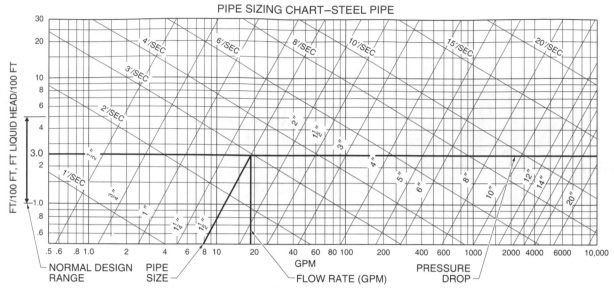

•ASHRAE 1989 Handbook — Fundamentals

Figure 19-10. Pressure drop through the piping and the pipe size are found on a pump sizing chart to find the water flow rate in a hydronic distribution system.

Adjusting Water Flow Rate. If the water flow rate in a piping system is greater than or less than the water flow rate required, adjustments should be made to attain the proper water flow rate. A circulating pump cannot be adjusted to change the water flow rate, but the water flow rate in the distribution system can be adjusted by changing the total system pressure drop. The total system pressure drop is changed by opening or closing balancing valves in the system.

The water flow rate of a pump in a hydronic distribution system is measured when the balancing valves are wide open. If the pump is the right size for the application, the water flow rate will be greater than the water flow rate needed for the application. To reduce the water flow rate to the required water flow rate, a greater pressure drop is introduced into the system by partially closing the balancing valve closest to the pump.

Balancing valves are sized by pressure drop. The balancing valves for any part of a piping system are selected for the pressure drop required at that point. The pressure drop of a valve is indicated by the pressure drop coefficient of the valve.

Pressure drop coefficient is the water flow rate through a valve or any control mechanism in a piping system that causes a pressure drop of 1 psi

through the valve or mechanism. To find the flow rate through a valve apply the formula:

$$Q = C_v \times \sqrt{PD}$$

where

Q = water flow rate (in gpm)

C_v = pressure drop coefficient

PD = pressure drop across valve (in psi)

Example: Finding Water Flow Rate Through Valve

A valve has a pressure drop coefficient of 21. The pressure drop across the valve is 4.58 psi. Find the water flow rate through the valve.

$$Q = C_v \times \sqrt{PD}$$

$$Q = 21 \times \sqrt{4.58}$$

$$Q = 21 \times 2.14$$

$$Q = \textbf{44.94 gpm}$$

To adjust the water flow rate through a balancing valve, the valve is opened or closed by turning the valve wheel or lever. If several branches lead off a main supply line, the balancing valves in each branch should be adjusted to provide the proper flow in the branch.

Water Flow in a Main or Branch

The water flow rate in main supply or return pipes and in each branch of a hydronic distribution system is regulated by adjusting the balancing valves that feed the piping. The water flow rate in the trunks and branches should be set automatically when the total pressure drop of the system is adjusted for the proper water flow rate in the circulating pump. When pressure drop is adjusted for the pump, care should be taken to recheck the water flow rate in the branches. Changes in the total system pressure drop affect the flow in individual branches as adjustments are made.

Measuring Water Flow Rate. To measure the water flow rate in the branches of a piping system, check the pressure drop across the balancing valves in the system. The valves are rated with a pressure drop coefficient so pressure drop through the valves can be used to find the water flow rate. A calibrated set of pressure gauges is used to read the pressure on each side of the valve. The difference between the readings is the pressure drop. A table or graph is used to find the water flow rate for a given pressure drop.

Adjusting Water Flow Rate. The balancing valve used for checking water flow rate by pressure drop is also used to adjust the pressure drop for the proper water flow rate in each branch of a piping system. A balancing valve has a valve wheel or lever that is used for adjusting water flow rate. Most balancing valves contain a calibrated scale that shows the change in water flow when changes are made to the valve wheel. See Figure 19-11.

Terminal Devices and Water Flow

The required water flow rate in the various terminal devices in a hydronic system depends on the type of system and the type of terminal device used. In a one-pipe series system, the water flow rate through each terminal device is the total water flow rate for the loop. In a two-pipe system, the water flow rate may be different through each terminal device. The water flow rate through each terminal device is

based on the amount of heat required from the terminal device. The layout drawing of the system shows the water flow rate required for each unit.

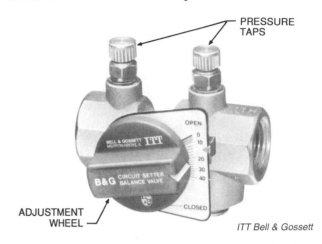

ITT Bell & Gossett

Figure 19-11. A balancing valve has a valve wheel or lever for adjusting water flow through the valve.

Measuring Water Flow Rate. The water flow rate through terminal devices is measured by the temperature difference method. The difference between the temperature of the water entering the system and the temperature of the water leaving the system is a function of the heating or cooling capacity of the terminal devices. The heating capacity of a terminal device is used to find the water flow rate by temperature difference.

The water flow rate through a terminal device is found by checking the temperature drop through each device. The inlet and outlet water temperatures are read with a thermometer. The water flow rate is calculated from the known heating capacity of the terminal device and the temperature difference.

Adjusting Water Flow Rate. The heating or cooling capacity of a terminal device is adjusted by adjusting the water flow rate through the device. The proper water flow rate for the application is selected for each terminal device. To adjust the distribution system properly, the water flow rate at each terminal device is set to the design water flow rate. Balancing valves located on the inlet or outlet piping to a terminal device are used for balancing. The water flow rate is found by pressure readings taken at pressure taps located on the valves, and the valves are adjusted to maintain the proper water flow rate.

Review Questions

1. Name the three major parts of a forced-air or hydronic distribution system that are adjusted during balancing.

2. What is the first component that is adjusted when balancing a forced-air distribution system?

3. Explain why a distribution system may not be installed exactly as it is designed.

4. List and describe the two types of forced-air distribution systems.

5. How is the total air flow rate determined for an application?

6. For a blower with a given wheel diameter, what two variables determine the air flow rate of the blower?

7. Where is the air flow rate data found for a blower?

8. What instrument measures the air flow rate in a duct?

9. What two air flow values are measured by a pitot tube?

10. When the temperature change method is used to determine the air flow rate in a duct, what two variables are used?

11. What two points in a piping system are used when pressures are taken for finding the total pressure drop in the system?

12. List and explain the formula used for finding the air flow rate through an air conditioning evaporator coil.

13. What is the relationship between air flow rate and blower wheel speed?

14. In what two ways is the air flow rate from a blower changed?

15. List and explain the formula used to determine the speed of a blower wheel when the motor speed and sheave sizes are known.

16. What are the main differences between belt drive blowers and direct drive blowers?

17. Where are balancing dampers normally located in a duct system?

18. List and explain the formula used for finding air flow rate from a register when the register size and air velocity from the register are known.

19. Name three valves used for balancing water flow in a hydronic piping system.

20. Where is the temperature difference measured to find the total water flow rate required in a hydronic distribution system?

Maintenance and Troubleshooting

Chapter 20

All components in heating and air conditioning systems require maintenance, which provides trouble-free operation, prolongs equipment life, and helps prevent breakdowns. If a component of a heating or air conditioning system fails, the operation of the entire system will be affected. Troubleshooting is the process of isolating a malfunctioning component or part.

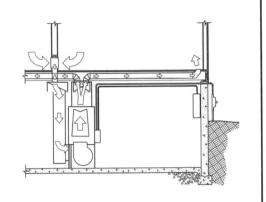

MAINTENANCE PROGRAM

Maintenance is the periodic upkeep of equipment. Heating and air conditioning systems should be inspected regularly on a maintenance program. A maintenance program includes maintenance calls scheduled at regular intervals. The length of time between maintenance calls depends on the type of system. Heating systems require inspection in the fall. Air conditioning systems require inspection in the spring. Combination heating and air conditioning systems require inspection in the spring and fall.

Maintenance may be performed by a maintenance technician or qualified service technician. A *maintenance technician* is a person trained in the maintenance tasks of heating and air conditioning systems. A maintenance technician must know when a system is operating properly and recognize signs of improper operation. A maintenance technician monitors the operation of the system, cleans components, inspects belts, and lubricates machinery. A *service technician* is a person trained to troubleshoot and repair equipment that is not operating properly.

Maintenance Call

A *maintenance call* is a scheduled visit in which a maintenance technician performs general and specific inspections of equipment, components, and operation of a heating or air conditioning system. A maintenance call may include visual inspection, mechanical inspection, and instrument tests.

On a general maintenance call, a maintenance technician will:

Look at all heating or air conditioning components for accumulations of dirt or dust that may affect operation of the components.

Determine if components are aligned properly.

Inspect important parts such as burner ports, blower belts, coils, and filters for wear or damage.

Inspect electrical wiring and contacts on electrical relays or starters for burned areas.

Specific inspections of the equipment are made at the same time as the general inspection. During a specific maintenance call, a maintenance technician will:

Measure electrical voltage and amperage.

Measure heating and/or cooling output.

Measure temperature control during normal heating or air conditioning system operation.

A visual inspection of the equipment includes inspecting the condition of the equipment, the air and fuel filters, blower belts, motors and shafts, and all visible operating parts. A visual check often shows if parts need cleaning, replacement, or lubrication.

A mechanical check of the parts of a system requires mechanical manipulation of the parts to verify the results of a visual inspection, or testing of parts that cannot be inspected visually. A mechanical check of parts includes physically checking belt tension, motor and blower shafts for play, and parts for excessive heat.

Instrument tests on equipment are made to ensure that electrical power, fuel supply, and combustion characteristics are all within acceptable ranges. Motors are checked with an ammeter to ensure they are not overloaded. An *ammeter* is a device used to measure current flow in an electric device. Voltage checks determine that proper voltage is applied to each component and ensure that excessive voltage drop does not affect operation. Pressure tests on fuel, and combustion tests may be used to determine firing characteristics.

Filters. All forced-air distribution systems have filters. Filters may be located in the return ductwork, return air plenum, or in the cabinet of the furnace or air conditioner. Filters should be inspected and cleaned or replaced on every maintenance call. An access panel allows easy access to the filter for inspection and replacement. See Figure 20-1.

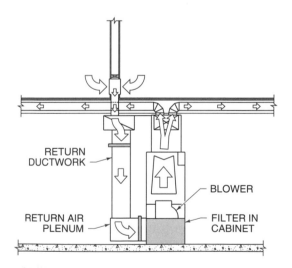

Figure 20-1. Filters may be located in the return ductwork, return air plenum, or in the cabinet of a furnace or air conditioner.

To inspect the filters, the blower is turned OFF by opening the electric disconnect switch. The access panel is removed and the filters are inspected.

The filters may be cleaned in place with a vacuum cleaner, removed for cleaning, or removed and replaced with new filters. A visual inspection is usually all that is required to determine if filters are dirty. Dust and dirt collect on the air inlet side of the filter media. If filters accumulate enough dust and dirt to reduce the air flow across the filter, the filter must be cleaned or replaced.

Large air distribution systems may have a pressure gauge located at the filter with pressure taps located on the inlet and outlet side of the filter. See Figure 20-2. Pressure drop across a filter is directly related to the accumulation of dust and dirt on the filter. A large amount of dust and dirt causes a large pressure drop across a filter. A manometer is connected to the pressure taps to measure air pressure drop across the filter.

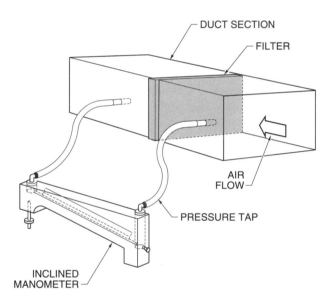

Figure 20-2. Large air distribution systems may have a manometer connected to pressure taps to measure the air pressure drop across the filter.

A *fuel filter* is a filter installed in fuel supply line to remove dirt and other contaminants from the fuel. Contaminants in the fuel may plug burner nozzles or pumps. Most fuel filters consist of a canister that holds a replaceable filter cartridge. The filter cartridge is removed for cleaning or replacement. A fuel filter cartridge should be inspected during a maintenance call at the beginning of each heating season. The filter cartridge should be cleaned or replaced. See Figure 20-3.

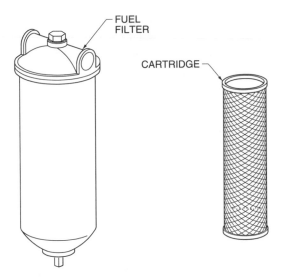

Figure 20-3. Fuel filters are installed in fuel supply lines to remove dirt and other contaminants in the fuel.

Blower Belts. During each maintenance call, all blower belts should be inspected for wear, tension, and alignment. Blower belts become worn and frayed if they are too tight, too loose, or if the blower and motor sheaves are not aligned properly. Blower belts should be replaced when they appear worn. A worn belt usually shows wear on the inside surfaces that fit in the sheaves. A new belt has slightly concave inside surfaces so the belt fits snugly at the top and bottom of the sheave. A worn belt has flat sides, which may show the fabric cord in the belt. See Figure 20-4.

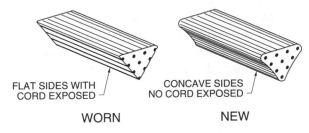

Figure 20-4. Blower belts become worn and frayed if they are too tight, too loose, or if the blower sheaves are not properly aligned.

For proper operation and maximum wear, a blower belt must be properly tensioned. A belt is properly tensioned when it can be deflected approximately 1″. To check blower belt tension, the blower is turned OFF by opening the electric disconnect switch. A straightedge is placed on top of the belt.

The belt is pulled away from the straightedge. When the resistance to deflection becomes noticeable, the distance between the straightedge and the belt is measured. The distance between the straightedge and the belt should be about 1″. If there is more or less than 1″ of deflection, the tension on the belt should be adjusted by loosening the motor mount bolts and moving the motor until the tension is correct. When the tension is correct, the motor mount bolts are tightened.

To check a blower belt for proper alignment, the furnace or air conditioner blower is turned OFF by opening the electric disconnect switch. A straightedge is placed along the top of the belt across the blower and motor sheaves. If the sheaves are aligned, the straightedge will be parallel with both sheaves. If the sheaves are not aligned, one sheave should be moved on its shaft until the sheave is aligned with the other. See Figure 20-5.

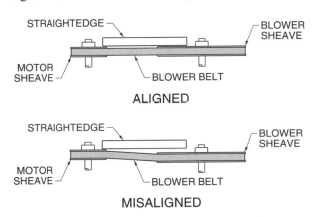

Figure 20-5. A straightedge is placed along the top of the belt across the blower and motor sheaves to check a blower belt for proper alignment.

Most sheaves are held in place on their shafts by a set screw or cap screw. To move a sheave on its shaft, the belt is removed and the set screw or cap screw is loosened. The sheave is moved along the shaft until it is aligned with the other sheave. When the sheaves are aligned, the set screw or cap screw is tightened and the belt is replaced.

Lubrication. During each maintenance call, all shaft bearings should be inspected for proper lubrication. Blower bearings are located on the ends of the blower shaft at the sides of the blower housing.

Bearings are inspected for signs of wear and excessive play.

To check a bearing, the furnace or air conditioner blower is turned OFF by opening the electric disconnect switch. An attempt is made to move the bearing on the shaft. Excessive play in any direction indicates wear between the bearing and the shaft. If excessive play is present, the bearing and shaft should be dismantled and inspected for signs of excessive wear. Excessive wear affects the rotation of the shaft in the bearing. If signs of excessive wear are present, the shaft, the bearings, or both should be replaced.

If a bearing is wearing excessively, the bearing will overheat during operation. A temperature check can indicate an excessively worn bearing. To check the temperature of a bearing, carefully touch the bearing while the blower is operating. If the bearing is warmer than the ambient air temperature or if the bearing is too hot to hold, the bearing should be lubricated.

To lubricate a bearing, find the lubrication port or oil or grease cup. An oil or grease cup is a small reservoir with a lid located on the outside of the bearing. A tube connects an oil or grease cup to the bearing surface. Oil cups have a hinged lid that closes over the oil reservoir. Grease cups have a threaded cap that covers the grease reservoir. See Figure 20-6.

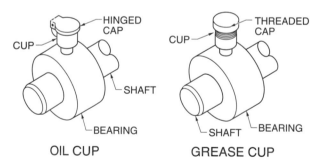

Figure 20-6. Oil and grease cups are small reservoirs located on the outside of a bearing that lubricate the bearing.

To lubricate a bearing that has an oil cup, open the lid on the oil cup and add a drop or two of oil inside the cup. The oil flows down the tube from the cup to the bearings. Most motors require lubrication once a year. Only one or two drops of oil are required.

To lubricate a bearing that has a grease cup, fill the cup with grease and turn the cap on the cup down two turns. Grease cups are found on older or very large blowers.

Compressors or other large operating components are lubricated as part of an annual maintenance program. If a compressor has an oil sight glass on it, the level of the oil is shown in the sight glass. If the oil level appears low, a service technician should be notified. A maintenance technician normally does not add oil to an air conditioning compressor.

If a condition that affects the proper operation of the equipment is discovered during a maintenance call, the condition should be corrected by the maintenance technician. If the problem cannot be resolved by the maintenance technician, the building owner should be notified that a service technician is required to make the repairs.

Maintenance Records

A record is kept of all work performed on the equipment during each maintenance call. This record becomes part of a job record. A job record is used for reference during future calls and during emergency service.

Job Sheet. A *job sheet* is a form where data related to inspections is recorded. A job sheet contains information about the job and the work performed on the equipment. A job sheet is filled out by a maintenance technician during a call. A job sheet is used for scheduling future maintenance calls and for recording information that may be needed during an emergency service call. See Figure 20-7.

Maintenance Checklist. A maintenance checklist is used on each maintenance call to ensure that all necessary items are inspected. A *maintenance checklist* is a form, which may be part of the job sheet, that lists all components to be inspected and tests to be conducted. The checklist may have spaces for listing the results of various tests. See Figure 20-8. Some items on the list may require inspection on each maintenance call, while other items may only need periodic inspection.

JOB SHEET

ABC HEATING AND COOLING
123 Main Street
Chicago, IL 60646 – Phone: 123-4567

Name		Description of Work Performed	Work Performed		
Street	Date		**Furnace**		**Condensing Unit**
			Replaced belt		Refrigerant No.
City			Adjusted belt		Cleaned coil
			Replaced pulley		Repaired leak
Phone			Adjusted pulley		Checked charge
			Cleaned blower		Leveled
Technician			Replaced bearings		Checked motor
			Oiled motor		Changed motor
Authorized by			Oiled bearings		Replaced belt
			Cleaned heat exch.		Adjusted belt

Qty.	Material and Service	Unit Price	Amount		
				Replace heat exch.	Replaced contactor
				Cleaned pilot	Replaced starting relay
				Adjusted pilot	Replaced starting cont.
				Replaced valve	Replace run cap.
				Replaced thermocouple	Repaired wiring
				Repaired valve	Replaced fuse
				Cleaned burner	Replaced compressor
				Duct	**Evaporator Unit**
				Repaired	Cleaned coil
				Adjusted	Repaired leak
				Thermostat	Replaced exp. valve
				Replaced	Adjusted exp. valve
				Adjusted	Replaced cap. tube
				Pumps	Cleared cap. tube
				Greased	Leveled coil
				Repaired	**Cooling Tower**
				Condensate Drains	Cleaned
				Cleaned pan drain	**Electric Heater**
				Repaired pan drain	Repaired wire
				Cleaned main drain	Replaced contacts
Total Materials				Repair main drain	
				Filters	

Hrs.	Labor	Rate	Amount	Recommendations	Cleaned	
					Replaced	

Total Summary	
Total Materials	
Total Labor	
Tax	

Total Labor	**Total**

Figure 20-7. A job sheet is a form that has space for recording data related to the inspections performed on equipment during a maintenance call.

Job File. A *job file* is a complete record of work done on a job during maintenance calls and emergency service calls. A job file is kept to ensure that proper maintenance is provided during each maintenance call. A job file is used with the job sheet and maintenance checklist to determine when maintenance calls are due and what work has been done in the past.

Follow-up

A follow-up program ensures that the heating and air conditioning equipment on a job is properly maintained. The follow-up program includes checking each job to ensure that the maintenance inspections were satisfactory and that the equipment is operating properly. A follow-up program may include extra maintenance calls. A follow-up program indicates to the equipment owner that the company maintaining the equipment has a sincere interest in proper maintenance of the equipment.

MAINTENANCE CHECKLIST

Company Name _____

Address _____

Homeowner's Name _____

Address _____

Date_____ Technician_____Time In _____ Time Out _____

Furnace Make and Model_____

Comments by Homeowner_____

_____

AT THERMOSTAT
__ Record thermostat setpoint temperature. _____
__ Check thermostat for dust or dirt and level.
__ Turn thermostat to lowest cooling setting.

AT FURNACE
__ Check supply voltage and record.
   Time _____ Voltage _____
__ Clean or change filters.
__ Clean out blower wheel and blower compartment.

BELT-DRIVE BLOWERS
__ Check wiring in blower compartment for loose connections or bad insulation.
__ Remove blower belt. Check for wear.
__ Check blower belt and motor bearings.
__ Check pulley and drive alignment.
__ Check pulley and drive set screws for tightness.
__ Check motor bracket for tightness.
__ Check blower for free operation.
__ Lubricate blower and motor bearings.
__ Put belt back on blower and drive pulleys and check belt tension slippage.

DIRECT-DRIVE BLOWERS
__ Check wiring in blower cabinet for loose connections and bad insulation.
__ Check motor bearings.
__ Check for free blower operation.
__ Check blower set screws for tightness.
__ Lubricate motor bearings.

CONDENSING UNIT
__ Check and clean condenser coil.
__ Oil condenser fan motor.
__ Check supply voltage and record.
   Time _____ Voltage _____
__ Check all wiring for loose connections.
__ Check all wiring for damaged insulation.
__ Gauge refrigeration system and check operating pressures.
__ Check refrigerant charge.
__ Check amperage draw on condenser fan motor.
   Nameplate _____ Actual _____
__ Check amperage draw on compressor.
   Nameplate _____ Actual _____
__ Visually inspect connecting tubing and coils for evidence of oil leak.
__ Return thermostat to original setpoint temperature.

EVAPORATOR COIL
__ Check and clean cooling coil.
__ Check and clean condensate drain.
__ Check static pressure.
__ Check temperature difference over coil.
__ Check for proper voltage at transformer and evaporator blower.

If unit is not running, refer to Service Handbook for cause of trouble.

BE SURE TO LEAVE ALL AREAS NEAT
AND CLEAN!

Figure 20-8. A maintenance checklist is used by a maintenance technician to ensure that all necessary items are inspected during a maintenance call.

TROUBLESHOOTING

Troubleshooting is the systematic elimination of the various parts of a heating or air conditioning system to locate a malfunctioning part. A *malfunctioning part* is an element of a heating or air conditioning system that does not operate properly. For example, a faulty thermostat affects the operation of the equipment it controls.

Troubleshooting combines observation, reasoning, and testing. Observation determines if the system is operating improperly. Reasoning determines what major section or component of the system is affected. Testing determines which part(s) of the major section or component is malfunctioning.

The major sections of heating and air conditioning systems include the fuel supply system, electrical power supply system, furnace or heat producing plant, air conditioner or chiller, air or water distribution system, and control system. The components of heating and air conditioning systems include the parts that make up the major sections. For example, the components of a heating system include the burner assembly, blower assembly, or circulating pump. The components of an air conditioner include a compressor, evaporator section, and condenser section. The parts of heating and air conditioning systems are the elements that make up the components. Parts of heating and air conditioning systems are burners, valves, motors, and relays.

Troubleshooting is usually performed by a service technician who understands the function of the major sections, components, and parts of a heating and air conditioning system. This understanding should extend to the controls, distribution system, and application of the equipment. When a service technician identifies a problem in a system, a service person is normally called to fix the equipment. Fixing the equipment may require repair or replacement of the malfunctioning part. In some cases, the repairs may be performed by the service technician.

When a heating or cooling system is inspected, all specific information about the components in the system should be known. Some information about a particular system may be located in the installation or service records. Inspecting the equipment and system during installation is the best way to identify the specific system and equipment. Some components are used for heating only, while other components are used for heating and air conditioning. Some systems contain humidifiers or dehumidifiers. Some systems have elaborate filters. When troubleshooting a heating or air conditioning system, the interaction of the various components to produce a desired heating or cooling effect must be understood. A malfunction in one part of a system may affect other parts.

For example, dirty filters in a forced-air system may cause a furnace to cycle the limit control ON and OFF. The problem is a dirty filter, but the limit control appears to be malfunctioning. Inspecting the limit control verifies that it is operating correctly. Checking the air temperature increase through the furnace and the air flow rate through the furnace shows a problem in the air distribution system. An inspection of the filters would indicate the actual cause of the problem.

Heating and air conditioning systems include electrical, mechanical, and chemical components. Because some chemicals are hazardous, persons working on heating or air conditioning systems must take proper safety precautions to prevent personal injury. Equipment is often operating during troubleshooting. Safety precautions must be taken when working on operating equipment.

Procedure

A troubleshooting procedure is a systematic procedure to efficiently troubleshoot a heating or air conditioning system. A systematic troubleshooting procedure allows a service technician to locate a problem more easily than with a random approach. Time and money are saved in the process. A systematic process includes analyzing the complaint, observing the system operation, and isolating the malfunctioning section, component, and part.

To troubleshoot a heating or air conditioning system, the complaint must be analyzed to determine which major section contains the problem. The complaint may be general or specific. A person who is not familiar with heating or air conditioning equipment may give a general complaint such as not enough heat or not enough cooling. A person who is familiar with heating or air conditioning equipment may give a specific complaint such as the burner does not ignite on a call for heat.

The first step in troubleshooting a heating or air conditioning system is to observe the system in operation and isolate the major section that contains the malfunction. For example, a complaint that a heating system produces no heat could be a problem originating in the fuel supply, electrical power supply, heating unit, or controls of the heating unit.

The next step in troubleshooting a heating or air conditioning system is to identify the component in the major section that contains the malfunction. For example, if no air is coming out of the registers of a forced-air heating or air conditioning system, the component to be investigated is the blower assembly and the related controls.

The last step in troubleshooting a heating or air conditioning system is to find the malfunctioning part in the component. For example, when troubleshooting the blower assembly, the electrical power source is checked first. If the electrical power source is functioning normally, then the blower motor should be checked. If the blower motor is functioning normally, then the blower motor controls should be checked. By investigating each part in turn, the malfunctioning part is found.

To troubleshoot a system efficiently, a service technician may use a troubleshooting chart, which outlines the steps to be followed in the procedure. A troubleshooting chart indicates the shortest possible route to troubleshooting success.

Charts. A *troubleshooting chart* is a chart that shows the major sections, components, and parts of a heating or air conditioning system, and is organized so that the effect of a malfunctioning part in a system can be traced to the faulty part in the major section of the system. Most troubleshooting charts start with a general complaint about the heating or air conditioning system. By answering questions about system performance, the major section that contains the problem is identified. More questions identify the malfunctioning component in the major section. Another set of questions helps identify the malfunctioning part in the component.

A *heating troubleshooting chart* is a flow chart that identifies malfunctioning parts of a heating system with a series of steps. A heating troubleshooting chart starts with the most common complaints concerning heating system performance. See Appendix. The most common complaints are:

no heat

not enough heat

too much heat

varying heat output

noise

odors

high operating costs

On a typical heating troubleshooting chart, complaints are referenced to the major sections of the system. The malfunctioning component in the major section is found on the chart by checking the system operation. The chart indicates that the various parts in the malfunctioning component should be checked by observing the operation of the components. The malfunctioning part in the component is found by checking each part of the component.

An *air conditioning troubleshooting chart* is a flow chart that identifies malfunctioning parts of an operating air conditioning system with a series of steps. The steps on the chart are systematic and proceed from the major complaints concerning system operation to identification of the malfunctioning part in the system. The major complaints concerning air conditioning system operation are:

no cooling

not enough cooling

too cool

noise

high operating costs

On a typical air conditioning troubleshooting chart, the main complaints are identified first. Under the main complaint heading, the symptoms of a malfunctioning major section are listed. The general cause of the symptom is found within the components in the section. Following the chart leads to the probable malfunctioning component. The chart leads from the cause in the component to the malfunctioning part in the component. See Appendix.

Forced-air Heating System

A forced-air heating system heats air and circulates the air through ductwork to building spaces. A forced-air heating system consists of the fuel supply system, electrical power supply system, heating unit, air distribution system, and control system. See Figure 20-9.

When a complaint is received concerning a forced-air heating system, the various major sections of the system are observed during operation. If a system is completely inoperative, the various sections should be observed when the control system calls for heat. Begin with the electrical power supply. Make sure the disconnect switch is closed and the fuses or circuit breakers are closed. If the electrical power is ON, proceed to the next section of the system to see if it is malfunctioning. These checks indicate which major section of the forced-air heating system is malfunctioning.

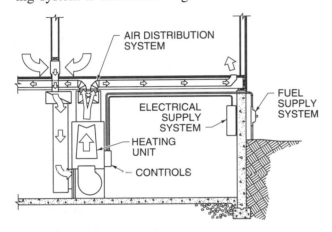

Figure 20-9. The major sections of a forced-air heating system are the fuel supply system, electrical supply system, heating unit, air distribution system, and controls.

For example, if cold air is flowing out of a register, the electrical power supply section must be operating correctly because the blower is running. The blower section must be operating correctly because air is flowing out of the register. The control section is inspected by turning the thermostat setpoint adjustor above the room temperature and observing if there is voltage at the heating unit relay. This is checked by placing the probes of a voltmeter set for the line voltage of the system on the terminals on each side of the coil connections of the relay. The voltage across these terminals should be the line voltage of the system. The fuel supply section is inspected by observing the fuel in the oil tank or in the oil or gas line at the unit. This process eliminates all major sections except the heating section. The next step is to inspect the components in the heating section.

Each of the major sections of a forced-air heating system is composed of several components. The heating section consists of the burner and heat exchanger. All combustion furnaces contain similar heat-producing components. A heating troubleshooting chart is used to determine which component in the heating section is malfunctioning. By following the complaint to the fault on a troubleshooting chart, the malfunctioning component should be identified. After the malfunctioning component is identified, the malfunctioning part in the component can be identified by eliminating each part.

Hydronic Heating System

A hydronic heating system heats water, which is circulated through a distribution system to terminal devices. The terminal devices heat air, which heats building spaces. The procedure for troubleshooting a forced-air heating system is used for troubleshooting a hydronic heating system. The main differences are in the boiler and water piping system. Many of the other components and parts of the two systems are alike. Similar methods are used for checking these parts.

For example, if all but one zone in a hydronic heating system is heating correctly, the major sections of the system are functioning properly. The problem lies in the components or parts of the section in the cold zone. Investigation reveals which zone is involved. The main components in a zone are the circulating pump, piping, terminal devices, and controls for the zone. See Figure 20-10. If there is no heat in a zone, an inspection may show that no water is circulating in the zone. The lack of water indicates a problem with the circulating pump in the zone. The pump and controls that operate the pump are inspected to locate the problem.

A water circulating pump is inspected by checking the electrical power at the pump motor. The controls that turn the pump ON and OFF are checked for proper operation. If electrical power is present at the pump and the controls are operating properly, the circulating pump may be malfunctioning. A circulating pump consists of a motor and pump assembly, which must be malfunctioning if all other components in the system are operating properly.

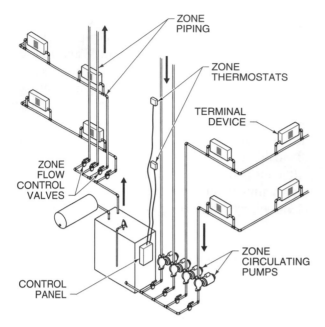

Figure 20-10. The main components in a zone of a hydronic system are the circulating pump, piping, terminal devices, and controls for that zone.

Forced Air Conditioning System

Major sections of a forced air conditioning system are the condenser, evaporator, distribution system, and control system. See Figure 20-11. A systematic process for troubleshooting a forced air conditioning system includes the same steps used for troubleshooting heating systems.

Inadequate cooling in a forced air conditioning system can be caused by a malfunctioning control system, distribution system, compressor, condenser, or evaporator. A combination heating and cooling system has a common air distribution system. The problem may be in one part of the system, but the effect of the problem is seen in the entire system.

For example, if a forced air conditioning system has stopped cooling, the warm air coming out of the registers indicates that the electrical supply system is operating properly and that the blower is functioning. The malfunction must be in the components that create the cooling effect. An inspection will show if the compressor is operating. If the compressor is operating, a check with pressure gauges will show if the proper high-pressure side and low-pressure side pressures are being maintained. If the pressures are not what they should be, additional checks can be made to see if the valves in the compressor are malfunctioning.

Hydronic Air Conditioning System

The major sections of a hydronic air conditioning system include the chiller, distribution system, terminal devices, and control system. See Figure 20-12. The procedure used to troubleshoot a hydronic air conditioning system is similar to the procedure used to troubleshoot a forced air conditioning system. The difference is in the chiller, piping system, and terminal devices.

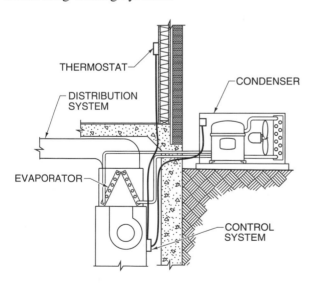

Figure 20-11. The major sections of a forced air conditioning system are the condenser, evaporator, distribution system, and control system.

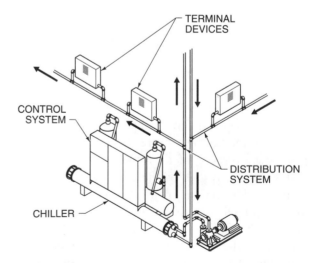

Figure 20-12. The major sections of a hydronic air conditioning system are the chiller, distribution system, terminal devices, and control system.

For example, no cooling is being produced from a hydronic distribution system. Warm air is coming out of the terminal devices and the electrical power is ON, which means that the blowers in the terminal devices are operating. This means that the problem is in the water distribution system or the chiller. The operation of the circulating pump and other components in the chiller are checked the same way the components of a forced air conditioning system are checked.

Follow-up

The follow-up to troubleshooting is service. When a heating or air conditioning system malfunctions and the malfunctioning part is identified, the malfunctioning part must be repaired or replaced. The parts may be repaired or replaced by the service technician who does the troubleshooting or by a service person trained for that job.

Review Questions

1. Define *maintenance*.

2. Explain how a maintenance technician differs from a service technician.

3. List and describe the four general tasks performed during a maintenance call.

4. List and describe the three specific tasks performed during a maintenance call.

5. When do fuel filters need to be checked?

6. Explain the procedure for tightening blower belts.

7. Explain the procedure for aligning blower belts.

8. Define *troubleshooting*.

9. If a blower bearing is not faulty, how should it feel to the touch?

10. Describe the systematic process used to efficiently troubleshoot a heating or air conditioning system.

11. What is the main difference between a maintenance checklist and a job sheet used by a maintenance technician?

12. Why is a good follow-up program an asset to a maintenance firm?

13. Name the three elements of troubleshooting.

14. List and describe the three steps of a systematic troubleshooting process.

15. Who normally fixes malfunctioning heating or air conditioning equipment?

16. How does a troubleshooting chart save time when checking a malfunctioning heating or air conditioning system?

17. What normally constitutes the follow-up to a troubleshooting call?

18. After troubleshooting, who services a heating or air conditioning unit?

19. What item is used by a maintenance technician to ensure that all necessary items are inspected during a maintenance call?

20. List the three places filters may be located in a forced-air heating or air conditioning system.

Appendices

Properties of Substances

Tables of Equivalents

Properties of Air and Psychrometrics

Refrigerants

Load Calculations

Duct Tables

Troubleshooting Charts

Symbols

PHYSICAL PROPERTIES OF VAPORS

Name	Normal Boiling Point*	Specific Heat§	Mass Density#	Thermal Conductivity**
Alcohol, Ethyl	173.3	.362		.0876
Alcohol, Methyl	148.9	.322		.2088
Ammonia	− 28	.525	.0482	.1536
Argon	−302.3	.125	.1114	.1128
Acetylene	−118.5	.377	.0732	.1296
Benzene	176.2	.31	.167	.0492
Bromine	137.8	.055	.38	.042
Butane	31.1	.377	.168	.0948
Carbon Dioxide	−109.3	.20	.123	.1008
Carbon Disulfide	115.2	.1431		
Carbon Monoxide	−312.7	.25	.078	.1596
Carbon Tetrachloride	169.8	.206		.45
Chlorine	− 30.3	.117	.201	.0552
Chloroform	143.1	.126		.0972
Ethyl Chloride	54.2	.426	.1793	.0605
Ethylene	−154.6	.352	.0783	.1224
Ethyl Ether	94.4	.589		.3276
Fluorine	−304.5	.194	.1022	.1764
Helium	−452.1	1.241	.0111	.9876
Hydrogen	−423.0	3.40	.00562	1.1664
Hydrogen Chloride	−120.8	.191	.1024	.0908
Hydrogen Sulfide	− 77.3	.238	.0961	.0901
Heptane	209.2	.476	.21	.1284
Hexane	154	.449	.21	.1165
Isobutane	− 10.9	.376	.154	.0972
Methyl Chloride	− 11.6	.184	.1440	.0648
Methane	−263.2	.520	.0448	.2136
Napthalene	−360.4	.313		
Neon	−412.6	.246		.3216
Nitric Oxide	−241.6	.238		
Nitrogen	−320.4	.248		.1656
Nitrous Oxide	−127.3	.203		.1201
Oxygen	−297.4	.218		.1692
Phenol	358.5	.34	.16	.1188
Propane	− 43.73	.3753	.126	.1044
Propylene	− 53.86	.349	.120	.0972
Sulfur Dioxide	14.0	.145	.183	.0588
Water Vapor	212.0	.489	.0373	.1716

* in °F
§ in Btu/lb × °F
# in lb/cu ft
** in Btu/hr × in. × °F

•*ASHRAE 1989 Handbook — Fundamentals*

PHYSICAL PROPERTIES OF LIQUIDS				
Name	Normal Boiling Point*	Specific Heat§	Mass Density#	Thermal Conductivity**
Benzene	176.2	.412	54.9	1.02
Calcium Chloride Brine		.744	73.8	3.984
Carbon Disulfide	115.3	.240	78.9	1.116
Carbon Tetrachloride	170.2	.201	99.5	.744
Chloroform	142.3	.234	92.96	.90
Ethyl Acetate	170.8	.468	52.3	1.212
Ethylene Glycol	388.4		69.22	1.2
Formic Acid	213.3	.526	76.16	1.248
Glycerine (Glycerol)	359		78.72	1.356
Heptane	209.2	.532	42.7	.8892
Hexane	154	.538	41.1	.864
Isobutyl Alcohol	226.4	.116	50.0	.984
Kerosene	400560	.50	51.2	1.032
Linseed Oil			58	
Methyl Acetate	134.6	.468	60.6	1.116
Nitric Acid	186.8	.42	94.45	1.92
Nitrobenzene	411.6	.348	75.2	11.52
Octane	258.3	.51	43.9	1.008
Petroleum		.4	40	
Sodium Chloride Brine	220.8	.745	71.8	4.044
Turpentine	303	.42	53.9	.876
Water	212	1.00	62.32	4.176
Zinc Sulfate and Water	.90	.90	69.2	4.044

* in °F at 14.7 psia
§ in Btu/lb × °F at 68°F
# in lb/cu ft at 68°F
** in Btu/hr × ft × °F

•*ASHRAE 1989 Handbook — Fundamentals*

PHYSICAL PROPERTIES OF SOLIDS			
Name	**Specific Heat***	**Mass Density**§	**Thermal Conductivity**#
Aluminum	.214	171	128
Ashes (wood)	.20	40	0.041
Asphalt	.22	132	0.43
Bakelite	.35	81	9.7
Bronze	.104	530	17
Cadmium	.055	540	53.7
Cast Iron	.12	450	27.6
Cement (Portland clinker)	.16	120	0.017
Chalk	.215	143	0.48
Charcoal (wood)	.20	15	0.03
Chrome Brick	.17	200	0.67
Clay	.22	63	
Coal	.3	90	0.098
Coal Tars	.35	75	0.07
Coke (petroleum, powdered)	.36	62	0.55
Concrete (stone)	.156	144	0.54
Copper (electrolytic)	.092	556	227
Cork (granulated)	.485	5.4	0.028
Cotton (fiber)	.319	95	0.024
Diamond	.147	151	27
Gold	.0312	1208	172
Graphite (powder)	.165		0.106
Gypsum	.259	78	0.25
Hemp (fiber)	.323	93	
Ice (32 °F)	.487	57.5	1.3
Lead	.0309	707	20.1
Leather (sole)		62.4	0.092
Limestone	.217	103	0.54
Magnesium	.241	108	91
Marble	.21	162	1.5
Nickel	.105	555	34.4
Paper	.32	58	0.075
Paraffin	.69	56	0.14
Plaster		132	0.43
Platinum	.032	1340	39.9
Porcelain	.18	162	1.3
Sand	.191	94.6	0.19
Sawdust		12	0.03
Silica	.316	140	0.83
Silver	.0560	654	245
Steel (mild)	.12	489	26.2
Stone (quarried)	.2	95	

* in Btu/lb × °F
§ in lb/cu ft
# in Btu/hr × ft × °F

•*ASHRAE 1989 Handbook — Fundamentals*

HEATING VALUES OF WOOD FUEL

Species	Btu/lb*	Lb/cord§	Btu/cord#
Alder	6720	2540	17,068,800
Apple	7890	4400	34,716,000
Ash	8300	3440	28,552,000
Aspen	5980	2160	12,916,800
White Birch	7470	3040	22,708,800
Western Cedar	6720	2060	13,843,200
Cherry	7470	3200	23,904,000
Cottonwood	5980	2160	12,916,800
Dogwood	8300	4230	35,109,000
Elm	7470	2260	16,882,200
Douglas Fir	7470	2970	22,185,900
Hemlock	6720	2700	18,144,000
Hickory	9500	4240	40,280,000
Juniper	7470	3150	23,530,500
Western Larch	7890	3330	26,273,700
Red Maple	7885	3200	25,232,000
Sugar Maple	7890	3680	29,035,200
Red Oak	8300	3680	30,544,000
White Oak	9500	4200	38,900,000
Ponderosa Pine	6720	2240	15,052,800
Yellow Pine	7890	2610	20,592,900
Poplar	5980	2080	12,438,400
Redwood	7470	2400	17,928,000
Norway Spruce	5980	2240	13,395,200

* approximate heating value, Btu/lb for air-seasoned wood (20% moisture)
§ approximate weight, lb/cord, for average density air-seasoned wood
# approximate lb/cu ft for air-seasoned wood (20% moisture), and for average density for species

•ASHRAE 1989 Handbook — Fundamentals

HEATING VALUES AND CHEMICAL COMPOSITION OF STANDARD GRADES OF COAL

Rank	Heating Value*	Chemical Composition§					
		Oxygen	Hydrogen	Carbon	Nitrogen	Sulfur	Ash
Anthracite	12,910	5.0	2.9	80.0	0.9	0.7	10.5
Semi-anthracite	13,770	5.0	3.9	80.4	1.1	1.1	8.5
Low-volatile Bituminous	14,340	5.0	4.7	81.7	1.4	1.2	6.0
Medium-volatile Bituminous	13,840	5.0	5.0	79.0	1.4	1.5	8.1
High-volatile Bituminous A	13,090	9.2	5.3	73.2	1.5	2.0	8.8
High-volatile Bituminous B	12,130	13.8	5.5	68.0	1.4	2.1	9.2
High-volatile Bituminous C	10,750	21.0	5.8	60.6	1.1	2.1	9.4
Subbituminous B	9150	29.5	6.2	52.5	1.0	1.0	9.8
Subbituminous C	8940	35.8	6.5	46.7	0.8	0.6	9.6
Lignite	6900	44.0	6.9	40.1	0.7	1.0	7.3

* in Btu/lb
§ in percentage

•ASHRAE 1989 Handbook — Fundamentals

CHEMICAL ELEMENTS

Name	Symbol	Atomic Weight*	Atomic Number	Name	Symbol	Atomic Weight*	Atomic Number
Actinium	Ac	[227]	89	Neon	Ne	20.183	10
Aluminum	Al	26.9815	13	Neptunium	Np	[237]	93
Americium	Am	[243]	95	Nickel	Ni	58.71	28
Antimony	Sb	121.75	51	Niobium	Nb	92.906	41
Argon	Ar	39.948	18	Nitrogen	N	14.0067	7
Arsenic	As	74.9216	33	Nobelium	No	[255]	102
Astatine	At	[210]	85	Osmium	Os	190.2	76
Barium	Ba	137.34	56	Oxygen	O	15.9994	8
Berkelium	Bk	[247]	97	Palladium	Pd	106.4	46
Beryllium	Be	9.0122	4	Phosphorus	P	30.9738	15
Bismuth	Bi	208.980	83	Platinum	Pt	195.09	78
Boron	B	10.811	5	Plutonium	Pu	[244]	94
Bromine	Br	79.909	35	Polonium	Po	[210]	84
Cadmium	Cd	112.40	48	Potassium	K	39.102	19
Calcium	Ca	40.08	20	Praseodymium	Pr	140.907	59
Californium	Cf	[251]	98	Promethium	Pm	[145]	61
Carbon	C	12.01115	6	Protactinium	Pa	[231]	91
Cerium	Ce	140.12	58	Radium	Ra	[226]	88
Cesium	Cs	132.905	55	Radon	Rn	[222]	86
Chlorine	Cl	35.453	17	Rhenium	Re	186.2	75
Chromium	Cr	51.996	24	Rhodium	Rh	102.905	45
Cobalt	Co	58.9332	27	Rubidium	Rb	85.47	37
Copper	Cu	63.54	29	Ruthenium	Ru	101.07	44
Curium	Cm	[247]	96	Samarium	Sm	150.35	62
Dysprosium	Dy	162.50	66	Scandium	Sc	44.956	21
Einsteinium	Es	[254]	99	Selenium	Se	78.96	34
Erbium	Er	167.26	68	Silicon	Si	28.086	14
Europium	Eu	151.96	63	Silver	Ag	107.870	47
Fermium	Fm	[257]	100	Sodium	Na	22.9898	11
Fluorine	F	18.9984	9	Strontium	Sr	87.62	38
Francium	Fr	[223]	87	Sulfur	S	32.064	16
Gadolinium	Gd	157.25	64	Tantalum	Ta	180.948	73
Gallium	Ga	69.72	31	Technetium	Tc	[97]	43
Germanium	Ge	72.59	32	Tellurium	Te	127.60	52
Gold	Au	196.967	79	Terbium	Tb	158.924	65
Hafnium	Hf	178.49	72	Thallium	Tl	204.37	81
Helium	He	4.0026	2	Thorium	Th	232.038	90
Holmium	Ho	164.930	67	Thulium	Tm	168.934	69
Hydrogen	H	1.00797	1	Tin	Sn	118.69	50
Indium	In	114.82	49	Titanium	Ti	47.90	22
Iodine	I	126.9044	53	Tungsten	W	183.85	74
Iridium	Ir	192.2	77	Unnilennium	Une	[266]	109
Iron	Fe	55.847	26	Unnilhexium	Unh	[263]	106
Krypton	Kr	83.80	36	Unniloctium	Uno	[265]	108
Lanthanum	La	138.91	57	Unnilpentium	Unp	[262]	105
Lawrencium	Lr	[256]	103	Unnilquadium	Unq	[261]	104
Lead	Pb	207.19	82	Unnilseptium	Uns	[262]	107
Lithium	Li	6.939	3	Uranium	U	238.03	92
Lutetium	Lu	174.97	71	Vanadium	V	50.942	23
Magnesium	Mg	24.312	12	Xenon	Xe	131.30	54
Manganese	Mn	54.9380	25	Ytterbium	Yb	173.04	70
Mendelevium	Md	[258]	101	Yttrium	Y	88.905	39
Mercury	Hg	200.59	80	Zinc	Zn	65.37	30
Molybdenum	Mo	95.94	42	Zirconium	Zr	91.22	40
Neodymium	Nd	144.24	60				

* A number in brackets indicates the mass number of the most stable isotope

PRESSURE EQUIVALENTS

psi*	psf§	in. Hg#	in. WC**	ft WC§§	atm##
1	144	2.042	27.7	2.31	.068
14.7	2116.3	30		33.95	1
.433	62.355			1	
.491		1	13.58	1.132	

EQUIVALENTS OF PSI

psi*	in. WC**	ft WC§§	in. Hg#
1	27.71	2.31	2.041
2	55.42	4.62	4.081
3	83.14	6.93	6.122
4	110.85	9.24	8.163
5	138.56	11.55	10.20
6	166.27	13.86	12.24
7	193.99	16.17	14.28
8	221.70	18.47	16.33
9	249.41	20.78	18.37
10	277.12	23.09	20.41
11	304.84	25.40	22.45
12	332.55	27.71	24.49
13	360.26	30.02	26.53
14	387.97	32.33	28.57
14.7	407.37	33.95	30.00
15	415.68	34.64	30.61
16	443.40	36.95	32.65
17	471.11	39.26	34.69
18	498.82	41.57	36.73
19	526.53	43.88	38.77
20	554.25	46.19	40.81
21	581.96	48.50	42.85
22	609.67	50.81	44.89
23	637.38	53.12	46.94
24	665.10	55.42	48.98
25	692.81	57.73	51.02

EQUIVALENTS OF IN. HG

in. Hg#	psi*	in. Hg#	psi*
30	14.730	15	7.365
29.939	14.7	14	6.874
29	14.239	13	6.383
28	13.748	12	5.892
27	13.257	11	5.401
26	12.776	10	4.910
25	12.275	9	4.419
24	11.784	8	3.928
23	11.293	7	3.437
22	10.802	6	2.946
21	10.311	5	2.455
20	9.820	4	1.964
19	9.329	3	1.473
18	8.838	2	.982
17	8.347	1	.491
16	7.856	0	0

* pounds per square inch
§ pounds per square foot
# inches of Mercury
** inches of water column
§§ feet of water column
atmospheres

POWERS OF 10

1×10^4	=	10,000	=	$10 \times 10 \times 10 \times 10$		Read ten to the fourth power
1×10^3	=	1000	=	$10 \times 10 \times 10$		Read ten to the third power or ten cubed
1×10^2	=	100	=	10×10		Read ten to the second power or ten squared
1×10^1	=	10	=	10		Read ten to the first power
1×10^0	=	1	=	1		Read ten to the zero power
1×10^{-1}	=	.1	=	1/10		Read ten to the minus first power
1×10^{-2}	=	.01	=	$1/(10 \times 10)$ or 1/100		Read ten to the minus second power
1×10^{-3}	=	.001	=	$1/(10 \times 10 \times 10)$ or 1/1000		Read ten to the minus third power
1×10^{-4}	=	.0001	=	$1/(10 \times 10 \times 10 \times 10)$ or 1/10,000		Read ten to the minus fourth power

UNITS OF ENERGY

Energy	Btu	ft lb	J	kcal	kWh
British thermal unit	1	777.9	1.056	0.252	2.930×10^{-4}
Foot-pound	1.285×10^{-3}	1	1.356	3.240×10^{-4}	3.766×10^{-7}
Joule	9.481×10^{-4}	0.7376	1	2.390×10^{-4}	2.778×10^{-7}
Kilocalorie	3.968	3.086	4.184	1	1.163×10^{-3}
Kilowatt-hour	3.413	2.655×10^6	3.6×10^6	860.2	1

UNITS OF POWER

Power	W	ft lb/s	HP	kW
Watt	1	0.7376	1.341×10^{-3}	0.001
Foot-pound/sec	1.356	1	1.818×10^{-3}	1.356×10^{-3}
Horsepower	745.7	550	1	0.7457
Kilowatt	1000	736.6	1.341	1

DRY BULB TEMPERATURE*

WET BULB DEPRESSION*

Dry Bulb	1	2	3	4	5	6	7	8	9	10	11	12	13	14	15	16	17	18	19	20	21	22	23	24	25	26	27	28	29	30	31	32	33	34	35	36	37	38	39	40	41	42	43	44	45
30	89	78	67	56	46	36	26	16	6																																				
32	89	79	69	59	49	39	30	20	11	2																																			
34	90	81	71	62	52	43	34	25	16	8																																			
36	91	82	73	64	55	46	38	29	21	13	5																																		
38	91	83	75	66	58	50	42	33	25	17	10	2																																	
40	92	83	75	68	60	52	45	37	29	22	15	7																																	
42	92	85	77	69	62	55	47	40	33	26	19	12	5																																
44	93	85	78	71	63	56	49	43	36	29	23	16	10	4																															
46	93	86	79	72	65	58	52	45	39	32	26	20	14	8	2																														
48	93	86	79	73	66	60	54	47	41	35	29	23	18	12	7	1																													
50	93	87	80	74	67	61	55	49	43	38	32	27	21	16	10	5																													
52	94	87	81	75	69	63	57	51	46	40	35	29	24	19	14	9	4																												
54	94	88	82	76	70	64	59	53	48	42	37	32	27	22	17	12	8	3																											
56	94	88	82	76	71	65	60	55	50	44	39	34	30	25	20	16	11	7	2																										
58	94	88	83	77	72	66	61	56	51	46	41	37	32	27	23	18	14	10	6	1																									
60	94	89	83	78	73	68	63	58	53	48	43	39	34	30	26	21	17	13	9	5	1																								
62	94	89	84	79	74	69	64	59	54	50	45	41	36	32	28	24	20	16	12	8	4	1																							
64	95	90	84	79	74	70	65	60	56	51	47	43	38	34	30	26	22	18	15	11	7	4	1																						
66	95	90	85	80	75	71	66	61	57	53	48	44	40	36	32	29	25	21	18	14	11	8	5	1																					
68	95	90	85	80	76	71	67	62	58	54	50	46	42	38	34	31	27	23	20	16	13	10	7	4	1																				
70	95	90	86	81	77	72	68	64	59	55	51	48	44	40	36	33	29	25	22	19	15	12	9	6	3																				
72	95	91	86	82	78	73	69	65	61	57	53	49	45	42	38	34	31	28	24	21	18	15	12	9	6	3																			
74	95	91	86	82	78	74	70	66	62	58	54	50	47	43	39	36	33	29	26	23	20	17	14	11	8	5	3																		
76	96	91	87	83	79	75	71	67	63	59	55	52	48	45	41	38	34	31	28	25	22	18	16	13	10	8	5	3																	
78	96	91	87	83	79	75	71	67	64	60	56	53	49	46	43	39	36	33	30	27	24	21	18	16	13	10	8	5	3																
80	96	92	87	83	80	76	72	68	65	61	57	54	50	47	44	41	38	35	32	29	26	23	20	18	15	12	10	7	5	2															
82	96	92	88	84	80	77	73	69	66	62	58	55	51	48	45	42	39	36	33	30	28	25	22	20	17	14	12	10	7	5	2														
84	96	92	88	84	80	77	73	70	66	63	59	55	52	49	46	44	41	38	35	32	30	27	25	22	20	17	15	13	10	8	5	3	1												
86	96	92	88	84	81	78	74	71	67	64	60	56	53	50	47	45	42	39	36	34	32	29	27	25	22	20	18	16	14	11	9	7	5	3	1										
88	96	92	88	85	81	78	75	71	68	64	61	57	54	51	48	46	43	40	37	35	33	30	28	26	24	21	19	17	15	13	11	9	7	5	3	1									
90	96	92	89	85	82	79	75	72	69	66	61	58	55	52	49	47	44	41	39	36	35	33	31	29	27	24	22	20	18	17	15	13	11	10	8	7	5	4	2	1					
92	96	93	89	85	82	79	76	73	69	66	62	59	56	53	50	48	45	42	40	37	36	34	32	30	28	26	24	22	20	18	16	14	13	11	9	7									
94	96	93	89	86	82	80	77	73	70	67	63	60	56	54	51	50	46	43	41	38	37	35	33	31	29	27	25	23	21	19	17	15	14	12	10	8	7	5							
96	96	93	90	86	83	80	77	74	71	67	64	60	57	54	52	51	48	45	42	40	39	37	35	33	31	28	26	24	22	20	19	17	15	14	12	10	8	7	5	4	2	1			
98	96	93	90	86	83	81	77	74	71	68	65	61	58	55	53	51	48	45	43	40	40	38	36	34	32	30	28	26	24	22	20	18	16	14	13	11	10	8	7	5	2	1			
100	96	93	90	87	83	81	78	75	72	69	65	62	59	56	54	52	49	46	44	41	41	39	37	35	33	31	29	27	25	23	21	19	17	16	14	12	11	9	7	5	2	1			
102	97	93	90	87	84	81	78	75	72	69	65	62	59	56	54	53	50	47	44	42	41	40	38	36	34	32	30	28	26	24	22	20	19	17	15	13	11	10	8	7	4	2	1		
104	97	93	90	87	84	82	78	75	72	69	66	63	60	57	55	53	50	48	45	43	42	41	39	37	35	33	31	29	27	25	24	22	20	18	16	14	13	11	10	8	5	4	2	1	
106	97	93	90	87	84	82	79	76	73	70	66	63	60	58	56	54	51	48	46	43	43	41	39	37	36	34	32	30	28	27	25	23	21	19	18	16	14	13	11	10	7	5	4	3	1
108	97	93	91	88	85	83	79	76	73	70	67	64	61	58	56	54	52	49	47	45	45	43	41	40	38	36	35	33	31	29	27	25	24	22	20	18	17	15	14	12	8	7	5	4	3

To find relative humidity, subtract the wet bulb temperature from the dry bulb temperature, which gives wet bulb depression. The relative humidity value is at the intersection of the wet bulb depression value and the dry bulb temperature.

* in °F

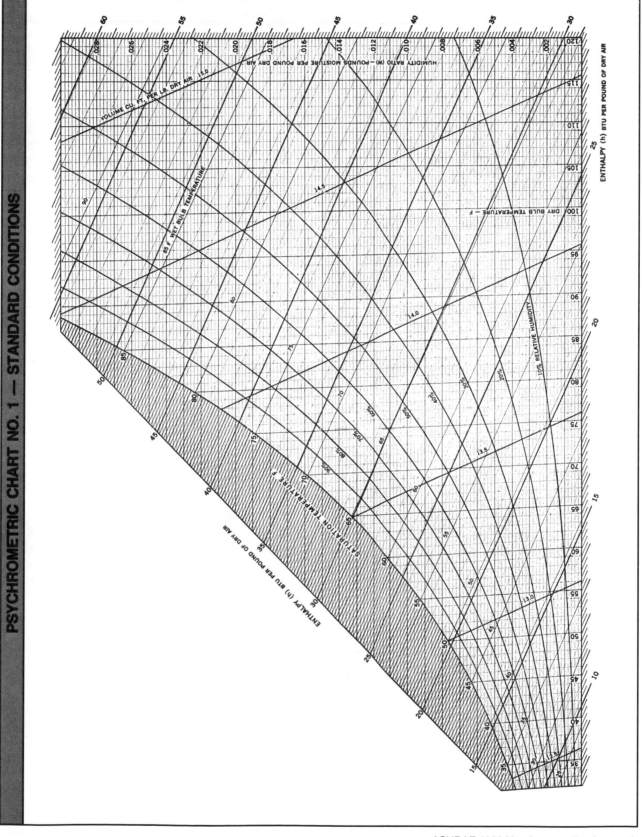

PSYCHROMETRIC CHART NO. 1 — STANDARD CONDITIONS

REFRIGERANT PROPERTIES

Refrigerant No.	Name	Chemical Formula	Molecular Mass	Normal Boiling Point*	Freezing Point*	Critical Tempera-ture*	Critical Pressure§	Critical Volume#
704	Helium	He	4.0026	−452.1	None	−450.3	33.21	.2311
702	Hydrogen (normal)	H₂	2.0159	−423.0	−434.5	−399.9	190.8	.5320
702	Hydrogen (para)	H₂	2.0159	−423.2	−434.8	−400.3	187.5	.5097
720	Neon	Ne	20.183	−410.9	−415.5	−379.7	493.1	.03316
728	Nitrogen	N₂	28.013	−320.4	−346.0	−232.4	492.9	.05092
729	Air	. . .	28.97	−317.8	. . .	−221.3	547.4	.04883
740	Argon	A	39.948	−302.6	−308.7	−188.1	710.4	.02990
732	Oxygen	O₂	31.9988	−297.3	−361.8	−181.1	736.9	.0375
50	Methane	CH₄	16.04	−258.7	−296	−116.5	673.1	.099
14	Tetrafluoromethane	CF₄	88.01	−198.3	−299	− 50.2	543	.0256
1150	Ethylene	C₂H₄	28.05	−154.7	−272	48.8	742.2	.070
170	Ethane	C₂H₆	30.07	−127.85	−297	90.0	709.8	.0830
744	Nitrous Oxide	N₂O	44.02	−129.1	−152	97.7	1048	.0355
23	Trifluoromethane	CHF₃	70.02	−115.7	−247	78.1	701.4	.0311
13	Chlorotrifluoromethane	CClF₃	104.47	−114.6	−294	83.9	561	.0277
744	Carbon Dioxide	CO₂	44.01	−109.2	− 69.9	87.9	1070.0	.0342
13	Bromotrifluoromethane	CBrF₃	148.93	− 71.95	−270	152.6	575	.0215
1270	Propylene	C₃H₆	42.09	− 53.86	−301	197.2	670.3	.0720
290	Propane	C₃H₈	44.10	− 43.73	−305.8	206.3	617.4	.0728
22	Chlorodifluoromethane	CHClF₂	86.48	− 41.36	−256	204.8	721.9	.0305
115	Chloropentafluoroethane	CClF₂CF₃	154.48	− 38.4	−159	175.9	457.6	.0261
717	Ammonia	NH₃	17.03	− 28.0	−107.9	271.4	1657	.068
12	Dichlorodifluoromethane	CCl₂F₂	120.93	− 21.62	−252	233.6	596.9	.0287
134	Tetrafluoroethane	CF₃CH₂F	102.03	− 15.08	−141.9	214.0	589.8	.029
152	Difluoroethane	CH₃CHF₂	66.05	− 13.0	−178.6	236.3	652	.0439
40	Methyl Chloride	CH₃Cl	50.49	− 11.6	−144	289.6	968.7	.0454
600	Isobutane	C₄H₁₀	58.13	10.89	−255.5	275.0	529.1	.0725
764	Sulfur Dioxide	SO₂	64.07	14.0	−103.9	315.5	1143	.0306
142	Chlorodifluoroethane	CH₃CClF₂	100.5	14.4	−204	278.8	598	.0368
630	Methyl Amine	CH₃NH₂	31.06	19.9	−134.5	314.4	1082	
318	Octafluorocyclobutane	C₄F₈	200.04	21.5	− 42.5	239.6	403.6	.0258
600	Butane	C₄H₁₀	58.13	31.1	−217.3	305.6	550.7	.0702
114	Dichlorotetrafluoroethane	CClF₂CClF₂	170.94	38.8	−137	294.3	473	.0275
21	Dichlorofluoromethane	CHCl₂F	102.92	47.8	−211	353.3	750	.0307
160	Ethyl Chloride	C₂H₅Cl	64.52	54.32	−216.9	369.0	764.4	.0485
631	Ethyl Amine	C₂H₅NH₂	45.08	61.88	−113	361.4	815.6	
11	Trichlorofluoromethane	CCl₃F	137.38	74.87	−168	388.4	639.5	.0289
611	Methyl Formate	C₂H₄O₂	60.05	89.22	−146	417.2	870	.0459
610	Ethyl Ether	C₄H₁₀O	74.12	94.3	−177.3	381.2	523	.0607
216	Dichlorohexafluoropropan	C₃Cl₂F₆	220.93	96.24	−193.7	356.0	399.5	.0279
30	Methylene Chloride	CH₂Cl₂	84.93	104.4	−142	458.6	882	
113	Trichlorotrifluoroethane	CCl₂FCClF₂	187.39	117.63	− 31	417.4	498.9	.0278
1130	Dichloroethylene	CHCl=CHCl	96.95	118	− 58	470	795	
1120	Trichloroethylene	CHCl=CCl₂	131.39	189.0	− 99	520	728	
718	Water	H₂O	18.02	212	32	705.6	3208	.0501

* in °F   # in cu ft/lb   •ASHRAE 1989 Handbook — Fundamentals
§ in psia

R-12—PROPERTIES OF SATURATED LIQUID AND SATURATED VAPOR

Temp. °F	Pressure		Volume§	Density#	Enthalpy**		Entropy§§	
	psia	psig	Vapor	Liquid	Liquid	Vapor	Liquid	Vapor
−140	0.25567	29.401*	110.72	104.03	−20.398	62.190	−.055433	.20292
−130	0.41131	29.084*	70.904	103.12	−18.380	63.258	−.049218	.19842
−120	0.64047	28.617*	46.858	102.21	−16.362	64.338	−.043187	.19440
−110	0.96829	27.950*	31.857	101.29	−14.341	65.430	−.037326	.19081
−100	1.4252	27.019*	22.220	100.36	−12.317	66.530	−.031620	.18760
− 90	2.0473	25.753*	15.861	99.429	−10.287	67.638	−.026058	.18474
− 85	2.4332	24.967*	13.509	98.958	−9.2699	68.194	−.023326	.18343
− 80	2.8765	24.065*	11.563	98.486	−8.2506	68.751	−.020626	.18219
− 75	3.3834	23.033*	9.9442	98.012	−7.2292	69.310	−.017956	.18102
− 70	3.9604	21.858*	8.5912	97.535	−6.2054	69.869	−.015314	.17991
− 65	4.6144	20.526*	7.4545	97.056	−5.1789	70.429	−.012700	.17887
− 60	5.3526	19.023*	6.4950	96.575	−4.1497	70.990	−.010112	.17789
− 55	6.1826	17.333*	5.6813	96.091	−3.1174	71.550	−.007549	.17697
− 50	7.1124	15.440*	4.9884	95.605	−2.0818	72.111	−.005010	.17609
− 45	8.1502	13.327*	4.3957	95.116	−1.0427	72.672	−.002494	.17527
− 40	9.3045	10.977*	3.8868	94.624	0.0	73.232	.0	.17450
− 38	9.8008	9.9666*	3.7035	94.426	0.4182	73.456	.000992	.17420
− 36	10.318	8.9139*	3.5307	94.228	0.8370	73.679	.001980	.17391
− 34	10.856	7.8180*	3.3677	94.029	1.2564	73.903	.002965	.17363
− 32	11.416	6.6777*	3.2138	93.830	1.6766	74.127	.003948	.17335
− 30	11.999	5.4916*	3.0684	93.631	2.0974	74.350	.004927	.17309
− 28	12.604	4.2586*	2.9310	93.431	2.5189	74.574	.005903	.17282
− 26	13.234	2.9773*	2.8011	93.230	2.9411	74.797	.006876	.17257
− 24	13.887	1.6466*	2.6782	93.029	3.3641	75.020	.007845	.17232
− 22	14.566	0.2651*	2.5618	92.827	3.7877	75.242	.008813	.17207
− 21.62	14.696	0.0	2.5407	92.789	3.8673	75.284	.008994	.17203
− 20	15.270	0.5739	2.4516	92.625	4.2121	75.465	.009777	.17184
− 18	16.000	1.3043	2.3471	92.422	4.6372	75.687	.010738	.17160
− 16	16.758	2.0615	2.2481	92.219	5.0631	75.909	.011697	.17138
− 14	17.542	2.8463	2.1541	92.015	5.4898	76.131	.012653	.17116
− 12	18.355	3.6592	2.0650	91.810	5.9173	76.353	.013606	.17094
− 10	19.197	4.5011	1.9803	91.605	6.3456	76.574	.014557	.17073
− 8	20.069	5.3725	1.8999	91.399	6.7747	76.795	.015505	.17053
− 6	20.970	6.2742	1.8234	91.192	7.2046	77.015	.016451	.17033
− 4	21.903	7.2067	1.7507	90.985	7.6354	77.236	.017394	.17014
− 2	22.867	8.1710	1.6815	90.777	8.0670	77.456	.018335	.16995

* in in. Hg. vacuum
§ in cu ft/lb
# in lb/cu ft
** in Btu/lb
§§ in Btu/lb × °R

continued

continued

R-12—PROPERTIES OF SATURATED LIQUID AND SATURATED VAPOR

Temp. °F	Pressure		Volume§	Density#	Enthalpy**		Entropy§§	
	psia	psig	Vapor	Liquid	Liquid	Vapor	Liquid	Vapor
0	23.863	9.1675	1.6157	90.569	8.4995	77.675	.019274	.16976
2	24.893	10.197	1.5530	90.360	8.9329	77.895	.020210	.16958
4	25.956	11.260	1.4933	90.150	9.3671	78.114	.021144	.16941
6	27.054	12.358	1.4364	89.939	9.8023	78.332	.022075	.16924
8	28.187	13.491	1.3821	89.728	10.238	78.550	.023005	.16907
10	29.356	14.660	1.3303	89.516	10.675	78.768	.023932	.16891
12	30.561	15.865	1.2809	89.303	11.113	78.985	.024857	.16875
14	31.804	17.108	1.2337	89.089	11.552	79.202	.025780	.16860
16	33.085	18.389	1.1887	88.875	11.992	79.418	.026701	.16845
18	34.405	19.709	1.1456	88.659	12.433	79.633	.027620	.16830
20	35.765	21.069	1.1045	88.443	12.874	79.849	.028537	.16816
25	39.341	24.645	1.0093	87.899	13.983	80.384	.030821	.16782
30	43.182	28.486	.92401	87.349	15.098	80.916	.033093	.16751
35	47.300	32.604	.84738	86.793	16.220	81.444	.035354	.16721
40	51.705	37.009	.77838	86.231	17.348	81.968	.037604	.16693
50	61.432	46.736	.65984	85.087	19.625	83.000	.042075	.16642
60	72.462	57.766	.56254	83.912	21.931	84.012	.046508	.16597
70	84.900	70.204	.48204	82.704	24.266	84.998	.050906	.16557
80	98.850	84.154	.41495	81.458	26.633	85.955	.055273	.16520
90	114.42	99.725	.35864	80.168	29.032	86.880	.059613	.16485
100	131.72	117.03	.31106	78.830	31.466	87.766	.063928	.16452
110	150.87	136.18	.27060	77.436	33.937	88.609	.068224	.16420
120	171.99	157.29	.23597	75.978	36.448	89.400	.072504	.16385
130	195.20	180.50	.20615	74.446	39.003	90.133	.076777	.16349
140	220.63	205.93	.18032	72.827	41.608	90.795	.081050	.16307
150	248.43	233.73	.15780	71.106	44.271	91.375	.085335	.16260
160	278.74	264.05	.13804	69.264	47.002	91.856	.089649	.16203
170	311.74	297.04	.12057	67.275	49.818	92.214	.094013	.16134
180	347.61	332.91	.10499	65.103	52.741	92.418	.098461	.16049
190	386.56	371.86	.090963	62.697	55.808	92.419	.10304	.15940
200	428.84	414.14	.078132	59.979	59.075	92.139	.10784	.15796
210	474.75	460.06	.066108	56.806	62.647	91.430	.11299	.15597
220	524.70	510.01	.054263	52.851	66.748	89.942	.11882	.15295
230	579.33	564.63	.040319	46.607	72.294	86.110	.12662	.14665
##233.2	598.3	583.6	.02871	34.83	79.40	79.40	.1368	.1368

* in in. Hg. vacuum
§ in cu ft/lb
# in lb/cu ft

** in Btu/lb
§§ in Btu/lb × °R
Critical point

continued

continued

	R-12—PROPERTIES OF SATURATED LIQUID AND SATURATED VAPOR									
	Viscosity*			Thermal Conductivity§			Specific Heat#			
Temp. °F	Sat. Liquid	Sat. Vapor	Gas**	Sat. Liquid	Sat. Vapor	Gas§§	Sat. Liquid	Sat. Vapor	Gas##	Temp. °F
−140	2.47			.0655			.199		.1085	−140
−120	1.97			.0631			.202		.1123	−120
−100	1.612			.0608			.204		.1160	−100
− 80	1.347			.0585			.207		.1196	− 80
− 60	1.146			.0561			.209	.126	.1230	− 60
− 40	.990			.0538			.212	.133	.1264	− 40
− 20	.866	.0249	2.49	.0514	.0040	4.00	.214	.139	.1296	− 20
0	.767	.0265	2.61	.0490	.0043	4.31	.217	.145	.1327	0
20	.687	.0279	2.72	.0467	.0046	4.63	.220	.150	.1356	20
40	.620	.0291	2.83	.0443	.0050	4.95	.224	.157	.1385	40
60	.564	.0301	2.94	.0420	.0053	5.28	.229	.164	.1413	60
80	.517	.0311	3.05	.0397	.0056	5.61	.234	.174	.1439	80
100	.477	.0324	3.15	.0373	.0060	5.94	.240	.185	.1465	100
120	.441	.0399	3.26	.0350	.0064	6.27	.251	.199	.1490	120
140	.409	.0359	3.36	.0326	.0068	6.60	.266	.216	.1513	140
160	.370	.0384	3.47	.0302	.0072	6.94	.288	.235	.1536	160
180	.329	.0417	3.57	.0276	.0076	7.28	.317	.290	.1558	180
200	.273	.0458	3.67	.0246	.0083	7.63	.356	.362	.1579	200
220	.200	.051	3.77	.0204	.0093	7.98	.406		.1599	220
230	.149	.060	3.82	.0161	.0107	8.16			.1609	230
234***	.075	.075	3.84	.0130	.0130	8.23			.1612	234***
240			3.87			8.34			.1618	240
260			3.96			8.71			.1637	260
280			4.06			9.08			.1654	280
300			4.15			9.45			.1671	300
320			4.25			9.82			.1687	320
340			4.34			10.1			.1703	340
360			4.43			10.5			.1718	360
380			4.53			10.8			.1732	380
400			4.62			11.2			.1746	400

* in lb/ft × hr
§ in Btu/hr × ft × °F
# in Btu/lb × °F
** at pressure of 1×10^{-2} atm
§§ at pressure of 1×10^{-3} atm
at pressure of 0 atm
*** Critical temperature

•*ASHRAE 1989 Handbook — Fundamentals*

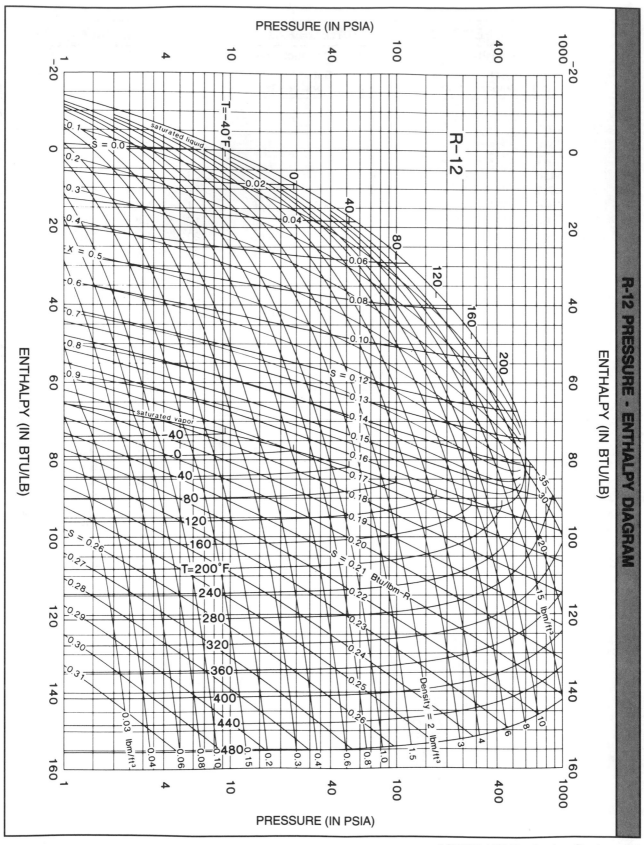

R-22—PROPERTIES OF SATURATED LIQUID AND SATURATED VAPOR

Temp. °F	Pressure		Volume§	Density#	Enthalpy**		Entropy§§	
	psia	psig	Vapor	Liquid	Liquid	Vapor	Liquid	Vapor
−130	0.68858	28.519*	59.170	96.313	−24.388	89.888	−.065456	.28118
−120	1.0725	27.738*	39.078	95.416	−21.538	91.049	−.056942	.27452
−110	1.6199	26.623*	26.578	94.509	−18.738	92.211	−.048818	.26848
−100	2.3802	25.075*	18.558	93.590	−15.980	93.371	−.041046	.26298
− 90	3.4111	22.976*	13.268	92.660	−13.259	94.527	−.033590	.25798
− 80	4.7793	20.191*	9.6902	91.717	−10.570	95.676	−.026418	.25342
− 75	5.6131	18.493^	8.3419	91.241	−9.2346	96.247	−.022929	.25128
− 70	6.5603	16.564*	7.2139	90.761	−7.9050	96.815	−.019500	.24924
− 65	7.6317	14.383*	6.2655	90.278	−6.5802	97.380	−.016128	.24728
− 60	8.8386	11.926*	5.4641	89.791	−5.2593	97.942	.012808	.24541
− 50	11.707	6.0851*	4.2039	88.807	−2.6263	99.055	−.006316	.24189
− 48	12.361	4.7548*	3.9962	88.608	−2.1007	99.275	−.005039	.24122
− 46	13.042	3.3666*	3.8007	88.408	−1.5753	99.495	−.003770	.24056
− 44	13.754	1.9186*	3.6168	88.208	−1.0501	99.714	−.002507	.23991
− 42	14.495	0.4090*	3.4437	88.007	−0.5250	99.932	−.001250	.23927
− 41.47	14.696	0.0	3.3997	87.954	−0.3865	99.990	−.000920	.23910
− 40	15.268	0.5717	3.2805	87.806	0.0	100.15	.0	.23864
− 38	16.072	1.3763	3.1267	87.604	0.5250	100.37	.001244	.23802
− 36	16.910	2.2138	2.9816	87.401	1.0500	100.58	.002482	.23741
− 34	17.781	3.0852	2.8446	87.197	1.5751	100.80	.003714	.23681
− 32	18.687	3.9914	2.7152	86.993	2.1003	101.01	.004940	.23622
− 30	19.629	4.9333	2.5930	86.788	2.6257	101.22	.006161	.23564
− 28	20.608	5.9119	2.4774	86.582	3.1512	101.44	.007377	.23506
− 26	21.624	6.9283	2.3680	86.375	3.6771	101.65	.008587	.23450
− 24	22.679	7.9832	2.2645	86.168	4.2032	101.86	.009792	.23394
− 22	23.774	9.0778	2.1664	85.960	4.7297	102.07	.010993	.23340
− 20	24.909	10.213	2.0735	85.751	5.2566	102.28	.012188	.23285
− 18	26.086	11.390	1.9854	85.542	5.7840	102.48	.013379	.23232
− 16	27.306	12.610	1.9018	85.331	6.3119	102.69	.014566	.23180
− 14	28.569	13.873	1.8225	85.120	6.8403	102.90	.015748	.23128
− 12	29.877	15.181	1.7472	84.908	7.3693	103.10	.016926	.23077
− 10	31.231	16.535	1.6757	84.695	7.8989	103.30	.018100	.23027
− 8	32.632	17.936	1.6077	84.481	8.4292	103.51	.019270	.22977
− 6	34.081	19.385	1.5430	84.266	8.9603	103.71	.020436	.22928
− 4	35.579	20.883	1.4815	84.051	9.4921	103.91	.021598	.22880
− 2	37.127	22.431	1.4230	83.834	10.025	104.10	.022757	.22832

* in in. Hg. vacuum
§ in cu ft/lb
# in lb/cu ft
** in Btu/lb
§§ in Btu/lb × °R

continued

continued

R-22—PROPERTIES OF SATURATED LIQUID AND SATURATED VAPOR

Temp. °F	Pressure		Volume§	Density#	Enthalpy**		Entropy§§	
	psia	psig	Vapor	Liquid	Liquid	Vapor	Liquid	Vapor
0	38.726	24.030	1.3672	83.617	10.558	104.30	.023912	.22785
2	40.378	25.682	1.3141	83.399	11.093	104.50	.025064	.22738
4	42.083	27.387	1.2635	83.179	11.628	104.69	.026213	.22693
6	43.843	29.147	1.2152	82.959	12.164	104.89	.027359	.22647
8	45.658	30.962	1.1692	82.738	12.702	105.08	.028502	.22602
10	47.530	32.834	1.1253	82.516	13.240	105.27	.029642	.22558
12	49.461	34.765	1.0833	82.292	13.779	105.46	.030779	.22515
14	51.450	36.754	1.0433	82.068	14.320	105.64	.031913	.22471
16	53.501	38.805	1.0050	81.843	14.862	105.83	.033045	.22429
18	55.612	40.916	.96841	81.616	15.405	106.02	.034175	.22387
20	57.786	43.090	.93343	81.389	15.950	106.20	.035302	.22345
25	63.505	48.809	.85246	80.815	17.317	106.65	.038110	.22243
30	69.641	54.945	.77984	80.234	18.693	107.09	.040905	.22143
35	76.215	61.519	.71454	79.645	20.078	107.52	.043689	.22046
40	83.246	68.550	.65571	79.049	21.474	107.94	.046464	.21951
45	90.754	76.058	.60258	78.443	22.880	108.35	.049229	.21858
50	98.758	84.062	.55451	77.829	24.298	108.74	.051987	.21767
55	107.28	92.583	.51093	77.206	25.728	109.12	.054739	.21677
60	116.34	101.64	.47134	76.572	27.170	109.49	.057486	.21589
65	125.95	111.26	.43531	75.928	28.626	109.84	.060228	.21502
70	136.15	121.45	.40245	75.273	30.095	110.18	.062968	.21416
80	158.36	143.66	.34497	73.926	33.077	110.80	.068441	.21246
90	183.14	168.44	.29668	72.525	36.121	111.35	.073911	.21077
100	210.67	195.97	.25582	71.061	39.233	111.81	.079400	.20907
110	241.13	226.44	.22102	69.524	42.422	112.17	.084906	.20734
120	274.73	260.03	.19118	67.901	45.694	112.42	.090444	.20554
130	311.66	296.96	.16542	66.174	49.064	112.52	.096033	.20365
140	352.14	337.45	.14300	64.319	52.550	112.47	.10170	.20161
150	396.42	381.72	.12334	62.301	56.177	112.20	.10749	.19938
160	444.75	430.06	.10590	60.068	59.989	111.67	.11345	.19684
170	497.46	482.76	.090228	57.532	64.055	110.76	.11970	.19386
180	554.89	540.19	.075819	54.533	68.504	109.30	.12640	.19018
190	617.52	602.82	.061991	50.703	73.617	106.88	.13399	.18518
200	686.02	671.32	.046923	44.671	80.406	101.99	.14394	.17666
##205.07	723.4	708.7	.03123	32.03	91.58	91.58	.1605	.1605

* in in. Hg. vacuum
§ in cu ft/lb
# in lb/cu ft

** in Btu/lb
§§ in Btu/lb × °R
Critical point

continued

continued

R-22—PROPERTIES OF SATURATED LIQUID AND SATURATED VAPOR

Temp. °F	Viscosity*			Thermal Conductivity§			Specific Heat#			Temp. °F
	Sat. Liquid	Sat. Vapor	Gas**	Sat. Liquid	Sat. Vapor	Gas§§	Sat. Liquid	Sat. Vapor	Gas##	
−100	1.167			.0789			.255		.1260	−100
− 80	1.014			.0757			.256		.1292	− 80
− 60	.894			.0725			.259	.139	.1324	− 60
− 40	.798	.0245	2.45	.0693	.0040	4.04	.262	.146	.1356	− 40
− 20	.719	.0257	2.57	.0661	.0044	4.43	.266	.152	.1388	− 20
0	.654	.0269	2.68	.0630	.0048	4.81	.271	.158	.1420	0
20	.599	.0282	2.80	.0598	.0052	5.20	.276	.165	.1452	20
40	.553	.0295	2.91	.0566	.0056	5.58	.283	.175	.1484	40
60	.513	.0309	3.03	.0534	.0060	5.97	.291	.187	.1515	60
80	.480	.0325	3.14	.0502	.0064	6.35	.300	.204	.1546	80
100	.449	.0343	3.25	.0471	.0068	6.74	.313	.226	.1577	100
120	.427	.0362	3.37	.0439	.0072	7.12	.332	.253	.1608	120
140	.392	.0383	3.48	.0407	.0077	7.51	.357	.288	.1638	140
160	.344	.0411	3.59	.0371	.0084	7.90	.390	.332	.1668	160
180	.285	.045	3.70	.0318	.0105	8.28	.433		.1697	180
190	.244	.049	3.75	.0288	.0119	8.48			.1712	190
200	.182	.058	3.81	.0238	.0140	8.67			.1726	200
205***	.074	.074	3.83	.0177	.0177	8.76			.1733	205***
220			3.92			9.05			.1754	220
240			4.02			9.44			.1782	240
300			4.34			10.59			.1863	300
400			4.86			12.5			.1983	400
440			5.06			13.3			.2026	440

* in lb/ft × hr
§ in Btu/hr × ft × °F
# in Btu/lb × °F
** at pressure of 1 × 10⁻² atm
§§ at pressure of 1 × 10⁻³ atm
at pressure of 0 atm
*** Critical temperature

•*ASHRAE 1989 Handbook — Fundamentals*

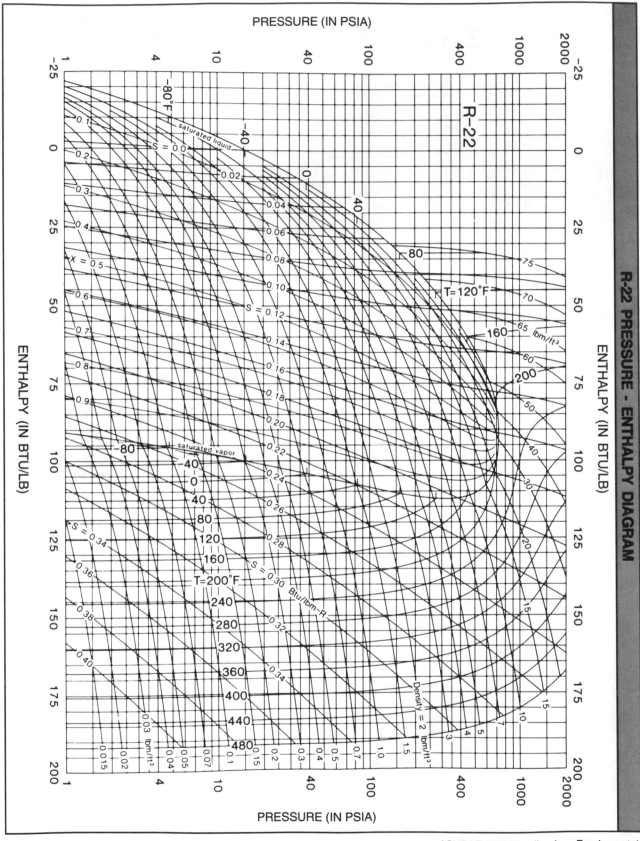

ASHRAE 1989 Handbook — Fundamentals

R-502—PROPERTIES OF SATURATED LIQUID AND SATURATED VAPOR

Temp. °F	Pressure		Volume§	Density#	Enthalpy**		Entropy§§	
	psia	psig	Vapor	Liquid	Liquid	Vapor	Liquid	Vapor
−75	7.2807	15.098*	4.9585	95.235	−7.6022	68.964	−.018844	.18020
−70	8.4343	12.749*	4.3241	94.699	−6.5643	69.570	−.016169	.17921
−65	9.7307	10.109*	3.7849	94.159	−5.5105	70.174	−.013488	.17828
−60	11.182	7.1539*	3.3248	93.615	−4.4406	70.777	−.010802	.17740
−55	12.802	3.8567*	2.9306	93.067	−3.3546	71.377	−.008110	.17656
−50	14.602	0.1906*	2.5915	92.514	−2.2525	71.975	−.005412	.17578
−49.75	14.696	0.0	2.5761	92.487	−2.1978	72.004	−.005279	.17574
−48	15.376	0.6804	2.4693	92.292	−1.8071	72.213	−.004331	.17547
−46	16.183	1.4865	2.3541	92.068	−1.3592	72.451	−.003249	.17518
−44	17.022	2.3256	2.2452	91.844	−0.9087	72.689	−.002167	.17489
−42	17.895	3.1986	2.1425	91.620	−0.4556	72.925	−.001084	.17461
−40	18.802	4.1064	2.0453	91.394	0.0	73.162	.0	.17433
−38	19.746	5.0500	1.9535	91.167	0.4582	73.398	.001085	.17406
−36	20.726	6.0303	1.8666	90.940	0.9189	73.633	.002170	.17380
−34	21.744	7.0482	1.7844	90.712	1.3822	73.867	.003256	.17354
−32	22.801	8.1048	1.7066	90.483	1.8481	74.101	.004343	.17329
−30	23.897	9.2010	1.6328	90.253	2.3165	74.335	.005430	.17304
−28	25.034	10.338	1.5629	90.022	2.7874	74.567	.006518	.17280
−26	26.212	11.516	1.4966	89.790	3.2608	74.799	.007607	.17257
−24	27.433	12.737	1.4336	89.557	3.7367	75.031	.008696	.17234
−22	28.697	14.001	1.3739	89.323	4.2152	75.261	.009786	.17211
−20	30.006	15.310	1.3172	89.088	4.6961	75.491	.010876	.17189
−18	31.361	16.665	1.2633	88.853	5.1795	75.720	.011966	.17168
−16	32.762	18.066	1.2120	88.616	5.6654	75.948	.013057	.17147
−14	34.211	19.515	1.1633	88.378	6.1537	76.175	.014149	.17126
−12	35.709	21.013	1.1169	88.139	6.6445	76.402	.015240	.17106
−10	37.256	22.560	1.0727	87.899	7.1377	76.627	.016332	.17087
− 8	38.854	24.158	1.0307	87.658	7.6333	76.852	.017425	.17067
− 6	40.504	25.808	.99066	87.416	8.1313	77.075	.018517	.17049
− 4	42.207	27.511	.95248	87.172	8.6318	77.298	.019610	.17030
− 2	43.964	29.268	.91608	86.928	9.1346	77.520	.020703	.17012
0	45.776	31.080	.88135	86.682	9.6397	77.741	.021796	.16995
2	47.644	32.948	.84821	86.435	10.147	77.960	.022889	.16978
4	49.569	34.873	.81657	86.187	10.657	78.179	.023982	.16961
6	51.552	36.856	.78636	85.937	11.169	78.397	.025075	.16944
8	53.594	38.898	.75749	85.686	11.684	78.613	.026168	.16928

* in in. Hg. vacuum
§ in cu ft/lb
# in lb/cu ft
** in Btu/lb
§§ in Btu/lb × °R

continued

continued

R-502—PROPERTIES OF SATURATED LIQUID AND SATURATED VAPOR

Temp. °F	Pressure		Volume§	Density#	Enthalpy**		Entropy§§	
	psia	psig	Vapor	Liquid	Liquid	Vapor	Liquid	Vapor
10	55.697	41.001	.72989	85.434	12.200	78.828	.027261	.16912
15	61.226	46.530	.66605	84.798	13.502	79.362	.029992	.16874
20	67.155	52.459	.60884	84.153	14.818	79.887	.032722	.16838
25	73.503	58.807	.55746	83.498	16.147	80.405	.035450	.16803
30	80.287	65.591	.51121	82.833	17.490	80.913	.038176	.16770
35	87.523	72.827	.46948	82.157	18.846	81.413	.040897	.16738
40	95.229	80.533	.43175	81.470	20.216	81.903	.043617	.16707
45	103.42	88.726	.39758	80.771	21.597	82.383	.046331	.16678
50	112.12	97.425	.36656	80.059	22.991	82.852	.049040	.16649
55	121.34	106.65	.33834	79.333	24.397	83.310	.051743	.16621
60	131.10	116.41	.31264	78.592	25.814	83.755	.054440	.16594
65	141.42	126.73	.28916	77.835	27.244	84.187	.057131	.16566
70	152.32	137.63	.26769	77.062	28.685	84.606	.059816	.16539
75	163.81	149.12	.24802	76.269	30.138	85.009	.062494	.16512
80	175.92	161.23	.22995	75.457	31.602	85.397	.065165	.16484
85	188.66	173.97	.21333	74.623	33.078	85.767	.067829	.16456
90	202.06	187.36	.19802	73.765	34.566	86.118	.070487	.16427
95	216.13	201.43	.18388	72.881	36.066	86.449	.073139	.16397
100	230.89	216.20	.17079	71.968	37.578	86.758	.075786	.16366
105	246.38	231.68	.15866	71.023	39.104	87.402	.078429	.16332
110	262.61	247.92	.14740	70.042	40.644	87.298	.081070	.16297
115	279.61	264.92	.13691	69.022	42.201	87.524	.083711	.16258
120	297.41	282.72	.12711	67.956	43.774	87.716	.086354	.16216
125	316.05	301.35	.11795	66.839	45.369	87.869	.089005	.16170
130	335.54	320.85	.10935	65.661	46.987	87.977	.091667	.16118
135	355.94	341.25	.10125	64.414	48.634	88.032	.094350	.16060
140	377.30	362.60	.093586	63.083	50.316	88.024	.097063	.15994
145	399.65	384.95	.086303	61.650	52.045	87.941	.099822	.15919
150	423.06	408.37	.079335	60.092	53.834	87.763	.10265	.15830
155	447.61	432.92	.072610	58.371	55.705	87.463	.10558	.15724
160	473.39	458.69	.066037	56.429	57.698	86.997	.10867	.15595
165	500.50	485.81	.059485	54.169	59.878	86.293	.11202	.15430
170	529.11	514.41	.052715	51.391	62.386	85.198	.11584	.15207
175	559.42	544.72	.045119	47.551	65.610	83.303	.12073	.14861
##179.9	591.0	576.3	.02857	35.00	74.81	74.81	.1346	.1346

* in in. Hg. vacuum
§ in cu ft/lb
# in lb/cu ft

** in Btu/lb
§§ in Btu/lb × °R
Critical point

continued

continued

R-502—PROPERTIES OF SATURATED LIQUID AND SATURATED VAPOR

Temp. °F	Viscosity*			Thermal Conductivity§			Specific Heat#			Temp. °F
	Sat. Liquid	Sat. Vapor	Gas**	Sat. Liquid	Sat. Vapor	Gas§§	Sat. Liquid	Sat. Vapor	Gas##	
−100	1.39			.0595			.244			−100
− 80	1.16			.0570			.248			− 80
− 60	1.00	.0228		.0545			.253	.138		− 60
− 40	.86	.0244	2.42	.0519	.0046	4.58	.259	.149	.148	− 40
− 20	.76	.0258	2.54	.0494	.0050	4.95	.264	.155	.151	− 20
0	.67	.0270	2.66	.0469	.0053	5.31	.271	.160	.154	0
20	.60	.0283	2.77	.0444	.0057	5.67	.277	.164	.157	20
40	.54	.0295	2.89	.0419	.0060	6.03	.285	.171	.160	40
60	.487	.0310	3.01	.0394	.0064	6.39	.292	.180	.164	60
80	.433	.0327	3.12	.0369	.0068	6.76	.300	.195	.167	80
100	.380	.0348	3.23	.0344	.0071	7.14	.308	.218	.170	100
120	.329	.0373	3.34	.0314	.0075	7.52	.316	.249	.173	120
140	.284	.039	3.45	.0281	.0083	7.91	.326	.310	.176	140
160	.243	.045	3.56	.0237	.0090	8.31	.335		.178	160
170	.207	.053	3.62	.021	.0103	8.52	.345		.179	170
180***	.074	.074	3.67	.014	.014	8.73			.181	180***
190			3.72			8.94			.182	190
200			3.78			9.16			.183	200
220			3.88			9.60			.186	220
240			3.99			10.07			.188	240
260			4.10			10.5			.190	260
280			4.20			11.0			.192	280
300			4.29			11.6			.193	300
320						12.1				320
340						12.7				340
360						13.3				360
380						13.9				380
440						16.0				440
460						16.7				460
500						18.4				500

* in lb/ft × hr
§ in Btu/hr × ft × °F
# in Btu/lb × °F
** at pressure of 1×10^{-2} atm
§§ at pressure of 1×10^{-3} atm
at pressure of 1 atm
*** Critical temperature

•*ASHRAE 1989 Handbook — Fundamentals*

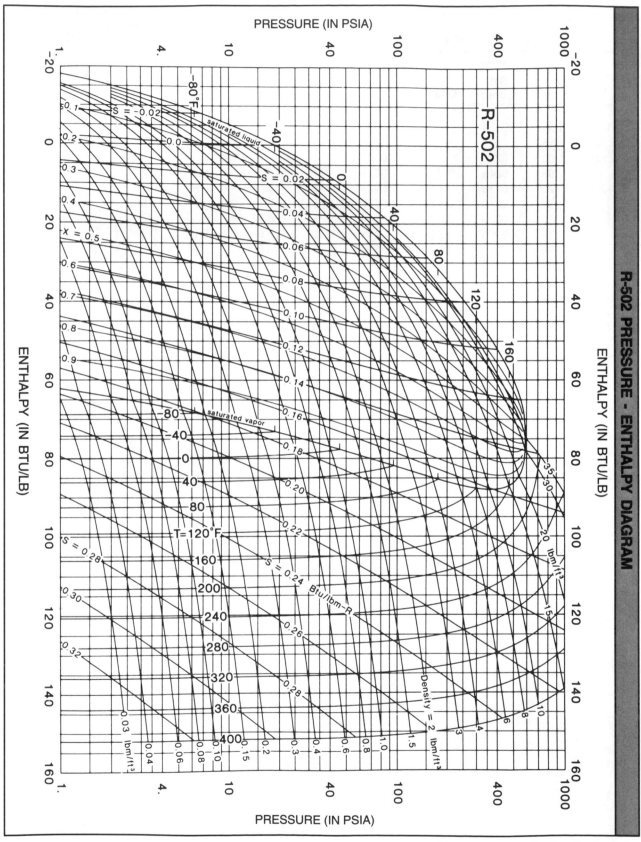

R-502 PRESSURE - ENTHALPY DIAGRAM

ASHRAE 1989 Handbook — Fundamentals

R-22 PRESSURE DROP CHART

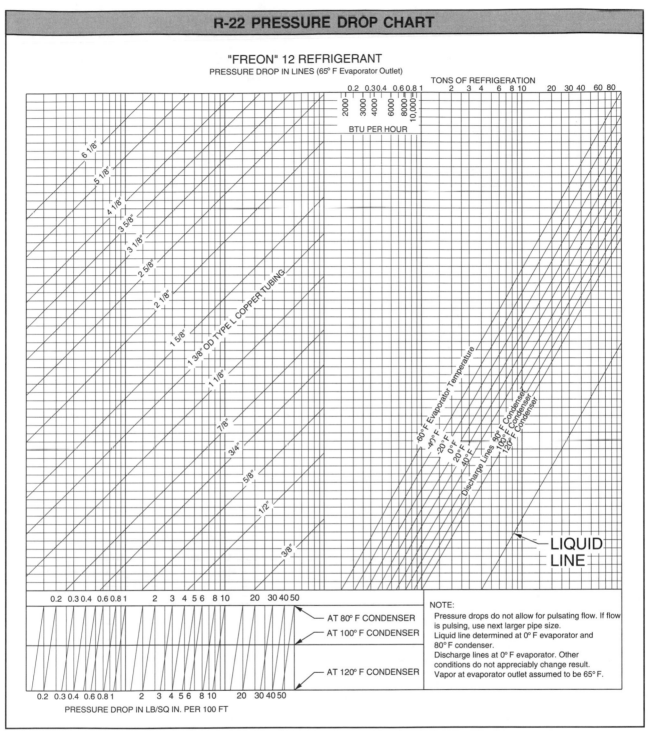

"FREON" 12 REFRIGERANT
PRESSURE DROP IN LINES (65° F Evaporator Outlet)

NOTE:
Pressure drops do not allow for pulsating flow. If flow is pulsating, use next larger pipe size.
Liquid line determined at 0° F evaporator and 80° F condenser.
Discharge lines at 0° F evaporator. Other conditions do not appreciably change result.
Vapor at evaporator outlet assumed to be 65° F.

PRESSURE DROP IN LB/SQ IN. PER 100 FT

•*Du Pont Co.*

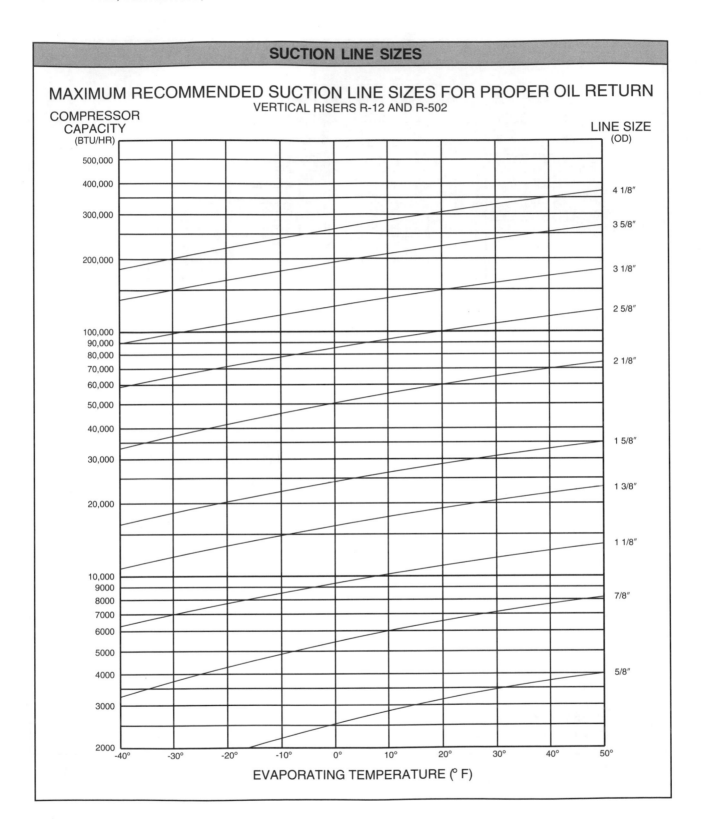

SUCTION LINE SIZES

MAXIMUM RECOMMENDED SUCTION LINE SIZES FOR PROPER OIL RETURN
VERTICAL RISERS R-12 AND R-502

OUTDOOR DESIGN TEMPERATURE

State and City	Lat.§	Winter DB*	Summer DB*	Daily Range*	WB*
ALABAMA					
Alexander City	32	18	96	21	79
Auburn	32	18	96	21	79
Birmingham	33	17	96	21	78
Huntsville	34	11	95	23	78
Mobile	30	25	95	18	80
Montgomery	32	22	96	21	79
Talladega	33	18	97	21	79
Tuscaloosa	33	20	98	22	79
ALASKA					
Anchorage	61	−23	71	15	60
Barrow	71	−45	57	12	54
Fairbanks	64	−51	82	24	64
Juneau	58	− 4	74	15	61
Kodiak	57	10	69	10	60
Nome	64	−31	66	10	58
ARIZONA					
Flagstaff	35	− 2	84	31	61
Phoenix	33	31	109	27	76
Tucson	32	28	104	26	72
Yuma	32	36	111	27	79
ARKANSAS					
Fort Smith	35	12	101	24	80
Hot Springs	34	17	101	22	80
Little Rock	34	15	99	22	80
Pine Bluff	34	16	100	22	81
CALIFORNIA					
Bakersfield	35	30	104	32	73
Burbank	34	37	95	25	71
Fresno	36	28	102	34	72
Laguna Beach	33	41	83	18	70
Long Beach	33	41	83	22	70
Los Angeles	33	41	83	15	70
Monterey	36	35	75	20	64
Napa	38	30	100	30	71
Oakland	37	34	85	19	66
Oceanside	33	41	83	13	70
Palm Springs	33	33	112	35	76
Pasadena	34	32	98	29	73
Sacramento	38	30	101	36	72
San Diego	32	42	83	12	71
San Fernando	34	37	95	38	71
San Francisco	37	38	74	14	64
San Jose	37	34	85	26	68
Santa Barbara	34	34	81	24	68
Santa Cruz	36	35	75	28	64
Santa Monica	34	41	83	16	70
Stockton	37	28	100	37	71
COLORADO					
Boulder	40	2	93	27	64

OUTDOOR DESIGN TEMPERATURE

State and City	Lat.§	Winter DB*	Summer DB*	Daily Range*	WB*
Colorado Spgs.	38	− 3	91	30	63
Denver	39	− 5	93	28	64
Pueblo	38	− 7	97	31	67
CONNECTICUT					
Bridgeport	41	6	86	18	75
Hartford	41	3	91	22	77
New Haven	41	3	88	17	76
Waterbury	41	− 4	88	21	75
DELAWARE					
Dover	39	11	92	18	79
Wilmington	39	10	92	20	77
DISTRICT OF COLUMBIA					
Washington	38	14	93	18	78
FLORIDA					
Cape Kennedy	28	35	90	15	80
Daytona Beach	29	32	92	15	80
Fort Lauderdale	26	42	92	15	80
Key West	24	55	90	9	80
Miami	25	44	91	15	79
Orlando	28	35	94	17	79
Pensacola	30	25	94	14	80
St. Petersburg	27	36	92	16	79
Sarasota	27	39	93	17	79
Tallahassee	30	27	94	19	79
Tampa	27	36	92	17	79
GEORGIA					
Athens	33	18	94	21	78
Atlanta	33	17	94	19	77
Augusta	33	20	97	19	80
Griffin	33	18	93	21	78
Macon	32	21	96	22	79
Savannah	32	24	96	20	80
HAWAII					
Honolulu	21	62	87	12	76
Kaneohe Bay	21	65	85	12	76
IDAHO					
Boise	43	3	96	31	68
Idaho Falls	43	−11	89	38	65
Lewiston	46	− 1	96	32	67
Twin Falls	42	− 3	99	34	64
ILLINOIS					
Aurora	41	− 6	93	20	79
Bloomington	40	− 6	92	21	78
Carbondale	37	2	95	21	80
Champaign	40	− 3	95	21	78
Chicago	41	− 9	90	15	79
Galesburg	40	− 7	93	22	78
Joliet	41	− 5	93	20	78
Kankakee	41	− 4	93	21	78
Macomb	40	− 5	95	22	79

* in °F   § in degrees

continued

continued

OUTDOOR DESIGN TEMPERATURE

State and City	Lat.§	Winter DB*	Summer DB*	Daily Range*	WB*
Peoria	40	− 8	91	22	78
Rantoul	40	− 4	94	21	78
Rockford	42	− 9	91	24	77
Springfield	39	− 3	94	21	79
Waukegan	42	− 6	92	21	78
INDIANA					
Fort Wayne	41	− 4	92	24	77
Hobart	41	− 4	91	21	77
Indianapolis	39	− 2	92	22	78
Kokomo	40	− 4	91	22	77
Lafayette	40	− 3	94	22	78
Muncie	40	− 3	92	22	76
South Bend	41	− 3	91	22	77
Terre Haute	39	− 2	95	22	79
Valparaiso	41	− 3	93	22	78
IOWA					
Ames	42	−11	93	23	78
Burlington	40	− 7	94	22	78
Cedar Rapids	41	−10	91	23	78
Des Moines	41	−10	94	23	78
Dubuque	42	−12	90	22	77
Iowa City	41	−11	92	22	80
Keokuk	40	− 5	95	22	79
Sioux City	42	−11	95	24	78
KANSAS					
Garden City	37	− 1	99	28	74
Liberal	37	2	99	28	73
Russell	38	0	101	29	78
Topeka	39	0	99	24	79
Wichita	37	3	101	23	77
KENTUCKY					
Bowling Green	35	4	94	21	79
Lexington	38	3	93	22	77
Louisville	38	5	95	23	79
LOUISIANA					
Alexandria	31	23	95	20	80
Baton Rouge	30	25	95	19	80
Lafayette	30	26	95	18	81
Monroe	32	20	99	20	79
New Orleans	29	29	93	16	81
Shreveport	32	20	99	20	79
MAINE					
Augusta	44	− 7	88	22	74
Bangor	44	−11	86	22	73
Caribou	46	−18	84	21	71
Portland	43	− 6	87	22	74
MARYLAND					
Baltimore	39	10	94	21	78
Frederick	39	8	94	22	78
Salisbury	38	12	93	18	79

OUTDOOR DESIGN TEMPERATURE

State and City	Lat.§	Winter DB*	Summer DB*	Daily Range*	WB*
MASSACHUSETTS					
Boston	42	6	91	16	75
Clinton	42	− 2	90	17	75
Lawrence	42	− 6	90	22	76
Lowell	42	− 4	91	21	76
New Bedford	41	5	85	19	74
Pittsfield	42	− 8	87	23	73
Worcester	42	0	87	18	73
MICHIGAN					
Battle Creek	42	1	92	23	76
Benton Harbor	42	1	91	20	75
Detroit	42	3	91	20	76
Flint	42	− 4	90	25	76
Grand Rapids	42	1	91	24	75
Holland	42	2	88	22	75
Kalamazoo	42	1	92	23	76
Lansing	42	− 3	90	24	75
Marquette	46	−12	84	18	72
Pontiac	42	0	90	21	76
Port Huron	42	0	90	21	76
Saginaw	43	0	91	23	76
Sault Ste. Marie	46	−12	84	23	72
MINNESOTA					
Alexandria	45	−22	91	24	76
Duluth	46	−21	85	22	72
International Falls	48	−29	85	26	71
Minneapolis	44	−16	92	22	77
Rochester	43	−17	90	24	77
MISSISSIPPI					
Biloxi	30	28	94	16	82
Clarksdale	34	14	96	21	80
Jackson	32	21	97	21	79
Laurel	31	24	96	21	81
Natchez	31	23	96	21	81
Vicksburg	32	22	97	21	81
MISSOURI					
Columbia	38	− 1	97	22	78
Hannibal	39	− 2	96	22	80
Jefferson City	38	2	98	23	78
Kansas City	39	2	99	20	78
Kirksville	40	− 5	96	24	78
Moberly	39	− 2	97	23	78
St. Joseph	39	− 3	96	23	81
St. Louis	38	2	97	21	78
Springfield	37	3	96	23	78
MONTANA					
Billings	45	−15	94	31	67
Butte	45	−24	86	35	60
Great Falls	47	−21	91	28	64
Lewiston	47	−22	90	30	65

* in °F   § in degrees

continued

continued

OUTDOOR DESIGN TEMPERATURE					
State and City	Lat.§	Winter	Summer		
		DB*	DB*	Daily Range*	WB*
Missoula	46	−13	92	36	65
NEBRASKA					
Columbus	41	− 6	98	25	77
Fremont	41	− 6	98	22	78
Grand Island	40	− 8	97	28	75
Lincoln	40	− 5	99	24	78
Norfolk	41	− 8	97	30	78
North Platte	41	− 8	97	28	74
Omaha	41	− 8	94	22	78
NEVADA					
Carson City	39	4	94	42	63
Las Vegas	36	25	108	30	71
Reno	39	6	96	45	64
NEW HAMPSHIRE					
Claremont	43	− 9	89	24	74
Concord	43	− 8	90	26	74
Manchester	42	− 8	91	24	75
Portsmouth	43	− 2	89	22	75
NEW JERSEY					
Atlantic City	39	10	92	18	78
Long Branch	40	10	93	18	78
Newark	40	10	94	20	77
New Brunswick	40	6	92	19	77
Trenton	40	11	91	19	78
NEW MEXICO					
Albuquerque	35	12	96	27	66
Carlsbad	32	13	103	28	72
Gallup	35	0	90	32	64
Los Alamos	35	5	89	32	62
Santa Fe	35	6	90	28	63
Silver City	32	5	95	30	66
NEW YORK					
Albany	42	− 6	91	23	75
Batavia	43	1	90	22	75
Buffalo	42	2	88	21	74
Geneva	42	− 3	90	22	75
Glens Falls	43	−11	88	23	74
Ithaca	42	− 5	88	24	74
Kingston	41	− 3	91	22	76
Lockport	43	4	89	21	76
New York City	40	11	92	17	76
Niagara Falls	43	4	89	20	76
Rochester	43	1	91	22	75
Syracuse	43	− 3	90	20	75
Utica	43	−12	88	22	75
NORTH CAROLINA					
Charlotte	35	18	95	20	77
Durham	35	16	94	20	78
Greensboro	36	14	93	21	77
Jacksonville	34	20	92	18	80

OUTDOOR DESIGN TEMPERATURE					
State and City	Lat.§	Winter	Summer		
		DB*	DB*	Daily Range*	WB*
Wilmington	34	23	93	18	81
Winston-Salem	36	16	94	20	76
NORTH DAKOTA					
Bismark	46	−23	95	27	73
Fargo	46	−22	92	25	76
Grand Forks	47	−26	91	25	74
Williston	48	−25	91	25	72
OHIO					
Akron-Canton	40	1	89	21	75
Athens	39	0	95	22	78
Bowling Green	41	− 2	92	23	76
Cambridge	40	1	93	23	78
Cincinnati	39	1	92	21	77
Cleveland	41	1	91	22	76
Columbus	40	0	92	24	77
Dayton	39	− 1	91	20	76
Fremont	41	− 3	90	24	76
Marion	40	0	93	23	77
Newark	40	− 1	94	23	77
Portsmouth	38	5	95	22	78
Toledo	41	− 3	90	25	76
Warren	41	0	89	23	74
OKLAHOMA					
Bartlesville	36	6	101	23	77
Chickasha	35	10	101	24	78
Lawton	34	12	101	24	78
McAlester	34	14	99	23	77
Norman	35	9	99	24	77
Oklahoma City	35	9	100	23	78
Seminole	35	11	99	23	77
Stillwater	36	8	100	24	77
Tulsa	36	8	101	22	79
Woodward	36	6	100	26	78
OREGON					
Albany	44	18	92	31	69
Astoria	46	25	75	16	65
Baker	44	− 1	92	30	65
Eugene	44	17	92	31	69
Grants Pass	42	20	99	33	71
Klamath Falls	42	4	90	36	63
Medford	42	19	98	35	70
Portland	45	17	89	23	69
Salem	44	18	92	31	69
PENNSYLVANIA					
Allentown	40	4	92	22	76
Altoona	40	0	90	23	74
Butler	40	1	90	22	75
Erie	42	4	88	18	75
Harrisburg	40	7	94	21	77
New Castle	41	2	91	23	75

* in °F § in degrees

continued

continued

OUTDOOR DESIGN TEMPERATURE

State and City	Lat.§	Winter DB*	Summer DB*	Summer Daily Range*	Summer WB*
Philadelphia	39	10	93	21	77
Pittsburgh	40	1	89	22	74
Reading	40	9	92	19	76
West Chester	39	9	92	20	77
Williamsport	41	2	92	23	75
RHODE ISLAND					
Newport	41	5	88	16	76
Providence	41	5	89	19	75
SOUTH CAROLINA					
Charleston	32	25	94	13	81
Columbia	33	20	97	22	79
Florence	34	22	94	21	80
Sumter	33	22	95	21	79
SOUTH DAKOTA					
Aberdeen	45	−19	94	27	77
Brookings	44	−17	95	25	77
Huron	44	−18	96	28	77
Rapid City	44	−11	95	28	71
Sioux Falls	43	−15	94	24	76
TENNESSEE					
Athens	35	13	95	22	77
Chattanooga	35	13	96	22	78
Dyersburg	36	10	96	21	81
Knoxville	35	13	94	21	77
Memphis	35	13	98	21	80
Murfreesboro	34	9	97	22	78
Nashville	36	9	97	21	78
TEXAS					
Abilene	32	15	101	22	75
Alice	27	31	100	20	82
Amarillo	35	6	98	26	71
Austin	30	24	100	22	78
Beaumont	29	27	95	19	81
Big Spring	32	16	100	26	74
Brownsville	25	35	94	18	80
Corpus Christi	27	31	95	19	80
Dallas	32	18	102	20	78
El Paso	31	20	100	27	69
Forth Worth	32	17	101	22	78
Galveston	29	31	90	10	81
Houston	29	27	96	18	80
Huntsville	30	22	100	20	78
Laredo	27	32	102	23	78
Lubbock	33	10	98	26	73
Mcallen	26	35	97	21	80
Midland	31	16	100	26	73
Pecos	31	16	100	27	73
San Antonio	29	25	99	19	77
Temple	31	22	100	22	78
Tyler	32	19	99	21	80

OUTDOOR DESIGN TEMPERATURE

State and City	Lat.§	Winter DB*	Summer DB*	Summer Daily Range*	Summer WB*
Victoria	28	29	98	18	82
Waco	31	21	101	22	78
Wichita Falls	33	14	103	24	77
UTAH					
Cedar City	37	− 2	93	32	65
Logan	41	− 3	93	33	65
Provo	40	1	98	32	66
Salt Lake City	40	3	97	32	66
VERMONT					
Barre	44	−16	84	23	73
Burlington	44	−12	88	23	74
Rutland	43	−13	87	23	74
VIRGINIA					
Charlottesville	38	14	94	23	77
Fredericksburg	38	10	96	21	78
Harrisonburg	38	12	93	23	75
Lynchburg	37	12	93	21	77
Norfolk	36	20	93	18	79
Petersburg	37	14	95	20	79
Richmond	37	14	95	21	79
Roanoke	37	12	93	23	75
Winchester	39	6	93	21	77
WASHINGTON					
Aberdeen	46	25	80	16	65
Bellingham	48	10	81	19	68
Olympia	46	16	87	32	67
Seattle-Tacoma	47	17	84	22	66
Spokane	47	− 6	93	28	65
Walla Walla	46	0	97	27	69
WEST VIRGINIA					
Charleston	38	7	92	20	76
Clarksburg	39	6	92	21	76
Parkersburg	39	7	93	21	77
Wheeling	40	1	89	21	74
WISCONSIN					
Beloit	42	− 7	92	24	78
Fon Du Lac	43	−12	89	23	76
Green Bay	44	−13	88	23	76
La Crosse	43	−13	91	22	77
Madison	43	−11	91	22	77
Milwaukee	42	− 8	90	21	76
Racine	42	− 6	91	21	77
Sheboygan	43	−10	89	20	77
Wausau	44	−16	91	23	76
WYOMING					
Casper	42	−11	92	31	63
Cheyenne	41	− 9	89	30	63
Laramie	41	−14	84	28	61
Rock Springs	41	− 9	86	32	59
Sheridan	44	−14	94	32	66

* in °F   § in degrees

Air Conditioning Contractors of America

HEAT TRANSFER FACTORS

Item	Design Temperature Difference*														
	30	35	40	45	50	55	60	65	70	75	80	85	90	95	100
Windows§ — Wood or Metal Frame															
Double-hung, Horizontal-slide, Casement, or Awning															
Single glass	45	50	60	65	75	80	90	95	105	110	120	125	135	140	150
With double glass or insulating glass	30	35	40	45	50	55	60	65	70	75	80	80	85	90	95
With storm sash	25	30	35	40	45	50	55	60	60	65	70	75	80	85	90
Fixed or Picture															
Single glass	40	50	55	60	70	75	85	90	95	105	110	115	125	130	140
Double glass or with storm sash	25	30	35	40	45	45	50	55	60	65	70	75	75	80	85
Jalousie															
Single glass	225	265	300	340	375	415	450	490	525	565	600	640	675	715	750
With storm sash	65	75	90	100	110	120	135	145	155	165	175	190	200	210	220
Doors§															
Sliding Glass Doors															
Single glass	75	85	100	115	125	140	150	165	175	190	200	210	225	240	250
Double glass	60	70	80	90	100	110	120	130	140	150	160	170	180	190	200
Other Doors															
Weatherstripped and with storm door	40	45	55	60	65	75	80	85	90	100	105	110	120	125	130
Weatherstripped or with storm door	70	85	95	110	120	135	145	155	170	180	195	205	215	230	240
No weatherstripping or storm door	135	160	180	200	225	250	270	290	315	340	360	380	405	430	450
Walls and Partitions§ — Wood Frame with Sheathing and Siding															
No insulation	8	9	10	11	13	14	15	16	18	19	20	21	23	24	25
R-5 polystyrene sheathing	3	4	4	5	6	6	7	7	8	8	9	9	10	10	11
R-7 batt insulation (2″ – 2¾″)	3	4	4	5	5	6	6	7	7	8	8	9	9	10	10
R-11 batt insulation (3″ – 3½″)	2	2	3	3	4	4	4	5	5	5	6	6	6	7	7
Partition Between Conditioned and Unconditioned Spaces															
Finished one side only, no insulation	17	19	22	25	28	30	33	36	39	41	44	47	49	52	55
Finished both sides, no insulation	9	11	12	14	16	17	19	20	22	23	25	26	28	29	31
Partition with 1″ polystyrene board R-5	4	4	5	5	6	7	7	8	8	9	10	10	11	11	12
R-7 insulation finished both sides	3	4	4	5	5	6	6	7	7	8	8	9	9	10	10
R-11 insulation finished both sides	2	3	3	4	4	4	5	5	6	6	6	7	7	8	8
Solid Masonry, Block, or Brick															
Plastered or plain	14	16	18	20	22	25	27	29	32	34	36	38	40	43	45
Furred, no insulation	9	10	12	13	14	16	17	19	20	22	23	25	26	28	29
Furred, with R-5 insulation	4	5	5	6	6	7	8	8	9	10	10	11	12	12	13
Basement or Crawl Space															
Above grade, no insulation	15	18	20	23	26	28	31	33	36	38	41	43	46	48	51
R-3.57 insulation (molded bead bd.)	5	6	7	8	9	10	11	12	13	14	14	15	16	17	18
R-5 insulation (ext. polystrene bd.)	4	5	6	6	7	8	9	9	10	11	12	12	13	14	14

* in °F
§ in Btu/hr per sq ft (Factors include heat loss for transmission and infiltration.)
Note: R values on this chart refer to thermal resistance value.

continued

continued

HEAT TRANSFER FACTORS

Item	Design Temperature Difference*														
	30	35	40	45	50	55	60	65	70	75	80	85	90	95	100
Basement or Crawl Space															
Wall of crawl space used as supply plenum, R-3.57 insulation	11	12	13	14	15	16	16	17	18	19	20	21	22	23	24
Wall of crawl space used as supply plenum, R-5 insulation	9	10	10	11	12	12	13	14	15	15	16	17	17	18	19
Below grade wall	2	2	2	3	3	3	4	4	4	5	5	5	5	6	6
Ceilings and Roofs§ — Ceiling Under Unconditioned Space or Vented Roof															
No insulation	18	21	24	27	30	33	36	39	42	45	48	51	54	57	60
R-11 insulation (3″ – 3¼″)	2	3	3	4	4	4	5	5	6	6	6	7	7	8	8
R-19 insulation (5¼″ – 6½″)	2	2	2	2	2	3	3	3	4	4	4	4	4	5	5
R-22 insulation (6″ – 7″)	1	1	2	2	2	2	2	3	3	3	3	3	4	4	4
R-30 insulation	1	1	1	2	2	2	2	2	2	3	3	3	3	3	3
R-38 insulation	1	1	1	1	1	1	2	2	2	2	2	2	2	3	3
R-44 insulation	1	1	1	1	1	1	1	2	2	2	2	2	2	2	2
Roof on Exposed Beams or Rafters															
Roofing on 1½ ″ wood decking no ins.	10	12	14	15	17	19	20	22	24	26	27	29	31	32	34
Roofing on 1½ ″ wood decking 1″ insulation between roofing and decking	5	6	7	8	8	9	10	11	12	13	13	14	15	16	17
Roofing on 1½ ″ wood decking 1½″ insulation between roofing and decking	4	5	5	6	7	7	8	9	10	10	11	12	12	13	14
Roofing on 2″ wood plank	6	7	8	9	10	11	12	14	15	16	17	18	19	20	21
Roofing on 3″ wood plank	5	5	6	7	8	8	9	10	11	11	12	13	14	14	15
Roofing on 1½ ″ fiberboard decking	6	7	8	9	10	10	11	12	13	14	15	16	17	18	19
Roofing on 2″ fiberboard decking	4	5	6	7	8	8	9	10	11	11	12	13	14	14	15
Roofing on 3″ firberboard decking	3	4	4	5	6	6	7	7	8	8	9	9	10	10	11
Roofing-Ceiling Combination															
No insulation	9	11	12	14	16	17	19	20	22	23	25	26	28	29	31
R-11 insulation	2	2	3	3	4	4	4	5	5	5	6	6	6	7	7
R-19 insulation	1	2	2	2	2	2	3	3	3	3	4	4	4	4	5
R-22 insulation (6″ – 7″)	1	1	2	2	2	2	2	3	3	3	3	3	4	4	4
Floors§ — Floors Over Unconditioned Space															
Over unconditioned room	4	5	6	6	7	8	8	9	10	11	11	12	13	13	14
No insulation	8	10	11	13	14	15	17	18	20	21	22	24	25	27	28
R-7 insulation (2″ – 2¾″)	3	3	4	4	5	5	6	6	7	7	8	8	9	9	9
R-11 insulation (3″ – 3½″)	2	2	3	3	4	4	4	5	5	5	6	6	6	7	7
R-19 insulation (5¼″ – 6½″)	1	2	2	2	2	2	3	3	3	3	4	4	4	4	4
Floor of Room Over Heated Crawl Space															
Less than 18″ below grade	35	40	40	45	45	50	50	55	55	60	60	65	65	70	75
18″ or more below grade	15	20	20	25	25	30	30	35	35	40	40	45	45	50	50

* in °F
§ in Btu/hr per sq ft (Factors include heat loss for transmission and infiltration.)
Note: R values on this chart refer to thermal resistance value.

•*Air Conditioning Contractors of America*

COOLING HEAT TRANSFER FACTORS

Type of Construction	Cooling Factor*		
	15°	20°	25°
Walls			
Wood Frame with Sheeting, Siding, and Veneer or Other Finish			
No insulation, ½″ gypsum board	5.0	6.4	7.8
R-11 cavity insulation + ½″ gypsum board	1.7	2.1	2.6
R-13 cavity insulation + ½″ gypsum board	1.5	1.9	2.3
R-13 cavity insulation + ¾″ bead board (R-2.7)	1.3	1.7	2.0
R-19 cavity insulation + ½″ gypsum board	1.1	1.4	1.7
R-19 cavity insulation + ¾″ extruded poly	0.9	1.2	1.4
Masonry			
Above grade – No insulation	5.8	8.3	10.9
Above grade + R-5	1.6	2.3	3.1
Above grade + R-11	0.9	1.3	1.6
Below grade – No insulation	0.0	0.0	0.0
Below grade + R-5	0.0	0.0	0.0
Below grade + R-11	0.0	0.0	0.0
Ceilings			
No insulation	17.0	19.2	21.4
2″ – 2½″ insulation R-7	4.4	4.9	5.5
3″ – 3½″ insulation R-11	3.2	3.7	4.1
5¼″ – 6½″ insulation R-19	2.1	2.3	2.6
6″ – 7″ insulation R-22	1.9	2.1	2.4
10″ – 12″ insulation R-38	1.0	1.1	1.3
12″ – 13″ insulation R-44	0.9	1.0	1.1
Cathedral type (roof/ceiling combination)			
No insulation	11.2	12.6	14.1
R-11	2.8	3.2	3.5
R-19	1.9	2.2	2.4
R-22	1.8	2.0	2.2
Floors			
Over Unconditioned Space			
Over basement or enclosed crawl space (not vented)	0.0	0.0	0.0
Over vented space or garage	3.9	5.8	7.7
Over vented space or garage + R-11 insulation	0.8	1.3	1.7
Over vented space or garage + R-19 insulation	0.5	0.8	1.1
Basement Concrete Slab Floor Unheated			
No edge insulation	0.0	0.0	0.0
1″ edge insulation R-5	0.0	0.0	0.0
2″ edge insulation R-9	0.0	0.0	0.0
Basement Concrete Slab Floor Duct in Slab			
No edge insulation	0.0	0.0	0.0
1″ edge insulation R-5	0.0	0.0	0.0
2″ edge insulation R-9	0.0	0.0	0.0

* in °F

•*Air Conditioning Contractors of America*

Note: R values on this chart refer to thermal resistance value.

COOLING HEAT TRANSFER FACTORS — WINDOWS AND DOORS*									
	Single Glass			Double Glass			Triple Glass		
	Temperature Difference			Temperature Difference			Temperature Difference		
Exposure	15°	20°	25°	15°	20°	25°	15°	20°	25°
N	18	22	26	14	16	18	11	12	13
NE & NW	37	41	46	31	33	35	26	27	28
E & W	52	56	60	44	46	48	38	39	40
SE & SW	45	49	53	39	41	43	33	34	35
S	28	32	36	23	25	27	19	20	21
Wood	8.6	10.9	13.2	8.6	10.9	13.2	8.6	10.9	13.2
Metal	3.5	4.5	5.4	3.5	4.5	5.4	3.5	4.5	5.4

* Inside shading by venetian blinds or draperies.   •*Air Conditioning Contractors of America*

DUCT HEAT LOSS MULTIPLIERS

Duct Location and Insulation Value Exposed to Outdoor Ambient Air — Attic, Garage, Exterior Wall, Open Crawl Space	Duct Loss Multipliers	
	Winter Design Below 15° F	Winter Design Above 15° F
None	1.30	1.25
R-2	1.20	1.15
R-4	1.15	1.10
R-6	1.10	1.05
Enclosed in Unheated Space — Vented or Unvented Crawl Space or Basement		
None	1.20	1.15
R-2	1.15	1.10
R-4	1.10	1.05
R-6	1.05	1.00
Duct Buried in or Under Concrete Slab — Edge Insulation		
None	1.25	1.20
R value = 3 to 4	1.15	1.10
R value = 5 to 7	1.10	1.05
R value = 7 to 9	1.05	1.00

•*Air Conditioning Contractors of America*

DUCT HEAT GAIN MULTIPLIERS

Duct Location and Insulation Value Exposed to Outdoor Ambient Air — Attic, Garage, Exterior Wall, Open Crawl Space	Duct Gain Multiplier
None	1.30
R-2	1.20
R-4	1.15
R-6	1.10
Enclosed in Unconditioned Space — Vented or Unvented Crawl Space or Basement	
None	1.15
R-2	1.10
R-4	1.05
R-6	1.00
Duct Buried in or Under Concrete Slab — Edge Insulation	
None	1.10
R value = 3 to 4	1.05
R value = 5 to 7	1.00
R value = 7 to 9	1.00

•*Air Conditioning Contractors of America*

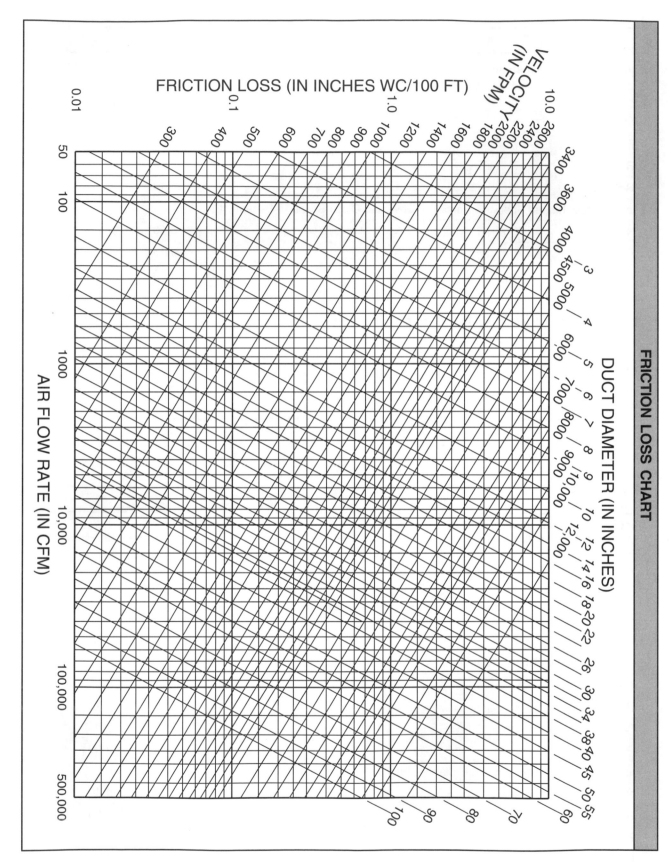

FRICTION LOSS CHART

ROUND-TO-RECTANGULAR DUCT CONVERSION CHART

Lgth. Adj.	Length of One Side of Rectangular Duct*																
	4.0	4.5	5.0	5.5	6.0	6.5	7.0	7.5	8.0	9.0	10.0	11.0	12.0	13.0	14.0	15.0	16.0
3.0	3.8	4.0	4.2	4.4	4.6	4.7	4.9	5.1	5.2	5.5	5.7	6.0	6.2	6.4	6.6	6.8	7.0
3.5	4.1	4.3	4.6	4.8	5.0	5.2	5.3	5.5	5.7	6.0	6.3	6.5	6.8	7.0	7.2	7.5	7.7
4.0	4.4	4.6	4.9	5.1	5.3	5.5	5.7	5.9	6.1	6.4	6.7	7.0	7.3	7.6	7.8	8.0	8.3
4.5	4.6	4.9	5.2	5.4	5.7	5.9	6.1	6.3	6.5	6.9	7.2	7.5	7.8	8.1	8.4	8.6	8.8
5.0	4.9	5.2	5.5	5.7	6.0	6.2	6.4	6.7	6.9	7.3	7.6	8.0	8.3	8.6	8.9	9.1	9.4
5.5	5.1	5.4	5.7	6.0	6.3	6.5	6.8	7.0	7.2	7.6	8.0	8.4	8.7	9.0	9.3	9.6	9.9

Lgth. Adj.	Length of One Side of Rectangular Duct*																				Lgth. Adj.
	6	7	8	9	10	11	12	13	14	15	16	17	18	19	20	22	24	26	28	30	
6	6.6																				6
7	7.1	7.7																			7
8	7.6	8.2	8.7																		8
9	8.0	8.7	9.3	9.8																	9
10	8.4	9.1	9.8	10.4	10.9																10
11	8.8	9.5	10.2	10.9	11.5	12.0															11
12	9.1	9.9	10.7	11.3	12.0	12.6	13.1														12
13	9.5	10.3	11.1	11.8	12.4	13.1	13.7	14.2													13
14	9.8	10.7	11.5	12.2	12.9	13.5	14.2	14.7	15.3												14
15	10.1	11.0	11.8	12.6	13.3	14.0	14.6	15.3	15.8	16.4											15
16	10.4	11.3	12.2	13.0	13.7	14.4	15.1	15.7	16.4	16.9	17.5										16
17	10.7	11.6	12.5	13.4	14.1	14.9	15.6	16.2	16.8	17.4	18.0	18.6									17
18	11.0	11.9	12.9	13.7	14.5	15.3	16.0	16.7	17.3	17.9	18.5	19.1	19.7								18
19	11.2	12.2	13.2	14.1	14.9	15.7	16.4	17.1	17.8	18.4	19.0	19.6	20.2	20.8							19
20	11.5	12.5	13.5	14.4	15.2	16.0	16.8	17.5	18.2	18.9	19.5	20.1	20.7	21.3	21.9						20
22	12.0	13.0	14.1	15.0	15.9	16.8	17.6	18.3	19.1	19.8	20.4	21.1	21.7	22.3	22.9	24.0					22
24	12.4	13.5	14.6	15.6	16.5	17.4	18.3	19.1	19.9	20.6	21.3	22.0	22.7	23.3	23.9	25.1	26.2				24
26	12.8	14.0	15.1	16.2	17.1	18.1	19.0	19.8	20.6	21.4	22.1	22.9	23.5	24.2	24.9	26.1	27.3	28.4			26
28	13.2	14.5	15.6	16.7	17.7	18.7	19.6	20.5	21.3	22.1	22.9	23.7	24.4	25.1	25.8	27.1	28.3	29.5	30.6		28
30	13.6	14.9	16.1	17.2	18.3	19.3	20.2	21.1	22.0	22.9	23.7	24.4	25.2	25.9	26.6	28.0	29.3	30.5	31.7	32.8	30
32	14.0	15.3	16.5	17.7	18.8	19.8	20.8	21.8	22.7	23.5	24.4	25.2	26.0	26.7	27.5	28.9	30.2	31.5	32.7	33.9	32
34	14.4	15.7	17.0	18.2	19.3	20.4	21.4	22.4	23.3	24.2	25.1	25.9	26.7	27.5	28.3	29.7	31.0	32.4	33.7	34.9	34
36	14.7	16.1	17.4	18.6	19.8	20.9	21.9	22.9	23.9	24.8	25.7	26.6	27.4	28.2	29.0	30.5	32.0	33.3	34.6	35.9	36
38	15.0	16.5	17.8	19.0	20.2	21.4	22.4	23.5	24.5	25.4	26.4	27.2	28.1	28.9	29.8	31.3	32.8	34.2	35.6	36.8	38
40	15.3	16.8	18.2	19.5	20.7	21.8	22.9	24.0	25.0	26.0	27.0	27.9	28.8	29.6	30.5	32.1	33.6	35.1	36.4	37.8	40
42	15.6	17.1	18.5	19.9	21.1	22.3	23.4	24.5	25.6	26.6	27.6	28.5	29.4	30.3	31.2	32.8	34.4	35.9	37.3	38.7	42
44	15.9	17.5	18.9	20.3	21.5	22.7	23.9	25.0	26.1	27.1	28.1	29.1	30.0	30.9	31.8	33.5	35.1	36.7	38.1	39.5	44
46	16.2	17.8	19.3	20.6	21.9	23.2	24.4	25.5	26.6	27.7	28.7	29.7	30.6	31.6	32.5	34.2	35.9	37.4	38.9	40.4	46
48	16.5	18.1	19.6	21.0	22.3	23.6	24.8	26.0	27.1	28.2	29.2	30.2	31.2	32.2	33.1	34.9	36.6	38.2	39.7	41.2	48
50	16.8	18.4	19.9	21.4	22.7	24.0	25.2	26.4	27.6	28.7	29.8	30.8	31.8	32.8	33.7	35.5	37.2	38.9	40.5	42.0	50
52	17.1	18.7	20.2	21.7	23.1	24.4	25.7	26.9	28.0	29.2	30.3	31.3	32.3	33.3	34.3	36.2	37.9	39.6	41.2	42.8	52
54	17.3	19.0	20.6	22.0	23.5	24.8	26.1	27.3	28.5	29.7	30.8	31.8	32.9	33.9	34.9	36.8	38.6	40.3	41.9	43.5	54
56	17.6	19.3	20.9	22.4	23.8	25.2	26.5	27.7	28.9	30.1	31.2	32.3	33.4	34.4	35.4	37.4	39.2	41.0	42.7	44.3	56
58	17.8	19.5	21.2	22.7	24.2	25.5	26.9	28.2	29.4	30.6	31.7	32.8	33.9	35.0	36.0	38.0	39.8	41.6	43.3	45.0	58
60	18.1	19.8	21.5	23.0	24.5	25.9	27.3	28.6	29.8	31.0	32.2	33.3	34.4	35.5	36.5	38.5	40.4	42.3	44.0	45.7	60

* in inches

ASHRAE 1989 Handbook — Fundamentals

HEATING TROUBLESHOOTING CHART

Complaint	Symptom	Cause	Description	Section, Component, or Part
No heating	Cool air out of supply registers	Controls	Burner not operating	Transformer Thermostat Heating relay Limit control
		Fuel supply	Burner not operating	Tanks empty Valves shut off
		Combustion controls	Burner not operating	Thermocouple Pilot safety valve Fuel valve
	No air out of registers	Power controls	No electricity	Disconnect switches Fuses blown Circuit breaker open
		Circulation system	Blower not operating	Blower motor Blower relay Blower belts
Insufficient heat	Air out of supply registers too cool	Controls	Burner not operating	Control transformer Thermostat Heating relay Limit control
		Combustion controls		Thermocouple Pilot safety valve Fuel valve
			Burner operating intermittently	Thermostat Limit control
Too much heat	Building overheats	Controls	Temperature exceeds setpoint	Thermostat Gas valve Heating relay
Fluctuating heat	Temperature changes in building	Controls	Burner cycles OFF and ON	Limit control Blower relay
		Circulation system	Burner cycles OFF and ON	Blower motor Blower relay
		Ductwork	Air supply	Sizing Dampers
System noisy	Equipment noisy	Controls	Humming	Transformer Relays
		Blower	Squeaks or rattles	Blower wheel Motor
		Burner	Whistle	Air adjustment
		Distribution	Pops and rattles	Ductwork Registers
Odor	Smell of fuel around unit	Burner connections	Smell fuel	Fuel line Fuel pump Fuel valves Filter
	Smell flue gas	Unit	Smell flue gas at unit	Draft diverter Barometric damper Flue
			Smell flue gas at supply registers	Heat exchanger
Cost of operation	Unit	Distribution system	Unit runs too long	Dirty filters Ductwork undersized Registers too small

AIR CONDITIONING TROUBLESHOOTING CHART

Complaint	Symptom	Cause	Description	Section, Component, or Part
No cooling	No air out of registers	Electric power	No power	Disconnect Fuses Circuit breakers
		Controls	Control voltage	Transformer Thermostat Limit control
		Blower	Not operating	Blower wheel Motor
		Evaporator	Obstructed	Coil
		Air distribution	Dampers	Closed
	Warm air out of registers	System	Equipment	Undersized
		Air distribution	Ductwork	Undersized
			Blower	Wheel Motor
			Dampers	Closed
		Controls	Calibration Setting	Thermostat Pressure switches
		Compressor	Not operating	Motor Valves
		Condenser	Insufficient air	Blower Motor Dirty coil
		Evaporator	Insufficient air	Dirty coil Dirty filter
Not enough cooling	Air supply too warm	Compressor	Cycling	Pressure switches Overloads
		Condenser	Not enough air	Blower wheel Motor Dirty coil Noncondensable gases
		Evaporator	Not enough air	Blower wheel Motor Dirty coil
			Expansion device	Adjustment Faulty
		Air distribution	Ductwork	Undersized Not insulated
			Supply registers	Dampers closed Too small
			Return air grills	Location Size
			Filters	Dirty
		Equipment	Runs all the time	Sized too small
Too cool	Air is cold out of registers	Controls	Calibration Anticipation	Thermostat
	Cold drafts	Air distribution	Air pattern	Supply registers location System balance

continued

continued

AIR CONDITIONING TROUBLESHOOTING CHART				
Complaint	**Symptom**	**Cause**	**Description**	**Section, Component, or Part**
Noisy system	Noisy units	Compressor	Knocking sounds	Mechanical problems
			Liquid slugs	Expansion device Refrigerant charge
		Condenser	Blower	Wheel Motor
		Evaporator	Blower	Wheel Motor
		Refrigerant lines	Vibrations	Lines too rigidly attached to supports
	Noise out of registers	Ductwork	Popping and ticking	Ducts need bracing
Too expensive to operate	Units runs constantly	Equipment	Runs constantly	Undersized
		Controls	Out of adjustment	Thermostat
			Welded contacts	Thermostat Compressor relay
		Air circulation	Motors run constantly	Shorted electrical circuits
			Filters	Dirty
		Evaporator	Too little air	Dirty coil Blower not operating
		Condenser	Too little air	Dirty coil Blower not operating
	Needs frequent service	System	Improperly designed	Undersized
			Improperly installed	Poor workmanship Inadequate maintenance

HVAC SYMBOLS

Equipment Symbols	Ductwork	Heating Piping
EXPOSED RADIATOR	DUCT (1ST FIGURE, WIDTH; 2ND FIGURE, DEPTH) — 12 × 20	HIGH PRESSURE STEAM — HPS —
RECESSED RADIATOR	DIRECTION OF FLOW	MEDIUM PRESSURE STEAM — MPS —
FLUSH ENCLOSED RADIATOR	FLEXIBLE CONNECTION	LOW PRESSURE STEAM — LPS —
PROJECTING ENCLOSED RADIATOR	DUCTWORK WITH ACOUSTICAL LINING	HIGH PRESSURE RETURN — HPR —
UNIT HEATER (PROPELLER)—PLAN	FIRE DAMPER WITH ACCESS DOOR FD AD	MEDIUM PRESSURE RETURN — MPR —
UNIT HEATER (CENTRIFUGAL)—PLAN	MANUAL VOLUME DAMPER VD	LOW PRESSURE RETURN — LPR —
UNIT VENTILATOR—PLAN	AUTOMATIC VOLUME DAMPER	BOILER BLOW OFF — BD —
STEAM	EXHAUST, RETURN OR OUTSIDE AIR DUCT—SECTION 20 × 12	CONDENSATE OR VACUUM PUMP DISCHARGE — VPD —
DUPLEX STRAINER	SUPPLY DUCT—SECTION 20 × 12	FEEDWATER PUMP DISCHARGE — PPD —
PRESSURE REDUCING VALVE	CEILING DIFFUSER SUPPLY OUTLET 20″ DIA. CD 1000 CFM	MAKE UP WATER — MU —
AIR LINE VALVE	CEILING DIFFUSER SUPPLY OUTLET 20 × 12 CD 700 CFM	AIR RELIEF LINE — V —
STRAINER		FUEL OIL SUCTION — FOS —
THERMOMETER	LINEAR DIFFUSER 96 ×6-LD 400 CFM	FUEL OIL RETURN — FOR —
PRESSURE GAUGE AND COCK	FLOOR REGISTER 20 × 12 FR 700 CFM	FUEL OIL VENT — FOV —
RELIEF VALVE	TURNING VANES	COMPRESSED AIR — A —
AUTOMATIC 3-WAY VALVE	FAN AND MOTOR WITH BELT GUARD	HOT WATER HEATING SUPPLY — HW —
AUTOMATIC 2-WAY VALVE		HOT WATER HEATING RETURN — HWR —
SOLENOID VALVE	LOUVER OPENING 20 × 12-L 700 CFM	

Air Conditioning Piping

REFRIGERANT LIQUID	— RL —
REFRIGERANT DISCHARGE	— RD —
REFRIGERANT SUCTION	— RS —
CONDENSER WATER SUPPLY	— CWS —
CONDENSER WATER RETURN	— CWR —
CHILLED WATER SUPPLY	— CHWS —
CHILLED WATER RETURN	— CHWR —
MAKE UP WATER	— MU —
HUMIDIFICATION LINE	— H —
DRAIN	— D —

REFRIGERATION SYMBOLS

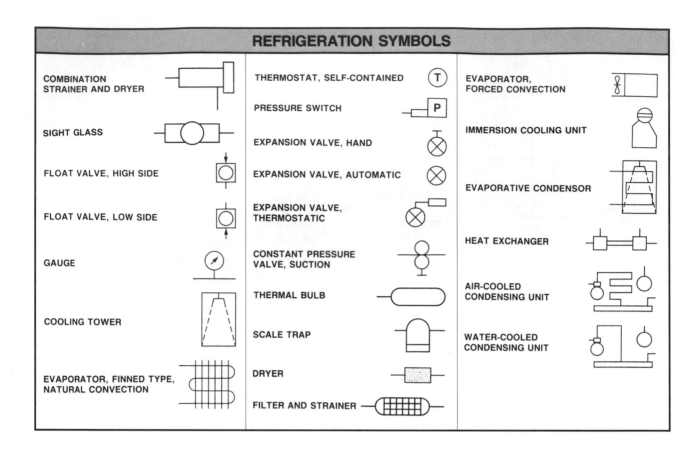

COMBINATION STRAINER AND DRYER	THERMOSTAT, SELF-CONTAINED	EVAPORATOR, FORCED CONVECTION
SIGHT GLASS	PRESSURE SWITCH	IMMERSION COOLING UNIT
FLOAT VALVE, HIGH SIDE	EXPANSION VALVE, HAND	
FLOAT VALVE, LOW SIDE	EXPANSION VALVE, AUTOMATIC	EVAPORATIVE CONDENSOR
GAUGE	EXPANSION VALVE, THERMOSTATIC	HEAT EXCHANGER
COOLING TOWER	CONSTANT PRESSURE VALVE, SUCTION	AIR-COOLED CONDENSING UNIT
	THERMAL BULB	WATER-COOLED CONDENSING UNIT
	SCALE TRAP	
EVAPORATOR, FINNED TYPE, NATURAL CONVECTION	DRYER	
	FILTER AND STRAINER	

Glossary

Note: Terms in this glossary are defined as they relate to heating, ventilating, and air conditioning equipment.

A

absolute pressure: (psia) Pressure above a perfect vacuum. Absolute pressure is the sum of gauge pressure plus atmospheric pressure. See *vacuum, gauge pressure,* and *atmospheric pressure.*

absolute zero: Theoretical condition at which no heat is present. Equal to 0°R, −460°F, 0°K, and −273°C.

absorbent: Fluid that has a strong attraction for another fluid. See *fluid.*

absorber: Component of an absorption refrigeration system in which a refrigerant is absorbed by the absorbent. See *absorption refrigeration system, refrigerant,* and *absorbent.*

absorption refrigeration system: Refrigeration system that uses the absorption of one chemical by another chemical and heat transfer to produce a refrigeration effect. See *refrigeration effect* and *heat transfer.*

AC electricity: See *alternating current.*

active solar heating system: System that uses mechanical components for bringing radiant heat from the sun into and storing heat in a building. See *solar heat.*

adiabatic change: Change in the pressure and temperature of a substance in a closed system that occurs without heat transfer.

air change factor: Value that represents the number of times per hour that the air in a building is completely replaced by outdoor air.

air circulation: See *circulation.*

air conditioner: Component in a forced air conditioning system that cools the air.

air conditioning troubleshooting chart: Flow chart that identifies malfunctioning parts of an operating air conditioning system with series of steps. See *malfunctioning part.*

air-cooled condenser: Condenser that uses air as the condensing medium. Heat is transferred from a refrigerant to the air. See *condenser, condensing medium,* and *refrigerant.*

air distribution system: System of ductwork in a forced-air heating or cooling system. Used to distribute conditioned air to building spaces.

air flow control: Control

air-cooled condenser

of the circulation of air through building spaces and the introduction of ventilation air into a building. See *circulation* and *ventilation.*

air-fuel mixture: Mixture of air and fuel that is necessary for efficient combustion. See *combustion.*

Airtrol®: Mechanical device that vents air from the water in a boiler and expansion tank. See *boiler* and *expansion tank.*

air velocity: Speed at which air moves from one point to another.

aldehyde: Chemical compound containing hydrogen, carbon, and oxygen that forms during incomplete combustion. See *incomplete combustion.*

Airtrol®

alternate fuel: Combustible material other than coal, oil, and gas.

alternate heat source: Natural source of heat, such as geothermal heat, that does not require combustion or electricity to produce heat. See *geothermal heat.*

alternating current: (AC) Electric current that continuously changes direction. See *current.*

ambient air: Unconditioned atmospheric air. See *atmospheric air.*

ammeter: Device used to measure current flow in an electric device. See *current.*

anemometer: Device that measures air velocity or air flow rate. See *air velocity.*

anthracite coal: Hard, high luster coal that yields little volatile matter as it burns. See *volatile matter.*

aquastat: Temperature-actuated electric switch used to limit the temperature of boiler water.

area: See *surface area.*

area of influence: Area from the front of a register to a point where the air velocity drops below 50 fpm. See *register* and *air velocity.*

aspect ratio: Ratio between the height and width of a rectangular duct.

atmospheric air: Mixture of dry air, moisture, and particles such as smoke and dust. See *dry air, moisture,* and *smoke.*

atmospheric burner: Burner that uses ambient air supplied at normal atmospheric pressure for combustion air. See *ambient air, atmospheric pressure,* and *combustion air.*

atmospheric pressure: Force exerted by the weight of the atmosphere on the earth's surface. Normal atmospheric pressure, which is a standard condition, equals 29.92″ Hg or 14.7 psi. See *standard conditions.*

atomization: Process of breaking a liquid into small droplets.

automatic expansion valve: Pressure-regulating valve that is opened and closed by the pressure in the refrigerant line ahead of the valve. Maintains a constant pressure in the evaporator of a refrigeration system. See *evaporator* and *refrigeration system.*

automatic flow control valve: Check valve in a hydronic heating system that opens or closes automatically by pressure from a circulating pump. See *check valve, hydronic heating system,* and *circulating pump.*

axial flow blower: Blower that contains a blower wheel, which works like a turbine wheel. See *blower* and *blower wheel.*

B

back pressure: Pressure produced by the ignition of an air-fuel mixture against the normal pressure of gas flow. See *air-fuel mixture.*

bag filter: High-efficiency filter made of filter paper and shaped like a large paper bag.

balancing: Adjusting the resistance to air or water flow through the distribution system of a heating or air conditioning system. Assures the proper amount of air or water flow through each terminal device for the amount of heat required at each terminal device. See *distribution system* and *terminal device.*

bag filter

balancing damper: Air flow damper located at a branch takeoff from the trunk duct in a forced-air distribution system. See *damper* and *trunk.*

ballast: Combination electric device that contains a transformer and an igniter, which generate heat. Used to start fluorescent lights. See *transformer.*

bare-tube coil: Coil of bare, copper tubes through which refrigerant flows. See *refrigerant* and *evaporator.*

barometric damper: Metal plate positioned in an opening in the flue so that atmospheric pressure can control the air flow through the combustion chamber and flue. Installed in the flue above a drum heat exchanger. See *atmospheric pressure* and *drum heat exchanger.*

baseboard convector: Convection heater enclosed in a low cabinet that fits along a baseboard. A finned tube transfers heat. See *finned tube.*

baseboard radiator: Baseboard-size section of vertical or horizontal tubes that is connected with headers at each end. Hot water from a boiler flows through the tubes. Has a lower heat output per lineal foot than a standing radiator. See *radiator, boiler,* and *standing radiator.*

bellows element: Accordion-like device that converts pressure variation into mechanical movement.

belt drive system: Motor-to-wheel connection that has a blower motor mounted on the scroll. The blower motor is connected to the blower wheel through a belt and sheave arrangement. See *scroll, blower wheel,* and *sheave.*

bidirectional motor: Motor that can turn in the forward and reverse direction.

bimetal element: Temperature sensor that consists of two different kinds of metal that are bonded together into a strip or coil.

bimetal overload relay: Relay that contains a set of contacts that are actuated by a bimetal element. See *relay, contacts,* and *bimetal element.*

bituminous coal: Coal, softer than anthracite coal, that yields a large amount of volatile matter when it burns. See *anthracite coal.*

blowdown: Rapid draining of boiler water that removes the minerals that lead to scale formation. See *scale.*

blowdown valve: Quick-opening manual valve located at the lowest part of a boiler. Used during blowdown. See *blowdown.*

blower: Mechanical device that consists of moving blades or vanes that force air through a venturi. See *venturi.*

blower control: Temperature-actuated switch that controls the blower motor in a furnace.

blower drive: Connection from an electric motor to a blower wheel. See *blower wheel.*

blower wheel: Sheet metal cylinder with curved vanes along its perimeter.

blue flame: Flame produced when combustion air mixes with fuel before the mixture reaches the burner face.

boiler: Pressure vessel that safely and efficiently transfers heat to water. See *pressure vessel.*

blower wheel

boiler combustion safety control: Control that detects flame. Prevents fuel from being sent to an unignited burner. See *pilot safety control, flame rod,* and *flame surveillance control.*

boiler control: Control that ensures that a boiler operates safely and that there is no danger to personnel or equipment. See *boiler operating control, boiler safety control,* and *combustion safety control.*

boiler gross unit output: See *gross unit output.*

boiler input: See *input rating.*

boiler net unit output: See *net unit output.*

boiler operating control: Control that automatically and safely energizes boiler burner(s) or resistance heating elements to maintain the temperature of the water at the setpoint temperature. See *burner, resistance heating element,* and *setpoint temperature.*

boiler output: See *output rating.*

boiler safety control: Control that monitors boiler operation and prevents excessive temperature and pressure by correcting a problem or by shutting the boiler down. See *boiler.*

boiling point: Temperature at which a substance changes from liquid state to gas state.

Bourdon tube: Circular stainless steel or bronze tube inside a mechanical pressure gauge that is flattened to make it more flexible. Tube moves according to the pressure in it, which changes the reading on the face of the gauge. See *pressure.*

Bourdon tube

British thermal unit: See *Btu.*

Btu: (British thermal unit) Unit of measure that expresses the quantity of heat required to change the temperature of 1 lb of water 1°F.

building component: Main part of a building structure such as the outside walls.

building data: Information that includes the name of each room, running feet of exposed wall, dimensions, ceiling height, and exposure of each room. Used for load calculations.

burner: Heat-producing element of a combustion furnace or boiler.

burner face: Point at which a flame is established. See *burner.*

burner port: Inlet for secondary air in a gas fuel-fired atmospheric burner. See *secondary air* and *gas fuel-fired atmospheric burner.*

burner tube: Tube used to mix fuel and air for combustion that has an opening on one end and burner ports located along the top. See *burner port.*

burner vestibule: Area where the burner(s) and burner controls are located in a furnace or boiler. See *furnace* and *boiler.*

burner tube

butane: Gas fuel containing 95% to 99% butane gas (C_4H_{10}) and up to 5% butylene (C_4H_8). A by-product of the oil refining process.

bypass circuit: Refrigerant line that contains a check valve on each side of an expansion device. Allows refrigerant flow when the refrigerant flow is opposite to the flow direction of the expansion device. See *check valve, expansion device,* and *refrigerant.*

C

cabinet: Sheet metal enclosure that completely covers and provides support for the components of a furnace or air conditioner. See *furnace* and *air conditioner.*

cabinet convector: Convection heating unit that has a sheet metal cabinet surrounding the hot water coil or tubes. See *convection heating.*

cabinet heater: Forced convection heater that has a blower, hot water coils, filter, and controls in one cabinet. Blower blows heated air from the hot water coil. See *forced convection heating* and *blower.*

cad cell: Light-sensitive device that detects flame. The cell is made of cadmium sulfide, which conducts electricity depending on the intensity of the light that strikes it. Used in flame surveillance controls. See *flame surveillance control.*

cad cell

calorie: (cal) Unit of measure that expresses the quantity of heat required to change the temperature of 1 g of water 1°C.

capillary tube: Long, thin tube that resists fluid flow, which causes a pressure decrease. Used as an expansion device in mechanical compression refrigeration systems. See *expansion device* and *mechanical compression refrigeration.*

cartridge fuse: Fibrous or plastic tube that contains a fuse wire that carries a specific amount of current. See *fuse.*

cascade system: Compression system that uses one refrigeration system to cool the refrigerant in another system. See *refrigerant.*

Celsius: (C) Scale used to express temperature in the metric system of measurements. Units of measure are degrees Celsius (°C).

central forced-air heating system: Heating system that uses a centrally located furnace to produce heat for a building.

central processing unit: (CPU) Control center of the computer system. Receives information through the input devices. See *input device.*

centrifugal blower: Blower that consists of a scroll, blower wheel, shaft, and inlet vanes. See *scroll, blower wheel,* and *inlet vanes.*

centrifugal compressor: Compressor that uses centrifugal force to move refrigerant vapor. See *compressor, centrifugal force,* and *refrigerant.*

centrifugal force: Force that pulls a body outward when it is spinning around a center.

centrifugal pump: Pump that has a rotating impeller wheel inside a cast iron or steel housing. See *impeller wheel.*

C factor: See *conductance factor.*

change of state: Process that occurs when a substance changes from one physical state to another, such as from ice to water and from water to steam.

centrifugal pump

check valve: Valve that allows flow in only one direction.

chiller: Hydronic air conditioning system that cools water, which cools air.

circuit breaker: Current-sensing device that is designed to open a circuit automatically if an overcurrent condition occurs.

circulating pump: Pump that moves water from a boiler and through the piping system and terminal devices of a hydronic heating system. See *boiler* and *terminal device.*

circulation: Movement of air or water.

clam shell heat exchanger: Heat exchanger that has multiple clam-shaped sections. See *heat exchanger.*

coal: Solid, black fuel that formed when organic material hardened in the earth over millions of years.

coefficient of performance: (COP) Theoretical operating efficiency of a refrigeration system. Ratio of the cooling effect achieved by a refrigeration system to energy used to achieve cooling effect.

clam shell heat exchanger

columnar form: Blank table that is divided into columns and rows by vertical and horizontal lines. Used for calculating heating and cooling loads. See *heating load* and *cooling load.*

combination system: Air conditioning system that contains the components for cooling and heating in one sheet metal cabinet.

combustion: Chemical reaction that occurs when oxygen reacts with the elements hydrogen (H) and carbon (C) in a fuel at an ignition temperature.

combustion air: Air provided at a burner for proper combustion of fuel. See *combustion.*

combustion air blower: Blower used to provide combustion air at a positive pressure at the burner face. See *combustion air* and *burner face.*

combustion chamber: Area in a heating unit where combustion takes place. See *combustion.*

combustion efficiency: Evaluation of how well chemical energy in a fuel is converted to thermal energy during combustion. Combustion efficiency is the efficiency at which a furnace burns fuel. See *combustion.*

combustion heat: Heat produced during combustion as the elements in a fuel react with oxygen at the ignition temperature. See *combustion, ignition temperature,* and *fuel.*

combustion safety control: Safety control that monitors firing for ignition and flame during a call for heat and shuts down the burner(s) if a malfunction occurs. See *safety control* and *burner.*

comfort: Condition that occurs when people cannot sense a difference between themselves and the surrounding air.

commercial building: Building that involves a large number of people.

complete combustion: Ideal combustion that occurs when enough oxygen is supplied by combustion air to combine with all of the elements in a fuel. Carbon dioxide and water vapor are produced. See *combustion.*

component: Major piece of equipment that makes up a furnace or air conditioner.

compression ratio: Ratio of the pressure on the high-pressure side (evaporator pressure) to the pressure on the low-pressure side (condenser pressure) of a refrigeration system.

compression stroke: Stroke that occurs after a piston completes its suction stroke and begins to move up in the cylinder toward the cylinder head. See *stroke* and *cylinder head.*

compressive force: Force that squeezes objects, such as air molecules, together.

compressor: Mechanical device that compresses refrigerant or other fluid. See *refrigerant.*

compressor performance: Cooling capacity produced by the amount of refrigerant the compressor moves through the refrigeration system. See *refrigerant.*

compressor suction pressure: Pressure produced by the compressor when refrigerant is drawn into the compressor. See *compressor* and *refrigerant.*

condensate: Liquid formed when a vapor cools below its dew point. See *dew point.*

condensation: Process that occurs when liquid (condensate) forms when moisture or other vapor cools below its dew point. See *condensate* and *dew point.*

condenser: Heat exchanger that removes heat from high-pressure refrigerant vapor. See *heat exchanger* and *refrigerant.*

condensing heat exchanger: Heat exchanger that reduces the temperature of the flue gas below the dew point of a heat exchanger. See *heat exchanger* and *dew point.*

condensing medium: Fluid (air or water) that has a lower temperature than the ambient temperature, which causes heat to flow to it.

condensing point: Temperature at which a vapor condenses to a liquid.

conductance factor: Amount of heat transferred through 1 sq ft of surface area of a material of given thickness. Given in Btu per hour per 1°F temperature difference through the material. Also known as *C factor*.

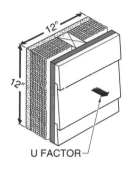

conductance factor

conduction: Heat transfer that occurs when molecules in a material are heated and the heat is passed from molecule to molecule through the material.

conduction factor: (U) Factor that represents the amount of heat that flows through a building component because of a temperature difference. Expressed in Btu per hour per square foot of material per degree Fahrenheit temperature difference through the material. See *building component, temperature difference, Btu,* and *Fahrenheit.*

conductivity factor: Amount of heat transferred through 1 sq ft of material that is 1″ thick. Given in Btu per hour per 1°F temperature difference through the material. Also known as *k factor*.

conductor: Electrical wiring in an electrical power circuit between the power supply and the equipment.

constant entropy: Ratio of the amount of heat added to a substance to the absolute temperature of the substance at the time the heat is added. Used to indicate energy lost to the disorganization of the molecular structure of a substance when heat is transferred.

constant quality: Percentage that expresses the ratio of refrigerant vapor to liquid refrigerant as a refrigerant changes state. See *refrigerant.*

constant air volume system: (CAV) Air distribution system in which the air flow rate in each building space remains constant, but the temperature of the supply air is varied by cycling the heating or air conditioning equipment ON and OFF.

constant volume: Volume of the refrigerant vapor that remains constant because of the relationship between the pressure and enthalpy of the refrigerant. Expressed in cubic feet per pound. See *refrigerant* and *enthalpy.*

contactor: Heavy-duty relay that has a coil and contacts that are designed to operate with the higher electric current that is required to run large electric motors. See *relay* and *contacts.*

contacts: Two electric conductors that carry current when joined. The position of contacts are identified as normally open (NO) or normally closed (NC).

control point: Point at which the sensor for a control system measures conditions. See *sensor.*

control signal: Medium used to communicate between sensor and operator in a control system. See *sensor* and *operator.*

control zone: Part of a building that is controlled by one controlling device.

convection: Heat transfer that occurs when currents circulate between warm and cool regions of a fluid. See *current.*

convection heating: Method of heating that occurs when currents circulate between warm and cool regions of a fluid, which creates an air flow pattern. See *cabinet convector* and *baseboard convector.*

cooler: Evaporator that cools water. See *evaporator* and *evaporating medium.*

cooling anticipator: Small heating element that is wired into the control circuit inside a thermostat case. Actuates a thermostat before a call for cooling is actually necessary. See *thermostat.*

cooling load: Amount of heat gained by a building because of a difference between the indoor temperature and outdoor temperature and infiltration or ventilation in the building. See *infiltration* and *ventilation.*

cooling tower: Evaporative water cooler that uses natural evaporation to cool water. Air circulates through the tower by natural convection or is blown through the tower by fans located in the tower. See *evaporation.*

corrosion: Condition that occurs when oxygen in air reacts and breaks down metal. Corrosion causes early failure of metal parts such as the boiler, circulating pump, and piping. See *boiler, circulating pump,* and *piping.*

crackage: Openings around windows, doors, or other openings in a building.

critical point: Pressure and temperature above which a substance does not change state regardless of the absorption or rejection of heat.

crude oil: Mixture of semisolids, liquids, and gases formed from the remains of organic materials that have been changed by pressure and heat over millions of years.

current: Fluid, such as water, air, or electricity, moving continuously in a certain direction. See *fluid.*

cycle: Combination of one suction stroke and one compression stroke. Each piston completes a cycle for every revolution of the crankshaft. See *suction stroke* and *compression stroke.*

cylinder head: Top part of a cylinder that seals the upper end of the cylinder.

D

damper: Plate that controls air flow.

datum line: Point at which the pressure in a column of water is zero. See *pressure.*

DC electricity: See *direct current.*

defrost cycle: Procedure of reversing refrigerant flow in a heat pump to melt frost or ice that builds up on the coil in the outdoor unit. See *refrigerant, heat pump,* and *outdoor unit.*

dehumidifier: Device that removes moisture from the air by causing moisture to condense. See *condensation* and *moisture.*

dehumidifier

density: Weight of an amount of a substance per unit of volume.

desiccant: Substance that acts as a drying agent. Absorbs liquid.

design condition: Condition of the air at which heating or air conditioning equipment provides comfort in building spaces.

design static pressure drop: Pressure drop per unit length of duct for a given size of duct at a given air flow rate. See *pressure drop.*

design technician: Person who has the knowledge and skill to plan heating and/or air conditioning systems.

design temperature: Temperature of the air at a predetermined set of conditions.

design temperature difference: Difference between the desired indoor temperature and the outdoor temperature for a particular season.

desired conditions: Setpoint conditions. See *setpoint temperature.*

dew point: (dp) Temperature below which moisture in the air begins to condense.

differential: Difference between the temperature at which the switch in a thermostat will turn the burner ON and the temperature at which the thermostat will turn the burner OFF. Prevents rapid cycling of the burner. See *thermostat.*

digital control system: Two-position control system.

digital valve: Two position (ON/OFF) valve, which is either completely open or completely closed.

dilution air: Atmospheric air that mixes with, dilutes, and cools the products of combustion. See *atmospheric air* and *products of combustion.*

dimensional change hygrometer: Hygrometer that operates on the principle that some materials absorb moisture and change size and shape depending on the amount of moisture in the air. See *hygrometer.*

diode: Solid-state device that allows current flow in only one direction. See *current.*

direct current: (DC) Electric current that flows in the same direction. See *current.*

direct drive system: Motor-to-wheel connection that has a blower wheel mounted directly on the motor shaft. Blower wheel turns as the motor turns the shaft. See *blower wheel.*

direct-fired heater: Unit heater that does not have a heat exchanger. A blower in the heater blows air directly through the combustion chamber and out of the

supply end of the unit. See *heat exchanger, blower,* and *combustion chamber.*

discomfort: Condition that occurs when people can sense a difference between themselves and the surrounding air.

disconnect: Switch that controls the flow of electricity to HVAC equipment such as an electric heating or cooling unit.

disk drive: Device that stores data on and retrieves data from a computer disk.

distribution system: Part of an HVAC system through which a heated or cooled medium is delivered to the building spaces that require heating or cooling. See *air distribution system* and *water distribution system.*

distributor: Piping arrangement that splits the refrigerant flow into several separate return bends on the evaporator coil to evenly distribute the refrigerant into the coils. See *refrigerant* and *evaporator.*

distributor

diverter fitting: Tee that meters water flow from a primary loop, which can be adjusted for gallons per minute required in the secondary loop through the terminal device. See *one-pipe primary-secondary system* and *terminal device.*

double duct system: Air distribution system that consists of a supply duct that carries cool air and a supply duct that carries heated air. See *air distribution system.*

double line drawing: Drawing of building spaces and components in which walls and partitions are shown with double lines. Wall thickness is shown at actual scale and fittings and transitions are shown as they appear. See *building component.*

double pipe condenser: Condenser that contains a small tube that runs through the center of a larger tube. Also known as *tube-in-tube condenser.* See *condenser.*

downflow furnace: Furnace in which heated air flows downward as it leaves the furnace. See *furnace.*

draft: Movement of air across a fire and through a heat exchanger. See *heat exchanger.*

draft diverter: Box made of sheet metal that runs the width of a heat exchanger. See *heat exchanger.*

draft inducer: Blower installed in a flue pipe to provide positive pressure in the flue, which carries the products of combustion up the stack. See *positive pressure* and *products of combustion.*

drum heat exchanger: Round drum or tube located on a combustion chamber. Transfers heat from products of combustion to air. See *combustion chamber* and *products of combustion.*

dry air: Elements that make up atmospheric air with the

moisture and particles removed. See *atmospheric air* and *moisture.*

dry base boiler: Boiler that has no water below the combustion chamber. See *combustion chamber.*

dry bulb temperature: (db) Measurement of sensible heat. See *sensible heat.*

dry bulb thermometer: Mercury thermometer that measures dry bulb temperature. See *dry bulb temperature.*

duct chase: Space provided in a building for installing ductwork. See *ductwork.*

dry base boiler

duct coil: Terminal device that is located in a duct. Air is supplied to the duct from a blower that may be remotely located. See *terminal device.*

duct heater: Unit heater that is installed in a duct and supplied with air from a remote blower. See *unit heater.*

duct section: Section of ductwork between two fittings.

duct size: Size of a duct expressed in inches of diameter for round ducts and in inches of width and height for rectangular ducts.

ductwork: Distribution system for a forced-air heating or cooling system. See *distribution system.*

dynamic pressure drop: Pressure drop in a duct fitting or transition caused by air turbulence as air flows through the fitting or transition. See *pressure drop* and *fitting.*

E

economizer package: Package of damper controls that brings outdoor air indoors to cool building spaces.

efficiency rating: Comparison of the furnace input rating with the output rating. Evaluation of how well a furnace burns fuel. See *input rating* and *output rating.*

electric heat: Electrical energy that has changed to thermal energy. See *thermal energy.*

electric spark igniter: Device that produces an electric spark, which is used to ignite either a pilot burner or main burner. See *pilot burner.*

electrical control system: Control system that uses AC electricity as a control signal. See *alternating current* and *control signal.*

electrical impedance hygrometer: Hygrometer based on the principle that the electrical conductivity of a substance changes as the amount of moisture in the air changes. See *hygrometer.*

electromechanical relay: Electric device that uses a magnetic coil to open or close one or more sets of contacts. See *relay* and *contacts.*

electronic control system: Control system that uses low-voltage DC electricity as a control signal. See *direct current* and *control signal.*

electrostatic filter: Device that cleans air that passes through electrically charged plates and collector cells.

emissivity: The ability of a surface to emit or absorb heat by thermal radiation. See *thermal radiation.*

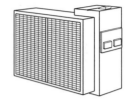

electrostatic filter

enthalpy: (h) Total heat contained in a substance, which is the sum of sensible heat and latent heat. See *sensible heat* and *latent heat.*

equal friction chart: Chart that shows the relationship between the air flow rate, static pressure drop, duct size, and air velocity. See *static pressure drop, duct size,* and *air velocity.*

equal friction method: Duct sizing method that considers that the static pressure is approximately equal at each branch takeoff in an air distribution system. See *static pressure.*

equal velocity method: Duct sizing method that considers that each duct section has the same air velocity. See *air velocity.*

equivalent temperature difference: Design temperature difference that is adjusted for solar gain. See *solar gain.*

evaporating medium: Fluid (air or water) at ambient temperature that is cooled when heat is transferred from the evaporating medium to the cold refrigerant. See *refrigerant.*

evaporation: Process that occurs when a liquid changes to a vapor by absorbing heat.

evaporative condenser: Condenser that uses water in the condenser coil, air blown past the coil, and evaporation of water from the outside surface of the condenser coil to remove heat from refrigerant. See *condenser, evaporation,* and *refrigerant.*

evaporator: Heat exchanger that adds heat to low-pressure refrigerant liquid. See *refrigerant.*

existing conditions: Conditions sensed by a sensor. See *sensor.*

expansion device: Valve or mechanical device that reduces the pressure on liquid refrigerant by allowing the refrigerant to expand. See *refrigerant.*

expansion tank: Boiler fitting that allows the water in a hydronic heating system to expand without raising the water pressure to dangerous levels. See *fitting* and *hydronic heating system.*

exposed surfaces: Building surfaces that are exposed to outdoor temperatures.

exposure: Geographic direction a wall faces. See *exposed surfaces.*

F

factors: Numerical values used for calculating heating and cooling loads that represent the heat produced or transferred under some specific condition. See *heating load* and *cooling load*.

Fahrenheit: (F) Scale used to express temperature in U.S. system of measurements. Units of measure are degrees Fahrenheit (°F).

feedback: Change in a signal. Feedback indicates that the conditions at a control point satisfy setpoint conditions. See *control point*.

feedwater valve: Valve that controls the flow of makeup water into a boiler to make up for losses. See *makeup water* and *boiler*.

feet of head: Unit of measure that expresses the height of a column of water that would be supported by a given pressure. See *pressure*.

filter: Porous material that removes particles from a moving fluid. See *fluid*.

filter-dryer: Combination filter and dryer located in the liquid line of a refrigeration system. Removes solid particles and moisture from a refrigerant. See *suction line* and *refrigeration system*.

filter-dryer

filter media: Material that makes up a filter in a forced-air system. See *filter*.

filtration: Process of removing particles and contaminants from air.

final condition: Properties of air after it goes through a process. See *properties of air*.

finned tube: Copper pipe or tube with aluminum fins, which provide a larger heat transfer surface area. Used in baseboard convectors. See *baseboard convector*.

finned-tube coil: Copper or aluminum tube with aluminum fins pressed on the tubing to increase the surface area of the coil. Part of an evaporator. See *surface area, evaporator,* and *condenser*.

first law of thermodynamics: See *thermodynamics, first law*.

fitting: Piece of equipment on a boiler or in the boiler piping that improves operation and efficiency of a hydronic heating system.

fitting loss coefficient: (coefficient C) Value that represents the ratio between the total static pressure loss through a fitting and the dynamic pressure at the fitting.

flame rod: Combustion safety control that conducts electricity through a flame for flame detection. See *combustion safety control*.

flame rod

flame surveillance control: Electronic combustion safety control that detects flame with a cad cell and controls a fuel valve according to the flame. See *combustion safety control* and *cad cell*.

flapper valve: Valve that opens to let the air-fuel mixture in but closes when a firing cycle begins because of back pressure. See *air-fuel mixture* and *back pressure*.

flat-plate coil: Coil pressed or buried in flat plates of metal. See *evaporator* and *condenser*.

float valve: Valve controlled by a hollow ball that floats in a liquid in a reservoir. Depending on the level of the liquid, the float ball and the attached linkage open or close a port in the valve. See *flooded evaporator coil* and *fuel oil-fired atmospheric burner*.

flooded evaporator coil: Evaporator coil that is full of liquid refrigerant during normal operation. See *refrigerant* and *float valve*.

flow control valve: Valve that regulates the flow of water in a hydronic heating system. May be a manual or automatic valve.

fluid: Any substance that takes the shape of its container. May be liquid or gas.

forced convection heating: Method of heating that occurs when convective currents are created by mechanically moving air past a hot water coil. Forced convection heaters use a blower or fan to move air. See *unit heater, cabinet heater,* and *unit ventilator*.

forced draft cooling tower: Cooling tower that has a fan located at the bottom of the tower that forces air through the tower. See *cooling tower* and *draft*.

free area: Total area of a register minus the area blocked by the frame or vanes.

freezing point: Temperature at which a substance freezes.

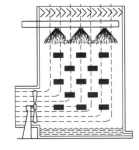

forced draft cooling tower

friction head: Effect of friction in a pipe, which occurs between the water moving through a pipe and the interior surfaces of the pipe.

friction loss: Decrease in air pressure due to friction as air moves through a duct.

fuel: Any material that is burned to produce heat.

fuel filter: Filter installed in a fuel supply line to remove dirt and other contaminants from the fuel. See *filter*.

fuel oil-fired atmospheric burner: Burner that consists of an open pot into which fuel oil flows at a controlled rate. See *burner*.

fuel oil power burner: Burner that atomizes fuel oil and provides combustion air under pressure.

furnace: Self-contained heating unit that includes a blower, burner, and heat exchanger or electric heating elements, and controls.

furnace heating capacity: See *output rating.*

furnace input rating: See *input rating.*

furnace output rating: See *output rating.*

fuse: Electrical overcurrent protection device located in an electrical power circuit.

G

gas fuel-fired atmospheric burner: Burner that mixes atmospheric air with a gas fuel to create a flame. See *burner.*

gas fuel power burner: Burner that uses natural or LP gas and contains a fan or blower to provide combustion air. See *combustion air.*

gas fuel valve: Shut off safety valve that controls the flow of fuel to the main burner and the pilot burner.

gas manifold: See *manifold.*

gate valve: Two position valve that has an internal gate that slides over the opening through which water flows.

gauge pressure: (psig) Pressure above atmospheric pressure that is used to express pressures inside a closed system. Expressed in pounds per square inch gauge. See *atmospheric pressure.*

geothermal heat: Heat that results when magma within the earth's crust comes in contact with groundwater, which results in steam. See *magma* and *groundwater.*

globe valve: Infinite position valve that has a gasket that is raised or lowered over a port through which water flows.

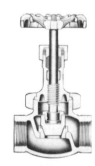

globe valve

grain: (gr) Unit of measure equal to $\frac{1}{7000}$ lb.

grill: Device that covers the opening of return ductwork.

gross unit output: Heat output of a boiler that is fired continuously. See *boiler* and *output rating.*

gross wall area: Total area of a wall including windows, doors, and other openings. See *surface area.*

groundwater: Water that sinks into the soil and subsurface rocks.

H

hand valve: Needle valve that has fine threads, which adjust the needle against the valve seat and control flow through the valve.

hardware: All physical units of a computer system, which includes the CPU, monitor, keyboard, disk drive, and printer. See *central processing unit, monitor, disk drive,* and *printer.*

header: Manifold that feeds several branch pipes or takes in fluid from several smaller pipes. See *fluid.*

heat: Form of energy identified by temperature difference or a change of state. Also known as *thermal energy.* See *temperature difference.*

heat anticipator: Small heating element located inside a thermostat that prevents the temperature from rising above the setpoint temperature by producing heat when the thermostat calls for heat. See *thermostat* and *setpoint temperature.*

heat anticipator

heat exchanger: Material that transfers heat from one substance to another without allowing the substances to mix.

heating capacity: See *output rating.*

heating control system: System that controls the temperature in a building by cycling heating equipment ON and OFF to maintain the temperature of a building within a few degrees of a setpoint temperature. See *setpoint temperature.*

heating load: Amount of heat lost by a building because of a difference between the indoor temperature and outdoor temperature, infiltration or ventilation in the building, and internal loads. See *infiltration, indoor design temperature,* and *ventilation.*

heating surface: Boiler metal that has heat from the combustion chamber on one side and water on the other side. See *combustion chamber.*

heating troubleshooting chart: Flow chart that identifies malfunctioning parts of a heating system with a series of steps. See *malfunctioning part.*

heating value: Quantity of heat in Btu that is released during combustion of one unit of an element. See *Btu.*

heat of compression: Thermal energy equivalent of mechanical energy expended by a motor that turns a compressor. Adds additional superheat to refrigerant.

heat pump: Mechanical compression refrigeration system that contains devices and controls that reverse the flow of refrigerant. Reversing the flow of refrigerant switches the relative position of the evaporator and condenser. See *mechanical compression refrigeration, refrigerant, evaporator,* and *condenser.*

heat pump thermostat: Component that incorporates a system switch, heating thermostat, and cooling thermostat. Switches the heat pump from the cooling mode to the heating mode and controls system operation in either mode. See *thermostat.*

heat rejection rate: Rate at which heat is transferred from the refrigerant in a condenser to the condensing medium. Function of the mass flow rate of the refrigerant and the heat of rejection. See *refrigerant, condenser,* and *condensing medium.*

heat sink: Heat that is contained in a substance that can be absorbed by a heat pump and used for heating. See *heat pump.*

heat transfer: Movement of heat from one material to another.

heat transfer coefficient: Amount of heat that will pass through a material per degree Fahrenheit temperature difference on each side of the material. See *conduction factor.*

heat transfer multiplier: Conduction factor used for calculating cooling or heating loads for residential or small commercial buildings. See *conduction factor, design temperature difference, heating load,* and *cooling load.*

hermetic compressor: Compressor in which the motor and compressor are sealed in the same housing. See *compressor* and *housing.*

hermetic compressor

high efficiency filter: See *bag filter.*

high-temperature limit control: Temperature-actuated electric switch that senses boiler water temperature. Shuts a burner OFF when the temperature rises above a setpoint.

horizontal furnace: Furnace in which heated air flows horizontally as it leaves the furnace.

hot gas: Hot, high-pressure refrigerant vapor that has been compressed and heated by a compressor. See *refrigerant* and *compressor.*

hot gas discharge line: Refrigerant line that connects the compressor to the condenser in a mechanical compression refrigeration system. See *refrigerant, compressor,* and *condenser.*

housing: Protective case or enclosure for moving parts.

humidifier: Device that adds moisture to air by causing water to evaporate into air. See *moisture.*

humidistat: Device that changes characteristics with changes in humidity.

humidity: Amount of moisture in the air.

humidity control system: System that controls the amount of moisture in the air to maintain comfort in building spaces. See *moisture* and *comfort.*

humidity ratio: (W) Ratio of the mass (weight) of the moisture in a quantity of air to the mass of the air and moisture together. Indicates the actual amount of moisture found in the air. Expressed in grains of moisture per pound of dry air or pounds of moisture per pound of dry air. See *moisture* and *grain.*

hydroelectric generating plant: Power plant where energy from moving water is converted to electrical energy. See *power.*

hydronic design static pressure drop: Pressure drop per unit length of pipe for a given size of pipe at a given water flow rate. See *pressure drop* and *water flow rate.*

hydronic heating system: Heating system that uses water, steam, or other fluid to carry heat from the point of generation to the point of use. See *fluid.*

hydronic radiant heater: Heater that has a radiant surface that is heated by hot water to a temperature high enough to radiate heat. See *radiant surface* and *thermal radiation.*

hygrometer: Instrument that measures humidity.

hyperbolic cooling tower: Cooling tower that has no fan. Natural draft moves air through a hyperbolic cooling tower. See *cooling tower.*

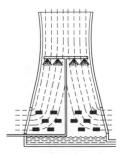
hyperbolic cooling tower

I

ignition temperature: Intensity of heat required to start the chemical reaction between the chemical energy in a fuel and oxygen in the air.

immersion heater: Electric heater that consists of copper rods enclosed in an insulated waterproof tube. Installed so that water surrounds the heater.

impeller wheel: Disk with blades that radiate from a central hub. Used in centrifugal pumps and centrifugal compressors. See *centrifugal pump* and *centrifugal compressor.*

inclined manometer: U-tube manometer designed so the bottom of the "U" is a long, inclined section of glass or plastic tubing. See *manometer.*

incomplete combustion: Improper combustion that occurs when not enough oxygen is supplied by combustion air to combine with all of the elements in a fuel. Carbon monoxide, aldehydes, and water vapor are produced. See *combustion air* and *aldehyde.*

individual duct section: Part of a distribution system between fittings in which the air flow, direction, or velocity changes due to the configuration of the duct.

indoor design temperature: Temperature selected for the inside of a building.

indoor unit: Package component in a heat pump that contains a coil heat exchanger and a blower. Depending on the direction of refrigerant flow, the indoor unit acts as the evaporator or condenser of the heat pump.

induced draft cooling tower: Cooling tower that has a fan located at the top of the tower that induces a draft by pulling the air through the tower. See *cooling tower* and *draft.*

industrial building: Building in which industrial processes are performed and heating or cooling processes are used for controlling the climate.

infiltration: Process that occurs when outdoor air leaks into a building.

infiltration air: Air that flows into a building when outer doors are open or when air leaks in through cracks around doors, windows, or other openings.

infrared radiant heater: Heating unit that heats by thermal radiation only. See *thermal radiation*.

initial condition: Properties of air before it goes through a process. See *properties of air*.

inlet vanes: Adjustable dampers that control the air flow to a blower. See *damper* and *blower*.

input device: Device, such as a keyboard, that allows an operator to enter information into a computer system.

input rating: Amount of heat produced as fuel burns. Found by

inlet vanes

multiplying the heating value of the fuel by the flow rate. Expressed in Btu/hr. Used for heating units such as forced-air furnaces and boilers. See *heating value*.

J

job file: Complete record of work done on a job during maintenance calls and emergency service calls. See *maintenance call*.

job sheet: Form used for equipment maintenance where data related to inspections is recorded. Contains information about the job and the work performed on the equipment. See *maintenance*.

Joule: (J) Unit of measure that expresses quantity of heat.

K

k factor: See *conductivity factor*.

kilocalorie: (kcal) Unit of measure equal to 1000 cal. See *calorie*.

kilojoule: (kJ) Unit of measure equal to 1000 J. See *Joule*.

king valve: Regular service valve located directly on the liquid line at the discharge side of the liquid receiver. See *service valve, liquid line,* and *liquid receiver*.

L

latent heat: Heat identified by a change of state and no temperature change. See *heat*.

latent heat of vaporization: Amount of heat required to change 1 lb of liquid refrigerant to 1 lb of vapor. See *latent heat*.

layout drawing: Drawing of the floor plan of a building that shows walls, partitions, windows, doors, fixtures, and other details that affect the location of the ductwork, registers, grills, piping, and terminal devices.

leaf valve: Valve that consists of a steel flapper, which is fastened at one end and is held in place by the tension of the flapper itself or by springs acting on the flapper.

limit switch: Electric switch that has a bimetal element, which senses the temperature of the surrounding air.

The switch shuts down a furnace if the furnace becomes overheated. See *bimetal element*.

liquid line: Refrigerant line that connects the condenser and the expansion device in a mechanical compression refrigeration system. See *refrigerant, condenser,* and *expansion device*.

liquid receiver: Storage tank for refrigerant that is located in the liquid line of a refrigeration system. See *refrigerant* and *liquid line*.

liquified petroleum gas: (LPG) By-product of the oil refining process used for heating. Also called *LP gas*.

liquid receiver

load calculation software program: Series of commands that electronically request data from an operator and manipulate data to determine heating and cooling loads. See *heating load* and *cooling load*.

load form: Document that is used by design technicians for arranging heating and cooling load variables and factors. See *heating load, cooling load, variables,* and *factors*.

low-efficiency filter: Filter made of fiberglass or other fibrous material treated with oil to help it hold dust and dirt. Often called slab filter. See *filter*.

low excess air burner: Burner that uses only the amount of air necessary for complete combustion. See *complete combustion*.

low-temperature limit control: Temperature-actuated electric switch that energizes a damper motor and shuts the damper if ventilation air temperature drops below a setpoint temperature, usually 32°F. See *damper, ventilation air,* and *setpoint temperature*.

LP gas: See *liquified petroleum gas*.

M

magma: Molten rock within the earth.

magnetic starter: Contactor with overload relays. See *overload relay*.

maintenance: Periodic upkeep of equipment.

maintenance call: Scheduled visit in which a maintenance technician performs general and specific inspections of equipment, components, and operation of a heating or air conditioning system.

maintenance checklist: Form, which may be part of the job sheet, that lists all components to

magnetic starter

be inspected and tests to be conducted during a maintenance call. See *job sheet* and *maintenance call.*

maintenance technician: Person trained in the maintenance tasks of heating and air conditioning systems.

major duct section: Independent part of an air distribution system through which all or part of the air supply from the blower flows. See *air distribution system.*

makeup air: Air that is used to replace air that is lost to exhaust.

makeup water: Water added to a boiler to replace water lost by leaks in a hydronic heating system. See *hydronic heating system.*

malfunctioning part: Element of a heating or air conditioning system that does not operate properly.

manifold: Pipe that has outlets for connecting other pipes.

manometer: Device that measures the pressures of vapors and gases. U-tube and inclined manometers are used to measure air pressure in ductwork. See *U-tube manometer* and *inclined manometer.*

manual disconnect: Protective metal box that contains fuses or circuit breakers and the disconnect, which is controlled with a lever that extends outside the box. See *fuse, circuit breaker,* and *disconnect.*

manual flow control valve: Globe valve that manually controls the flow of water in a hydronic heating system. See *globe valve.*

manufactured gas: Gas produced from coal, oil, liquid petroleum, natural gas or as a by-product of manufacturing processes.

mass: Quantity of matter held together that is considered one body.

mechanical action: Manner in which a compressor compresses a vapor. See *compressor.*

mechanical compression refrigeration: Refrigeration process that produces a refrigeration effect with mechanical equipment. See *refrigeration effect.*

medium-efficiency filter: Filter that contains filter media made of dense fibrous mats or filter paper. See *filter media.*

mercury barometer: Instrument consisting of mercury, a glass tube, and a dish that is used to measure atmospheric pressure. See *pressure.*

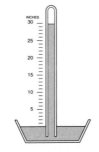

mercury barometer

mercury switch: Switch that uses the movement of mercury in a glass bulb to control flow of electricity in a circuit.

mil: Unit of measure equal to $1/1000$ of an inch.

miscibility: Ability of a substance to mix with other substances.

modular forced-air heating system: Heating system that uses more than one heating unit to produce heat for a building.

modulating control system: Control system in which the sensor regulates the operator in proportion to changes in existing conditions. See *existing conditions.*

modulating DC signal: Varying DC signal.

modulating valve: Infinite position valve, which may be completely open, completely closed, or at any intermediate position in response to the control signal the valve receives.

moist air: Mixture of dry air and moisture.

moisture: Gaseous form of water that is always present in atmospheric air. Also known as *water vapor.* See *atmospheric air.*

moisture indicator: Colored chemical patch located inside the glass window of a sight glass. Color of the chemical patch indicates whether moisture is present in the refrigerant. See *sight glass* and *moisture.*

monitor: Output device that displays information on a screen, which is similar to a television screen.

multibulb thermostat: Thermostat that contains more than one mercury bulb switch. See *thermostat* and *mercury bulb.*

multistage thermostat: Thermostat that contains several mercury bulb switches that make and break contacts in stages. See *thermostat.*

multistage centrifugal compressor: Centrifugal compressor that has more than one impeller wheel. See *centrifugal compressor* and *impeller wheel.*

N

natural gas: Colorless and odorless gas fuel. See *fuel.*

net unit output: Gross boiler output multiplied by a percentage of loss because of pickup in the piping and terminal devices. See *gross unit output, pickup,* and *terminal device.*

net wall area: Area of a wall after the area of windows, doors, and other openings have been subtracted.

O

one-pipe primary-secondary system: Hydronic piping system in which the terminal devices are on secondary loops, which are connected in parallel with the primary loop. Primary loop is the piping that connects the entire system. See *terminal device.*

one-pipe series system: Hydronic piping system that circulates water through one pipe and through each terminal device in turn. See *terminal device.*

opaque: Light cannot pass through.

open compressor: Compressor that has all of the components except for the motor inside one housing. See *compressor* and *housing.*

open pot burner: Burner used in space heaters and other small heating units. Uses fuel oil to produce flame.

open compressor

operating control: Control that cycles HVAC equipment ON or OFF as required. Operating controls for HVAC systems include transformers, thermostats, relays, contactors, magnetic starters, and solenoids. See *transformer, thermostat, relay, contactor, magnetic starter,* and *solenoid.*

operator: Mechanical device that switches heating, ventilating, and air conditioning equipment ON or OFF. Operators include relays, contactors, solenoids, and primary control systems. See *relay, contactor, solenoid,* and *primary control system.*

orifice: 1. In a burner, a precisely sized hole through which gas fuel flows. **2.** In an absorption refrigeration system, a restriction in the refrigerant line that leads from the condenser to the evaporator. See *absorption refrigeration system, condenser,* and *evaporator.*

outdoor design temperature: The expected outdoor temperature that a heating load or cooling load must balance.

outdoor design temperature tables: Tables of expected temperatures developed from records of temperatures that have occurred in an area over many years.

outdoor unit: Package component of a heat pump that contains a coil heat exchanger and a blower. Depending on the direction of refrigerant flow, the outdoor unit acts as the evaporator or condenser of the heat pump. See *heat pump, heat exchanger, refrigerant, evaporator,* and *condenser.*

output device: Device, such as a monitor or printer, that allows a CPU to output information. See *central processing unit.*

output rating: Amount of heat in Btu that a heating unit produces in one hour. Calculated after heat losses. Also known as *heating capacity.* See *Btu.*

outside diameter: Distance from outside edge to outside edge of a pipe or tube.

overload: Condition that occurs when a motor is connected to an excessive load.

overload relay: Electric switch controlled by electric current flow or the ambient air temperature.

P

package air conditioner: Self-contained air conditioning system that has all of the components contained in one sheet metal cabinet.

passive solar heating system: Solar heating system that uses no mechanical components for bringing heat into or for storing heat in a building.

perimeter loop system: Ductwork distribution system that consists of a single loop of ductwork with feeder branches that supply air to the loop from the supply plenum. See *supply plenum.*

physiological functions: Natural physical and chemical functions of an organism.

pickup: Additional heat needed to warm the water in a hydronic heating system after a period of off-time such as overnight.

pilot burner: Small burner located near the burner tubes that produces a pilot light. See *pilot light.*

pilot light: Small standing flame used to start combustion in a gas fuel-fired atmospheric burner. Produced by a pilot burner. See *gas fuel-fired atmospheric burner* and *pilot burner.*

pilot burner

pilot safety control: Safety control that determines if the pilot light is burning. See *pilot light.*

pipe section: Length of pipe that runs from the outlet of one fitting to the outlet of the previous fitting.

piping: Pipe used to carry water from a boiler to terminal devices.

piping loop: Secondary loop of pipe off a main supply and return pipe.

pitot tube: Device used to sense and measure static pressure and total pressure in a duct. See *static pressure.*

plan: Drawing of a building that shows dimensions, construction materials, location, and arrangement of the spaces within the building. Used for calculating heating load and cooling load. See *heating load* and *cooling load.*

plate valve: Valve that consists of a floating plate held in place by springs. Tension in the spring holds the valve closed until the pressure of refrigerant vapor opens the valve.

pneumatic control system: Low-pressure control system with a maximum air pressure of about 15 psi. Modulating system that uses compressed air pressure as a control signal.

pole: Number of load circuits that the contacts in a relay or contactor control at one time. See *contacts.*

positive displacement: Moving a fixed amount of a substance with every cycle. See *compressor* and *cycle.*

positive pressure: Pressure greater than atmospheric pressure. See *atmospheric pressure.*

potentiometer: Variable-resistance electric device that divides voltage proportionally between two circuits. Receives a signal and converts it to mechanical action in a valve motor. See *voltage.*

pounds per square inch: (psi) Unit of measure used to express pressure. Pressure exerted on 1 sq in. of surface. See *pressure.*

power: Energy used per unit of time.

power burner: Burner in which a fan or blower is used to supply and control combustion air. Air and fuel are introduced under pressure at the burner face. See *combustion air* and *burner face.*

power control: Device used to control the flow of electricity to HVAC equipment such as a furnace or a boiler. Power controls include disconnects, fuses, and circuit breakers. See *disconnect, fuse,* and *circuit breaker.*

power burner: See *gas fuel power burner* and *fuel oil power burner.*

prefilter: Filter that filters large particulate matter. Installed ahead of bag filters in the air stream. See *bag filter.*

prepared form: Preprinted form consisting of columns and rows that identify required information. Used for calculating heating and cooling loads. See *heating load* and *cooling load.*

pressure: (P) Force per unit area.

pressure atomizing burner: Burner that sprays oil into the combustion chamber through an atomizing nozzle. See *atomization.*

pressure drop: Decrease in water pressure caused by friction between water and the inside surface of a pipe as the water moves through the pipe.

pressure drop coefficient: Water flow rate through a valve or control mechanism in a piping system that causes a pressure drop of 1 psi through the valve or mechanism. See *pressure drop.*

pressure-enthalpy diagram: Graphic representation of the thermodynamic properties of a refrigerant.

pressure-reducing valve: Valve that reduces the pressure of makeup water to approximately 12 psi to 18 psi so that the water can be used in a boiler. See *makeup water* and *boiler.*

pressure switch: Electric switch that contains contacts and a spring-loaded lever arrangement. The lever opens and closes the contacts. See *contacts.*

pressure-temperature gauge: Gauge that measures the temperature and pressure of the water at the point on a boiler where the gauge is located.

Pressuretrol®: Pressure-actuated mercury switch that controls the burner on a high-pressure boiler by starting or stopping the boiler burner(s) based on the pressure inside the boiler. See *mercury switch.*

Pressuretrol®

pressure vessel: Tank or container that operates at a pressure greater than atmospheric pressure. See *atmospheric pressure.*

primary air: Air that mixes with a fuel to begin combustion. See *combustion.*

primary control system: System of operating controls for power burners combined with combustion safety controls.

printer: Output device that prints information from a CPU to paper. See *output device* and *central processing unit.*

products of combustion: Heat and gases produced when the chemicals in a fuel react with oxygen in the air and recombine to form new compounds.

propane: Gas fuel containing 90% to 95% propane gas (C_3H_8) and 5% to 10% propylene (C_3H_6). A by-product of the oil refining process.

propane-butane mixture: Mixture of propane and butane used when the properties of propane or butane alone are not acceptable. See *propane* and *butane.*

propeller fan: Mechanical device that consists of blades mounted on a central hub.

propeller fan

properties of air: Characteristics of air, which are temperature, humidity, enthalpy, and volume. See *temperature, humidity, enthalpy,* and *volume.*

proportional thermostat: Thermostat that contains a potentiometer that sends out an electric signal that varies as the temperature varies. See *potentiometer.*

psi: See *pounds per square inch.*

psia: See *absolute pressure.*

psig: See *gauge pressure.*

psychrometer: Hygrometer that measures humidity by comparing the wet bulb and dry bulb temperatures of the air. Consists of a wet bulb thermometer and a dry bulb thermometer mounted on a base. See *hygrometer, wet bulb thermometer,* and *dry bulb thermometer.*

psychrometric chart: Graph that defines the properties of the air at various conditions.

pulse burner: Low excess air burner that introduces the air-fuel mixture to the burner face in small amounts or pulses. See *low excess air burner, air-fuel mixture,* and *burner face.*

pumpdown control system: Control system that has a solenoid valve in the liquid line.

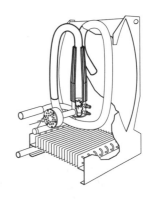
pulse burner

R

radial system: Ductwork distribution system that consists of branches that run out radially from the supply plenum of a furnace. See *supply plenum.*

radiant heating: Heat produced by thermal radiation.

radiant panel: Factory-built panel with a radiant surface that is heated by a piping coil or grid built into it. See *radiant surface.*

radiant surface: Heated surface from which radiant heat waves are generated.

radiation: See *thermal radiation.*

radiator: Heat-distributing device that consists of metal coil or tube units through which hot water passes. Used as terminal device. See *terminal device.*

reciprocating compressor: Compressor that uses mechanical energy to compress a vapor. See *compressor.*

refinery: Plant where crude oil is separated into petroleum products.

refrigerant: Fluid used for transferring heat in a refrigeration system. See *refrigeration system.*

refrigerant control valve: Combination expansion device and check valve used on a heat pump. A ball inside the check valve moves back and forth inside the valve housing to allow and stop flow.

refrigerant flow rate: Amount of refrigerant that flows through the refrigeration system per unit of time. Compressor controls the refrigerant flow rate. See *refrigerant* and *compressor.*

refrigerant property table: Table that contains values for and information about the properties of a refrigerant at saturation conditions or at other pressures and temperatures. See *saturation conditions.*

refrigeration: Process of moving heat from an area where it is undesirable to an area where it is not objectionable. See *heat.*

refrigeration effect: Cooling effect produced by a refrigeration system. Expressed in Btu per hour. See *Btu.*

refrigeration system: Closed system that controls the pressure and temperature of a refrigerant to regulate the absorption and rejection of heat by a refrigerant.

register: Device that covers the opening of the supply ductwork.

reheat: Heat supplied at the point of use while a ventilated air supply comes from a central location.

reheat unit: Heater that reheats air from a central air supply.

relative humidity: (rh) Amount of moisture in the air compared to the amount of moisture the air would hold if it were saturated. See *humidity* and *moisture.*

relay: Electric device that controls the flow of electric current in one circuit with another circuit. See *current.*

remote bulb: Sensor that consists of a small refrigerant-filled metal bulb connected to a thermostat by a thin tube. See *thermostat.*

residence: Dwelling where one family resides.

relay

resistance heating element: Heating element that consists of a grid of wires that have a known resistance to the flow of electricity.

reversing valve: Four-way valve that reverses the flow of refrigerant in a heat pump. Consists of a piston and a cylinder that has four refrigerant line connections.

rooftop air conditioner: Air-cooled package air conditioner that is located on the roof of a building.

rotary compressor: Positive-displacement compressor that uses a rolling piston or rotor that rotates in a cylinder to compress a fluid. See *positive displacement, compressor,* and *fluid.*

rotating vane compressor: Positive-displacement compressor that compresses refrigerant with multiple vanes that are located in the rotor. The vanes form a seal as they are forced against the cylinder wall by centrifugal force. See *positive displacement, compressor,* and *centrifugal force.*

S

safety control: Control that monitors the operation of HVAC equipment and turns the equipment OFF if the equipment becomes hazardous to personnel or damaging to equipment.

safety valve: Valve that prevents excessive pressure from building up in a boiler and hydronic heating system. See *boiler.*

saturated liquid: Liquid at a certain temperature and pressure that will vaporize if the pressure or temperature increases.

saturated vapor: Vapor at a certain temperature and pressure that will condense if the pressure or temperature decreases.

saturation conditions: Temperature and pressure of a refrigerant at which the refrigerant changes state. See *refrigerant.*

saturation line: Curve on a psychrometric chart where the wet bulb temperature and dew point scales begin. See *psychrometric chart, wet bulb temperature,* and *dew point.*

saturation pressure: Pressure at which a substance such as a refrigerant changes state. Both liquid and vapor are present at saturation pressure. Saturation pressure varies with the temperature of the substance. See *saturation conditions.*

saturation temperature: Temperature at which a substance such as a refrigerant changes state. Saturation temperature varies with the pressure on the substance. Both liquid and vapor are present at saturation temperature. See *saturation conditions.*

scale: Hard, brittle substance that forms when minerals and salts such as calcium carbonate and magnesium carbonate deposit on heating surfaces. Buildup reduces heat transfer through heating surfaces of a boiler. See *heating surface, heat transfer* and *boiler.*

schematic diagram: Diagram that uses lines and symbols to represent the electrical circuits and components in an electrical system.

screw compressor: Compressor that contains a pair of screw-like helical gears that interlock as they turn.

Also known as a *helical gear compressor*.

scroll: Sheet metal enclosure that surrounds the blower wheel of a centrifugal blower. See *blower wheel*.

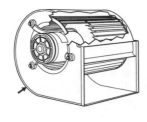

scroll

seasonal energy efficiency rating: (SEER) Cooling performance rating of a refrigeration system such as an air conditioner, which operates under normal conditions over a period of time.

secondary air: Air that enters combustion after a flame has been established.

second law of thermodynamics: See *thermodynamics, second law*.

seismograph: Device that measures and records vibrations in the earth.

self-contained hydraulic motor: Motor that contains an electronically controlled motor, a hydraulic fluid reservoir, small pump, and piston operator.

semi-hermetic compressor: Compressor that has all of the components and a motor located inside a housing. Also known as *serviceable compressor* because the housing can be opened and the components can be serviced on site. See *compressor* and *housing*.

semi-hermetic compressor

sensible heat: Heat that does not involve a change of state measured with a thermometer or sensed by a person.

sensitivity: Degree of control achieved by a control system compared to the degree of control desired.

sensor: Device that changes size, shape, or resistance due to a change in conditions.

service technician: Person trained to troubleshoot and repair HVAC equipment that is not operating properly.

service valve: Manually operated valve that contains two valve seats.

setpoint adjustor: Lever or dial on a thermostat that indicates the setpoint temperature on an exposed scale. See *thermostat* and *setpoint temperature*.

setpoint temperature: 1. In a forced-air heating system, the temperature at which the switch in a thermostat opens and closes. **2.** In a hydronic heating system, the temperature at which boiler water is maintained. See *thermostat* and *boiler*.

sheave: Grooved wheel or pulley. Used with V belts in belt drive systems. See *V belt* and *belt drive system*.

shell-and-coil cooler: Heat exchanger that contains a coil of copper tubing inside a shell. Water circulates through the shell and refrigerant circulates through the coil.

shell-and-tube cooler: Heat exchanger that contains tubes that run from one end of the shell to the other. Refrigerant flows through the tubes.

sight glass: Fitting for refrigerant lines that contains a small, glass window. Refrigerant flow can be observed through the window.

single duct system: Air distribution system that consists of a supply duct that carries both cool and warm air and a return duct that returns the air. See *air distribution system*.

single line drawing: Drawing in which walls and partitions are shown by a single line with no attempt to show actual wall thickness.

sling psychrometer: Psychrometer mounted on a handle so that it can be rotated rapidly. See *psychrometer*.

smoke: Unburned particles of carbon that are carried away from the flame by the convection currents generated by the heat of the flame. See *convection* and *current*.

software: Program that distributes information to a computer system and directs various functions.

solar energy: Energy transmitted from the sun by radiation. See *thermal radiation*.

solar gain: Heat gain caused by radiant energy from the sun that strikes opaque objects.

solar heat: Heat created by the visible (light) and invisible (ultraviolet) energy rays of the sun.

solenoid: Electric switch that controls one electrical circuit with another. Consists of a hollow coil that contains a metal rod, which closes or opens contacts. See *contacts*.

solenoid-operated pilot valve: Small valve in the refrigerant line that contains a solenoid. Solenoid is operated by a low-voltage electric signal from a thermostat.

soot: Unburned particles of carbon that collect on burners, in the combustion chamber, or in the flue when the surface temperature of these parts is lower than the ignition temperature of the unburned carbon.

specifications: Written supplements to plans that describe the materials used during construction of a building. See *plan*.

specific gravity: Ratio of the weight of a substance to the weight of an equal volume of water. Specific gravity of a gas is the weight of 1 cu ft of the gas compared to the weight of 1 cu ft of dry air at 29.92″ Hg (inches of mercury) and 68°F, which are standard conditions. See *standard conditions*.

specific heat: Ability of a material to hold heat. Expressed as the ratio of the quantity of heat required to raise the temperature of a substance 1°F to that required to raise the temperature of an equal mass of water 1°F.

specific volume: (v) Volume of air at any given temperature in cubic feet per pound.

split system: Air conditioning system that has separate sections for the evaporator and condenser. See *air conditioner, evaporator,* and *condenser.*

spot heating: Method of heating that provides heat at a particular area. Unused space is not heated by spot heating. Spot heating is done with radiant panels. See *radiant panel.*

spring safety valve: Safety valve that controls the pressure in a boiler by exhausting hot water and steam to the atmosphere when the pressure in the boiler rises above the normal working pressure of the boiler. See *safety valve.*

stack switch: Mechanical combustion safety control that contains a bimetal element that senses flue-gas temperature and converts it to mechanical motion. See *combustion safety control, bimetal element,* and *temperature.*

spring safety valve

stagnant air: Air that contains an excess of impurities and lacks the amount of oxygen required to provide comfort.

standard conditions: Values used as a reference for comparing properties of air at different elevations and pressures. One pound of dry air and its associated moisture at standard conditions has a pressure of 29.92″ Hg (14.7 psi), temperature of 68°F, volume of 13.33 cu ft/lb, and density of 0.0753 lb/cu ft. See *properties of air, dry air,* and *pounds per square inch.*

standard sample: Sample of coal from an area that consists of different kinds of coal. Represents all of the coal in that area.

standing radiator: Vertical or horizontal tubes that are connected with headers at each end. Hot water from a boiler flows through the tubes. See *radiator* and *boiler.*

static head: Weight of water in a vertical column above a datum line. See *datum line.*

static pressure: Pressure that acts through weight only with no motion. Static pressure is expressed in inches of water column. In a duct, air pressure measured at right angles to the direction of air flow. Tends to burst a duct. See *water column.*

standing radiator

static pressure drop: Decrease in air pressure caused by friction between the air moving through a duct and the internal surfaces of the duct.

static regain method: Duct sizing that considers that each duct section is sized so that the static pressure increase at each takeoff offsets the friction loss in the preceding duct section.

stationary vane compressor: Positive-displacement compressor that has a spring-loaded vane in the side of the cylinder wall. The vane rides against the rolling piston and helps direct the refrigerant. See *positive displacement* and *refrigerant.*

stop valve: Valve that stops water flow. Located in a hydronic heating system where water may have to be completely shut OFF. Gate valves and globe valves are used as stop valves. See *gate valve* and *globe valve.*

strobe light tachometer: Tachometer that contains a strobe light, which flashes a beam of light at specified time intervals. See *tachometer.*

stroke: Travel distance of a piston inside a cylinder.

subcooling: Process of cooling of a substance such as a refrigerant to a temperature that is lower than the saturated temperature of the substance at a particular pressure. See *refrigerant* and *saturation temperature.*

suction accumulator: Metal container located in the suction line between the evaporator and the compressor that catches liquid refrigerant before the refrigerant reaches the compressor. See *suction line, evaporator,* and *compressor.*

suction line: Refrigerant line that connects the evaporator and the compressor in a mechanical compression refrigeration system. See *refrigerant, evaporator,* and *compressor.*

suction accumulator

suction stroke: Stroke in a cylinder that occurs when the piston moves down. Decreases the pressure in the cylinder. See *stroke.*

summer dry bulb temperature: Warmest dry bulb temperature expected to occur in an area while disregarding the highest temperature that occurs in from 1% to 5% of the total hours in the three hottest months of the year. See *dry bulb temperature.*

summer wet bulb temperature: Wet bulb temperature that occurs concurrently with the summer dry bulb temperature. Considered in cooling loads because of the effect of humidity on comfort at higher temperatures. See *wet bulb temperature, summer dry bulb temperature,* and *comfort.*

superheat: Sensible heat that is added to a substance

after the substance has turned to vapor. See *sensible heat.*

superheated temperature: Temperature higher than the saturated vapor temperature for the pressure of the refrigerant.

supply plenum: Sealed sheet metal chamber that connects the furnace supply air opening to the supply ductwork.

surface radiation system: Heating system that uses the interior surfaces of a room as radiant surfaces, which heat the air in the room. See *radiant surface.*

surface area: Amount of area of a material in square feet.

surface temperature: Temperature of a radiant surface.

system graph: Graphic representation of the operation of a refrigeration system on a pressure-enthalpy diagram. See *pressure-enthalpy diagram.*

system pressure drop: Total static pressure drop in a forced-air or hydronic distribution system. Difference in air or water pressure between the point at the beginning of a distribution system and the point at the end of the system. See *pressure drop.*

T

table of equivalent lengths: Table used to convert dynamic pressure drop to friction loss in feet of duct. Contains the length of duct that gives the same static pressure drop as a particular fitting or transition. See *dynamic pressure drop* and *static pressure.*

tachometer: Device used for measuring the speed of a shaft or wheel.

tapping valve: Valve that pierces a refrigerant line.

temperature: Measurement of the intensity of heat.

temperature-actuated defrost control: Control that consists of a remote bulb thermostat and an electric control switch.

temperature difference: Difference between the temperatures of two materials, the temperatures on both sides of a material, or the initial and final temperatures of a material through which heat has been transferred.

temperature gradient: Variation in temperature in a substance. See *temperature.*

temperature stratification: Variation of air temperature in a building that occurs when warm air rises to the ceiling and cold air drops to the floor.

temperature swing: Difference between the setpoint temperature and the actual temperature. Lowers the temperature in a building. See *setpoint temperature.*

terminal device: Device that transfers heat from the water in a piping system to the air in building spaces.

terminal device control: Control that regulates the temperature of the air in a building space by regulating the air and water flow through a terminal device. See *circulating pump, zone valve,* and *blower.*

therm: Quantity of gas required to produce 100,000 Btu of heat.

thermal energy: See *heat.*

thermal radiation: Transfer of energy in the form of radiant (electromagnetic) waves.

thermal transmission factor: Numerical value that represents the amount of heat that passes through a material when there is a temperature difference across the material.

thermistor: Electronic device that changes resistance in response to a temperature change.

thermocouple: Pair of electrical conductors that have different current-carrying characteristics welded together at one end.

thermodynamics: Science of thermal energy and how it transforms to and from other forms of energy.

thermodynamics, first law: Energy cannot be created or destroyed but may be changed from one form to another.

thermodynamics, second law: Heat always flows from a material at a high temperature to a material at a low temperature.

thermoelectric expansion valve: Temperature-actuated expansion device that controls the flow of refrigerant with a solid-state sensor, electric current, and a control signal. See *expansion device.*

thermostat: Temperature-actuated electric switch that operates and controls burner(s) or heating elements in a heating unit.

thermostatic expansion valve: Expansion device that controls the amount of superheat the refrigerant absorbs in the evaporator by pressure generated in a remote bulb. See *expansion device, superheat, evaporator,* and *remote bulb.*

thermostatic expansion valve

throw: The number of different control circuits that each individual pole in a relay or contactor controls. Circuits are identified as single-throw (ST) circuits or double-throw (DT) circuits. See *pole.*

ton of cooling: (tons) Amount of heat required to melt a ton of ice over a 24-hour period. Equal to 288,000 Btu/24 hours or 12,000 Btu/hr. See *Btu.*

total head: Sum of friction head and static head. See *friction head* and *static head.*

transformer: Electric device that changes the voltage in an electrical circuit. See *voltage.*

transistor: Solid-state device that allows current to flow through a primary circuit when a secondary circuit is energized.

troubleshooting: Systematic elimination of the various parts of a heating or air conditioning system to locate a malfunctioning part. See *malfunctioning part.*

troubleshooting chart: Chart that shows the major sections, components, and parts of a heating or air conditioning system. Organized so that the effect of a malfunctioning part in a system can be traced to the effect the part has on the major sections of the system.

trunk: Main supply duct that extends from the supply plenum. See *supply plenum.*

trunk and branch system: Ductwork distribution system that consists of one or more trunks that run out from the supply plenum of a furnace. See *trunk* and *supply plenum.*

two-pipe direct return hydronic piping system: Hydronic piping system that circulates supply water in the opposite direction of the circulation of return water.

two-pipe reverse return system: Hydronic piping system that circulates return water in the same direction as supply water.

two-stage compression system: Compression system that uses more than one compressor to raise the pressure of a refrigerant. See *compressor* and *refrigerant.*

U

unit air conditioner: Self-contained air conditioner that contains a coil and a blower in one cabinet.

unitary system: Air conditioning system that has all components enclosed in one cabinet.

unit heater: Self-contained forced convection heater used in forced-air or hydronic heating systems that contains a fan or blower, hot water coil, and controls in one cabinet. Unit heaters are normally used as space heaters and are not connected to ductwork. See *forced convection heating.*

unit ventilator: Forced convection heater that has a blower, hot water coil, filters, controls, and a cabinet that has an opening for outdoor ventilation. See *forced convection heating.*

upflow furnace: Furnace in which heated air flows upward as it leaves the furnace.

U-tube manometer: U-shaped section of glass or plastic tubing that is partially filled with water or mercury. Used to measure pressure. See *manometer.*

V

vacuum: Pressure lower than atmospheric pressure. Vacuum is expressed in inches of mercury. See *pressure* and *atmospheric pressure.*

valve wrench: Ratchet wrench that fits common valve stems. Usually $\frac{1}{4}''$ to $\frac{3}{8}''$ in size.

variable air volume system: (VAV) Air distribution system in which the air flow rate in the building spaces is varied by mixing dampers, but the temperature of the supply air remains constant.

variables: Data unique to a building that relates to the specific location and the specifications of a building.

variable-volume damper: Damper that controls the air at a terminal device in a VAV distribution system. See *variable air volume system.*

variable-volume duct system: Control system that has variable-volume dampers on a primary cooling duct.

V belt: Closed-looped belts made of rubber, nylon, polyester, and rayon. Used for belt drive blower motors. See *belt drive system.*

velocity pressure: Air pressure in a duct that is measured parallel to the direction of air flow.

velocity reduction method: Duct sizing that considers that the air velocity is reduced at each branch takeoff.

velometer: Device that measures the velocity of air flowing out of a register.

ventilation: Process of introducing fresh air into a building.

ventilation air: Air that is brought into a building to keep fresh air in building spaces.

venturi: Restriction that causes increased air pressure as air moves through it.

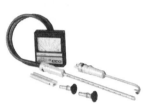

velometer

viscosity: Ability of a liquid or semiliquid to resist flow.

volatile matter: Organic gases or vapors (usually oils and tars) given off when coal burns.

voltage: Electric potential that causes current to flow.

volume: (V) Amount of space occupied by a three-dimensional figure. Expressed in cubic units.

volumetric capacity: Volume of refrigerant that a compressor moves. Also called *volumetric displacement.*

volumetric flow rate: Rate of refrigerant flow expressed in cubic feet per minute. Mass flow rate expressed as volume.

W

water column: (WC) Pressure required to raise a column of water a given height. Expressed in inches.

water-cooled condenser: Condenser that uses water as the condensing medium. Heat is transferred from the refrigerant to the water.

water distribution system: System of piping in a hydronic heating or cooling system. Used to distribute hot or cold water to building spaces to condition the air.

water flow rate: Volumetric flow rate of water as it moves through a pipe section expressed in gallons per minute. See *volumetric flow rate.*

water hammer: Condition that occurs when water pounds against the inside of a pipe. Can cause damage to valves and other boiler fittings. See *fitting.*

water pressure: Force exerted by water per unit of area. Expressed in pounds per square inch or in feet of head. See *pressure.*

water temperature control system: Group of components that controls the temperature of water in a hydronic heating system.

water vapor: See *moisture.*

watt: (W) Unit of electrical power. One watt equals 3.414 Btu. See *Btu.*

weight: Force with which a body is pulled downward by gravity.

wet base boiler: Boiler that has water on all sides of the combustion chamber. See *combustion chamber*.

wet bulb depression: Difference between wet bulb and dry bulb temperature readings. See *wet bulb temperature* and *dry bulb temperature*.

wet bulb temperature: Measurement of the amount of moisture in the air.

wet bulb thermometer: Mercury thermometer that has a small cotton sock placed over the bulb.

wet leg boiler: Boiler that has water along the sides of the combustion chamber. See *combustion chamber*.

wind power: Energy created by the movement of wind.

winter dry bulb temperature: Coldest temperature expected to occur in an area while disregarding the lowest temperatures that occur in from 1% to 2½% of the total hours in the three coldest months of the year. See *dry bulb temperature*.

Y

yellow flame: Flame produced when combustion air mixes with a fuel at the burner face. See *burner face*.

Z

zone: Specific section of a building that requires separate temperature control.

zone valve: Valve that regulates the flow of water in a control zone or terminal device of a building. Valves used in the piping of a hydronic heating or cooling system to regulate the flow of water. See *control zone*.

zone valve

Index